MAKING TRANSCENDENCE TRANSPARENT

T0155697

MAKING TRANSCENDENCE
TRANSPARENT

An intuitive approach to classical transcendental number theory

Edward B. Burger   **Robert Tubbs**

 Springer

Edward B. Burger
Department of Mathematics
Williams College
Williamstown, MA 01267
USA

Robert Tubbs
Department of Mathematics
University of Colorado at Boulder
Boulder, CO 80309
USA

Cover illustration: The cover image was rendered by Steve Barber and Joseph Piliero.

ISBN   978-1-4419-1948-9   Printed on acid-free paper.

9 8 7 6 5 4 3 2 1

Springer-Verlag is a part of Springer Science+Business Media
springeronline.com

Contents

Preface: *The Journey Ahead*   ix

Number 0.   1.41421356237309504880016887242...
A Prequel to Transcendence:
The nature of numbers and the irrationality of $\sqrt{2}$ 1

  0.1   A natural beginning   1
  0.2   The problem of addition   1
  0.3   The problem of multiplication   2
  0.4   A search for missing numbers   4
  0.5   Is i a number?   5
  0.6   Transcending the algebraic numbers   6
       Postscript: Tools of the transcendence trade   7

Number 1.   0.110001000000000000000000010000...
Incredible Numbers Incredibly Close to Modest Rationals:
Liouville's theorem and the transcendence of $\sum_{n=1}^{\infty} 10^{-n!}$ 9

  1.1   Turning a mountain of transcendence into a molehill   9
  1.2   One small rational for Liouville, one giant step for
       transcendence   10
  1.3   Our first transcendental number   11
  1.4   The proof of Liouville's Theorem   13
       *Algebraic Excursion: Irreducible polynomials evaluated at
       rational values*   14
  1.5   Liouville numbers   18
  1.6   The number 0.12345678910111213141516...   20
  1.7   Roth's Theorem: The ultimate Liouville result   24
  1.8   Life after Liouville   25

Number 2.   2.718281828459045235360287471.3 . . .
The Powerful Power Series for e:
Polynomial vanishing and the transcendence of *e*   27

2.1   Fourier's proof of Euler's slick result   27
2.2   A first attempt at a proof   29
2.3   The classic vanishing polynomial trick   32
2.4   The first part of the proof of Theorem 2.2—The elusive estimate   34
2.5   The dramatic conclusion of the proof Theorem 2.2—Arithmetic
      conquers all   36
2.6   The transcendence of *e*   37
2.7   Foreshadowing algebraic exponents—The irrationality
      of $e^{\sqrt{n}}$ and π 40

Number 3.   4.113250378782927517173581.8151 . . .
Conjugation and Symmetry as a Means Towards Transcendence:
The Lindemann–Weierstrass Theorem and the transcendence of $e^{\sqrt{2}}$ 43

3.1   Algebraic exponents through the looking glass—A wonderland of
      transcendence   43
3.2   Heading towards Hermite—A partial result casts
      some foreshadowing   45
3.3   A surprisingly non-special, special case of the
      Lindemann–Weierstrass Theorem   52
      Algebraic Excursion: Symmetric functions and conjugates   52
3.4   A conjugate appetizer—The delicate but surprisingly non-special,
      special case   58
3.5   The main dish—Serving up a spaghetti of symmetry   62
3.6   Algebraic Independence—Freeing ourselves of unhealthy
      dependencies   70

Number 4.   23.14069263277926900572908.6367 . . .
The Analytic Adventures of e^z:
Siegel's Lemma and the transcendence of e^{π} 77

4.1   Giving e^z a complex by taking away its power series   77
4.2   Throwing in our two bits as a warm-up to the irrationality of e^{π} 78
4.3   Our first ill-fated attempt at a proof—A star-crossed relationship
      between the order of vanishing and the degree   83
4.4   Polynomials in two variables and the power of *i*   87
4.5   The ever-popular polynomial construction   88
      *Algebraic Excursion: Solving a system of linear
      equations in integers*   92
4.6   At long last, a proof that e^{π} is irrational   96

4.7   A first glance at the transcendence of e^{π}—Some algebraic obstacles   101
      Algebraic Excursion: Re-expressing the powers of an algebraic
      number   102
4.8   The transcendence of e^{π} 104

Number 5.   2.6651441426902251886502972498 . . .
Debunking Conspiracy Theories for Independent Functions:
The Gelfond–Schneider Theorem and the transcendence of $2^{\sqrt{2}}$ **113**

5.1   Algebraic exponents and bases—Moving beyond e by focusing on e 113
5.2   Some sketchy thoughts on the proof of Theorem 5.2   115
5.3   Distilling three algebraic numbers down to one primitive element   118
      Algebraic Excursion: Number fields and primitive elements   119
5.4   Beating the (linear) system—Constructing a polynomial
      via linear equations   122
5.5   The proof of a real special case   127
5.6   Moving out to the vast complex plane   133
5.7   Widening our perspective and extending our results   136
5.8   Algebraic values of algebraically independent functions   140

Number 6.   2.718281828459 . . . + 0.11000100000 . . .
Class Distinctions Among Complex Numbers:
Mahler's classification and the transcendence of $e + \sum_{n=1}^{\infty} 10^{-n!}$ **147**

6.1   The power of making polynomials nearly vanish   147
6.2   A rational approach to irrationality   149
6.3   An algebraic approach to transcendence   152
6.4   Detecting subtle distinctions among the transcendent   158
6.5   A critical consequence of the classification   163
6.6   Which is the most popular class? Granting "Most favored
      number status"   167
6.7   The proofs of Lemmas 6.14 and 6.15   174
      Algebraic Excursion: Algebraic approximations and polynomials
      with small moduli   175
6.8   Declassified quantities: e, π, and the elusive T-numbers   181

Number 7.   7.4162987092054876737354013887 . . .
Extending Our Reach Through Periodic Functions:
The Weierstrass $\wp$-function and the transcendence of $\frac{\Gamma(1/4)^2}{\sqrt{\pi}}$ **183**

7.1   Transcending our beloved e^z and challenging its centrality   183
7.2   A circle of ideas behind elliptic curves   184
7.3   Entire periodic functions   191
7.4   Pinning down $\wp(z)$ and uncovering a group homomorphism   194

7.5   A new transcendence result   200
7.6   Exploring the gamma function and infinite products   203
7.7   The proof of Theorem 7.3   208
7.8   The transcendence of gamma values and the
       Schneider–Lang Theorem revisited   216
       *Algebraic Excursion: An introduction to the theory of complex
       multiplication*   216

Number 8. $1 + \frac{1}{T^2-T} + \frac{1}{(T^4-T)(T^2-T)^2} + \frac{1}{(T^8-T)(T^4-T)^2(T^2-T)^4} + \cdots$

Transcending Numbers and Discovering a More Formal e:
Function fields and the transcendence of $e_C(1)$ **223**

8.1   Moving beyond numbers   223
8.2   An intimate interlude—How to get close in function fields   225
8.3   A formal search for *e*   228
8.4   Finding the factorial in $F_q[T]$ 229
       Algebraic Excursion: Irreducible polynomials over F_q   232
8.5   The transcendence of $e_C(1)$ 234
8.6   A repeated look at $e_C(z)$ through periodicity   240
8.7   The transcendence of π_C 249
8.8   Revisiting $\wp(z)$ and moving beyond Carlitz   250

Appendix: *Selected highlights from complex analysis*   **255**

A.1   Analytic and entire functions   255
A.2   Contour integrals   256
A.3   Cauchy's integral formula   256
A.4   The Maximum Modulus Principle   257

Acknowledgments   **259**

Index   **261**

Preface

The Journey Ahead

At the heart of transcendental number theory lies an intriguing paradox: While essentially all numbers are transcendental, establishing the transcendence of a particular number is a monumental task. Thus transcendental numbers are an enigmatic species of number: We know they are all around us and yet it requires enormous effort to catch one. More often than not, they slip through our fingers and dissappear back into the dense jungle of numbers. Here we will venture to tame a few of these incredible creatures. In the pages ahead we offer an approach to transcendence that not only includes the intricate analysis but also the beautiful ideas behind the technical details.

The phrase "classical transcendental number theory" in the title of this book refers to the most widely known results that were obtained in the nineteenth and early twentieth centuries. The reason for this focus is threefold. Firstly, this body of work requires only the mathematical techniques and tools familiar to advanced undergraduate mathematics students, and thus this area can be appreciated by a wide range of readers. Secondly, the ideas behind modern transcendence results are almost always an elaboration of the classical arguments we will explore here. And finally, and perhaps more importantly, this early work yields the transcendence of such admired and well-known numbers as e, π, and even $2^{\sqrt{2}}$.

While the theory of transcendental numbers is a fundamental and important area of number theory, it is not widely known. Part of the reason for its relative obscurity is that upon first, or even second or third inspection, the arguments and techniques employed appear to be formidable if not impenetrable. In truth, however, the underlying principles upon which the entire discipline is based are both straightforward and central to all of number theory.

Our quest is to see beyond the complexity and intricacies of the subject and develop some intuition into the delicate ideas that are at the center of transcendental number theory. Thus we endeavor not only to present the proofs of the transcendence results we will encounter, but more importantly to make those results and their justifications natural and intuitively sound. Our desire is to bring some of the beautiful ideas from one of the richest areas of mathematics to life, and to inspire further explorations.

Number 0

1.41421356237309504880168887242...

A Prequel to Transcendence:
The nature of numbers and the irrationality of $\sqrt{2}$

0.1 A natural beginning

The study of the nature of numbers is one of the most ancient and fundamental pursuits in all of mathematics. Before we embark upon our journey into the world of transcendental number theory, we offer this brisk panoramic tour of the notion of number in which we not only introduce some basic background, but, more importantly, attempt to provide a framework within which the theory of transcendence will find its rightful place in our quest for an understanding of the intrinsic properties of numbers.

Naïvely, at the most basic level, when we think of numbers, we think of enumeration; that is, we use numbers in our daily lives to count objects. Given this point of view, the most natural of numbers are the *natural numbers*: The set $\mathbb{N} = \{1, 2, 3, \ldots\}$. The moment we conjure up algebraic and analytic thoughts, the story quickly becomes more interesting.

0.2 The problem of addition

While the natural numbers empower us to count, they are not robust enough to enable us to answer even incredibly simple questions about the most elementary of all arithmetic operations—addition. For example, given a fixed natural number $n > 1$, it is easy to find natural numbers x and y that satisfy $x + y = m$. However, the issue becomes much more challenging if we also fix the natural number $y = n$. That is, given fixed natural numbers m and n, find a natural number x satisfying $x + n = m$. If $n \geq m$, then no such a natural number exists.

Since any reasonable concept of number should be rich enough to contain solutions to all such simple algebraic equations, we are inspired to expand our notion of number so as to include all solutions to all such linear equations. Hence we are led to consider the set of *integers*, denoted as $\mathbb{Z} = \{\ldots, -3, -2, -1, 0, 1, 2, 3, \ldots\}$.

But even the collection of all integers is not sufficient to solve all simple arithmetic problems. If we allow the variable x in our linear equation to appear repeatedly, say for example, $x + x = m$ or $x + x + x = m$, then we discover that for a given integer m, there may not exist an integer x satisfying the algebraic equation

$$\underbrace{x + x + \cdots + x}_{n \text{ times}} = m,$$

where n is a natural number. In order to solve these linear equations, we again must relax our view of number and define the *rational numbers*, denoted by $\mathbb{Q}$, to be the set of all quotients m/n, where $m \in \mathbb{Z}$ and $n \in \mathbb{N}$.

0.3 The problem of multiplication

If we view $\mathbb{Q}$ as our entire world of numbers and move beyond addition to multiplication, then we encounter difficulties analogous to those presented by addition within the context of the natural numbers. While given any rational number r/s it is easy to find rational numbers x and y such that $x \times y = \frac{r}{s}$, even if we specify $y = \frac{u}{v}$, for some nonzero rational number u/v, if we now consider repeated multiplication, the issue quickly becomes far more complicated. That is, if we attempt to solve an equation

$$\underbrace{x \times x \times x \times \cdots \times x}_{n \text{ times}} = \frac{r}{s},$$

then we often encounter equations that are impossible to solve over $\mathbb{Q}$.

We begin by considering the simplest repeated multiplication problem: Solving the equation $x \times x = m$, where m is a natural number. The equation $x \times x = 1$ clearly has integer solutions, for example, $x = 1$. However by simple trial and error we see that the equation $x \times x = 2$ does not have any integer solutions. Furthermore, if we experiment with rational numbers $x = \frac{r}{s}$, then we never quite locate an *exact* solution. Our desire to develop a notion of number that allows us to solve elementary addition and multiplication problems together with our frustration in not being able to find a rational solution to $x \times x = 2$, leads us to expand our world of number and denote a solution to this equation by a special symbol, $\sqrt{2}$.

The "number" $\sqrt{2}$ is not as easily understood as were the solutions to the equations we first considered, in part because we can prove that $\sqrt{2}$ is *not* a rational number. There are many proofs of this fact, some of them possibly 2500 years old, but the one we offer here foreshadows and highlights almost every transcendence argument we will encounter in the chapters to come. In fact, this strategy of first exploring a more basic irrationality result that mirrors the essence of the more challenging transcendence theorem will become a recurring theme throughout our journey. Thus the proof below not only reflects the structure of all the arguments we will explore, but also exemplifies, in a larger sense, the basic philosophy of our approach.

THEOREM 0.1   *The number $\sqrt{2}$ is not a rational number.*

Proof. We assume that $\sqrt{2}$ is rational, say $\sqrt{2} = \frac{r}{s}$, where r and s are natural numbers. We now construct a sequence of polynomials $\mathcal{P}_n(z)$, called *auxiliary polynomials*, that produce nonzero integers when evaluated at $z = \sqrt{2}$. To build these polynomials, we first introduce two sequences of integers, $\{a_n\}$ and $\{b_n\}$, which we define recursively as follows: We declare that $a_0 = 0, a_1 = 1; b_0 = 1, b_1 = 1;$ and for any $n \geq 2$,

$$a_n = 2a_{n-1} + a_{n-2} \quad \text{and} \quad b_n = 2b_{n-1} + b_{n-2}. \qquad (0.1)$$

Next we let $\mathcal{P}_n(z) = a_n^2 z^2 - b_n^2$ and observe that $\mathcal{P}_n(\sqrt{2}) = 2a_n^2 - b_n^2$ is an integer. We now claim that $\mathcal{P}_n(\sqrt{2}) \neq 0$. To establish this claim, we will actually prove more; in particular, we will show by induction that for all $n \geq 1$, the following two identities hold:

$$2a_n^2 - b_n^2 = (-1)^{n+1} \quad \text{and} \quad 2a_{n-1}a_n - b_{n-1}b_n = (-1)^n. \qquad (0.2)$$

It is easy to establish these identities in the case when $n = 1$. We now assume that both of the identities in (0.2) hold for all indices up through a particular choice of n. To establish the first identity for the index $n + 1$, we apply the recurrence relations in (0.1) and deduce that

$$2a_{n+1}^2 - b_{n+1}^2 = 2(2a_n + a_{n-1})^2 - (2b_n + b_{n-1})^2$$
$$= 4\left(2a_n^2 - b_n^2\right) + 4(2a_{n-1}a_n - b_{n-1}b_n) + \left(2a_{n-1}^2 - b_{n-1}^2\right),$$

which, in view of our inductive hypotheses, implies that

$$2a_{n+1}^2 - b_{n+1}^2 = 4(-1)^{n+1} + 4(-1)^n + (-1)^n = (-1)^n = (-1)^{n+2}.$$

The previous equality establishes the first identity of (0.2) for the index $n + 1$. Similarly we have that

$$2a_n a_{n+1} - b_n b_{n+1} = 2a_n(2a_n + a_{n-1}) - b_n(2b_n + b_{n-1})$$
$$= 2\left(2a_n^2 - b_n^2\right) + (2a_{n-1}a_n - b_{n-1}b_n)$$
$$= 2(-1)^{n+1} + (-1)^n = (-1)^{n+1},$$

which establishes our claim. In particular, this little argument reveals that for all $n \geq 1, \left|\mathcal{P}_n(\sqrt{2})\right| = 1$.

In view of our assumption that $\sqrt{2} = \frac{r}{s}$, we have another expression for $\left| \mathscr{P}_n(\sqrt{2}) \right|$, namely

$$1 = \left| \mathscr{P}_n(\sqrt{2}) \right| = \left| \left(a_n\sqrt{2} - b_n \right) \left(a_n\sqrt{2} + b_n \right) \right|$$

$$= \left| \left(a_n\frac{r}{s} - b_n \right) \left(a_n\frac{r}{s} + b_n \right) \right|$$

$$= |a_nr - b_ns| \left(\frac{a_nr + b_ns}{s^2} \right).$$

Clearly $a_nr - b_ns$ is an integer and we have just established that it is nonzero and satisfies

$$0 < |a_nr - b_ns| = \frac{s^2}{a_nr + b_ns}.$$

By (0.1) we see that the sequences $\{a_n\}$ and $\{b_n\}$ are strictly increasing and thus it follows that the sequence $\{a_nr + b_ns\}$ is strictly increasing as well. Therefore there exists an index $\tilde{n}$ satisfying $s^2 < a_{\tilde{n}}r + b_{\tilde{n}}s$. Our auxiliary polynomial of interest is $\mathscr{P}_{\tilde{n}}(z)$. Using it we have constructed an integer $a_{\tilde{n}}r - b_{\tilde{n}}s$ satisfying

$$0 < |a_{\tilde{n}}r - b_{\tilde{n}}s| < 1,$$

which plainly is impossible since there are no integers between 0 and 1. Thus $\sqrt{2}$ is irrational.   ∎

A natural, and healthy, inclination is to now expand the concept of number to include $\sqrt{2}$ and all other solutions to analogous algebraic equations. Moreover, as there is no reason to confine our attention to only quadratic equations, we now boldly extend our notion of number further still and accept as "numbers" all solutions to algebraic equations having rational coeffficients. This collection is known as the set of *algebraic numbers* and is denoted by $\overline{\mathbb{Q}}$. While this sweeping declaration allows us to solve all algebraic equations, there are still gaps. To uncover these holes and fill them in, we now turn our attention to analytical issues.

0.4 A search for missing numbers

In our failed attempt to find a rational solution to the equation $x \times x = 2$, we may have observed that some rationals are very *nearly* a solution. For example, the rational numbers

$$\frac{3}{2}, \frac{7}{5}, \frac{17}{12}, \frac{41}{29}, \frac{99}{70}, \cdots$$

appear to be approaching an exact solution. In particular, we note that

$$\left(\frac{99}{70} \right)^2 - 2 = \frac{1}{4900},$$

which is surprisingly small. As an aside, we remark that these mysterious rational numbers are simply the b_n/a_n's from the proof of Theorem 0.1. By calculating b_n/a_n for an ever-growing list of n's, we obtain increasingly better rational approximations to $\sqrt{2}$. This observation hints at a method for expanding our notion of number to include $\sqrt{2}$ without referring to a solution to an algebraic equation: A "number" is a quantity that can be approximated arbitrarily well by rational numbers.

While what is perhaps the most unsettling feature about the above notion of number is its utter vagueness, we can make it precise by importing a fundamental idea from calculus: A "number" is any quantity that is represented as a limit of a converging sequence of rational numbers. Alas this definition may appear to be a bit circular since we might believe that we can only determine that a sequence of rational numbers converges if we already know its limit. Fortunately, elementary analysis comes to our rescue: The Cauchy Criterion for the convergence of a sequence allows us to confirm that a sequence converges without specifying its limit. Thus we now adopt as an operative definiton of number: A "number" is that which is the limit of a Cauchy sequence of rational numbers. In some sense we widen our view of number to include not only the rational numbers but those for which rational numbers approach, such as $\sqrt{2}$. These newly induced "numbers" are known as *irrational numbers* and the totality of all such numbers comprise the set of *real numbers*, denoted as $\mathbb{R}$.

0.5 Is *i* a number?

We now have arrived at two concepts of number, those elements that are solutions to polynomial equations having rational coefficients and those elements that are limits of Cauchy sequences of rational numbers. While both notions of number appear to be reasonable, the next theorem asserts that they are not the same.

THEOREM 0.2 *The number $i = \sqrt{-1}$ is not the limit of a sequence of rational numbers.*

Challenge 0.1 *Provide a proof of Theorem 0.2.*

Faced with the realization that $\overline{\mathbb{Q}}$ is not equal to $\mathbb{R}$, we develop an even more expansive sense of number by considering objects of the form $r \times \alpha$, where $r \in \mathbb{R}$ and $\alpha \in \overline{\mathbb{Q}}$. Since 1 is both a real and an algebraic number, we immediately see that this new collection contains all the algebraic and all the real numbers. However, to ensure that we can "do the math," that is, perform arithmetic, we construct an even larger set from this vast one. Specifically we consider our set of "numbers" to be that set obtained by taking all possible sums, products and quotients of terms of the form $r \times \alpha$. Thus we have necessarily constructed a collection within which we can solve all basic addition and multiplication problems.

Unless we uncover some underlying and unifying scheme, the set we just constructed will be an unwieldly and perhaps unmanageable notion of number; after all it includes wild expressions such as

$$\left(e^6 - \frac{16}{19} + \pi^{29} + \sqrt[4]{-1209}\right) \Big/ \left(\sqrt{-\frac{8}{9}} + \sqrt[5]{i - \sqrt{403}}\right).$$

In the nineteenth century, Carl Friedrich Gauss proved a remarkable theorem, one consequence of which is that every element in the above collection may be expressed in an incredibly simple form, specifically as $r_1 + r_2 i$, where r_1 and r_2 are real numbers. In view of our first description of this set it is perhaps properly known as the set of *complex numbers*, denoted as $\mathbb{C}$.

In fact, Gauss actually showed that not only does $\mathbb{C}$ contain both the real and algebraic numbers, but $\mathbb{C}$ is sufficient for solving *all* algebraic equations, not simply equations with rational coefficients. That is, any polynomial equation with complex coefficients has all of its solutions in $\mathbb{C}$. Thus we say that the field of complex numbers is *algebraically closed*. Moreover, is it a straightforward exercise to verify that, just as the limit of a Cauchy sequence of real numbers is a real number, the limit of a Cauchy sequence of complex numbers is a complex number. Thus we say the field of complex numbers is *complete*. At long-last we have arrived at a reasonable notion of *number*.

0.6 Transcending the algebraic numbers

By our construction of the set of complex numbers we see that every algebraic number is a complex number. We now face a natural question: Does the converse of the previous sentence hold? That is, is every complex number an algebraic number? This question is where our story begins. While the mathematical community believed the answer was "no," this question remained opened until 1844 when Joseph Liouville proved a simple result that allowed him to explicitly show that a specific class of real numbers are demonstrable not algebraic. A real or complex number that is not algebraic is said to be a *transcendental number.*

In the second half of the nineteenth century the naturally-occuring numbers e and π were also shown to be transcendental. Establishing the transcendence of these values required herculian efforts, which might seem paradoxical given another important nineteenth century theorem: Almost all complex numbers are transcendental. This result was established by Georg Cantor in 1872 and implies that if we select a complex number (or even a real number) at random, then with probability 1, we have chosen a transcendental number. We suddenly see that the more familiar set of algebraic numbers is essentially invisible.

Even though the past century was one of spectacular growth in the subtlety and power of mathematical ideas and techniques, it is still the case that establishing the transcendence of a particular number is a rare achievement. In fact, our knowledge of transcendental numbers is extraordinarily limited, and thus these commonplace numbers remain safely shrouded in a veil of mystery. We now will attempt to lift a piece of that veil and develop an intuitive understanding of the transcendental.

Postscript. **Tools of the transcendence trade.** The arguments we explore in this book require several ideas from algebra, some results from complex analysis, and always in the background the fundamental properties of power series. While we develop much of the necessary background as we require it along the way, here we offer a few highlights of some basic notions and notation.

Some basic algebraic notions. Given that a transcendental number is one that is not algebraic, in order to understand transcendental numbers we must also explore the algebraic ones. Thus we will find ourselves studying polynomials in great depth. Toward that end we recall that given a set of numbers S, we define the collection $S[z]$ to be

$$S[z] = \left\{ s_N z^N + s_{N-1} z^{N-1} + \cdots + s_1 z + s_0 : s_n \in S \right\},$$

that is, $S[z]$ is the set of all polynomials having coefficients from the set S. For a polynomial $p(z) = \sum_{n=0}^{N} s_n z^n \in S[z]$, with $s_N \neq 0$, we define its *degree*, $\deg(p)$, to be N, and call s_N its *leading coefficient*. If $s_N = 1$, then we say $p(z)$ is a *monic polynomial*. The polynomial having all its coefficients equal to 0 is called the *zero polynomial*. We define the *degree* of the zero polynomial to be $-\infty$.

We now offer a more precise definition of an algebraic number. A number $\alpha \in \mathbb{C}$ is called *algebraic* if it is the zero of some nonzero polynomial $p(z) \in \mathbb{Z}[z]$. A polynomial $p(z) \in \mathbb{Z}[z]$ is called *irreducible* if it cannot be factored into two polynomials in $\mathbb{Q}[z]$ each having smaller degree than $\deg(p)$. If α is algebraic, then there exists a unique irreducible polynomial $p(z) \in \mathbb{Z}[z]$ with the properties that α is a zero of $p(z)$; the leading coefficient of $p(z)$ is positive; and the coefficients of $p(z)$ are relatively prime integers. We call this polynomial $p(z)$ the *minimal polynomial associated with* α. We define the *degree of* α, $\deg(\alpha)$, to be the degree of its minimal polynomial. The zeros of the minimal polynomial for α are called the *conjugates of* α.

Some basic complex notions. Given a complex number $z = a + bi$, we call a the *real part of z*, denoted by $\mathrm{Re}(z)$, and view b as the *imaginary part of z*, written as $\mathrm{Im}(z)$. If we consider z as the the point (a, b) in the Cartesian plane, then it is intuitively plain that we should define the absolute value (also known as the *modulus*) of z to be $|z| = \sqrt{a^2 + b^2}$.

Either prominently in the foreground or lurking in the background will be the ever-present power series $e^z = \sum_{n=0}^{\infty} z^n/n!$. Of course we are able to view the exponential function as a function of a complex variable by first observing that formally, for real θ,

$$e^{i\theta} = \sum_{n=0}^{\infty} \frac{(i\theta)^n}{n!} = \sum_{n=0}^{\infty} (-1)^n \frac{\theta^{2n}}{(2n)!} + i \sum_{n=0}^{\infty} (-1)^n \frac{\theta^{2n+1}}{(2n+1)!} = \cos\theta + i\sin\theta.$$

$$(0.3)$$

Thus for real r and θ, the complex number $re^{i\theta}$ can be viewed as the point in the Cartesian plane given by the polar coordinates (r, θ). It also follows that

$\left|re^{i\theta}\right| = |r|$ and $e^{a+bi} = e^a e^{bi}$. Furthermore, if we let $\theta = 2\pi$ in (0.3), then we see that $e^{2\pi i} = \cos(2\pi) + i\sin(2\pi) = 1$, and thus are immediately led to the famous and incredibly important identity $e^{2\pi i} - 1 = 0$. In the *Appendix* we include some additional highlights from complex analysis.

With these modest preliminaries behind us, we are ready to embark upon our transcendental adventure.

Number 1

0.11000100000000000000000010000 ...

Incredible Numbers Incredibly Close to Modest Rationals:
Liouville's Theorem and the transcendence of $\sum_{n=1}^{\infty} 10^{-n!}$

1.1 Turning a mountain of transcendence into a molehill

As we begin our journey into the theory of transcendental numbers, we are immediately faced with a nearly insurmountable obstacle: A transcendental number is defined not by what it *is* but rather by what it is *not*. What will become apparent as we develop the classical theory of transcendental numbers is that every demonstration of the transcendence of a particular number is indirect—a number is shown to be transcendental by showing that it is not algebraic.

Since there are infinitely many polynomials with integer coefficients, it is not possible to verify that a number α is not algebraic by considering each polynomial individually and verifying that it does not vanish at α. Instead, we assume there exists a nonzero polynomial $P(x)$, with integer coefficients, satisfying $P(\alpha) = 0$, and then hope for the worst, specifically, that this assumption violates some more basic principle of mathematics. Fortunately, as we will discover, this strategy for verifying the transcendence of certain special numbers is a fruitful one. Unfortunately, as we will also discover, this approach involves extremely technical and intricate mathematical arguments. One of the central goals of this book is to make those elaborate ideas understandable and ideally even intuitive.

While the theory of transcendence is complex, the overarching principles are simple. In fact, a key ingredient used throughout transcendental number theory is the following critical observation that we already employed in the proof of the irrationality of $\sqrt{2}$ in Chapter 0.

THE FUNDAMENTAL PRINCIPLE OF NUMBER THEORY. *There are no integers between 0 and 1.*

Given this principle, we now offer a ridiculously general outline for the typical proof in transcendental number theory.

How to prove that a number α is transcendental in two steps

1. **Assume α is algebraic.**   In particular, assume there exists a nonzero polynomial $P(z)$ with integer coefficients such that $P(\alpha) = 0$.
2. **Produce an integer violating the Fundamental Principle of Number Theory.** Using the integer coefficients from the proposed polynomial in Step 1, and some distinguishing features about α, produce an integer $\mathcal{N}$ between 0 and 1.

All of the interesting mathematics in this outline is hidden in Step 2, so the outline is not especially illuminating. An elaboration of this approach better conveys the structure of the typical transcendence proof.

How to prove that a number α is transcendental in six steps

1. **Assume α is algebraic.**   In particular, assume there exists a nonzero polynomial $P(z)$ with integer coefficients such that $P(\alpha) = 0$.
2. **Build an integer.**   Using the integer coefficients from the proposed polynomial in Step 1, and some additional features of α, produce an integer $\mathcal{N}$.
   (This step usually involves some technical results from algebra.)
3. **Show that $\mathcal{N}$ is not zero.**   Prove that the integer $\mathcal{N}$ satisfies $0 < |\mathcal{N}|$.
   (Typically this step is the most difficult.)
4. **Give an upper bound for $\mathcal{N}$.**   Prove that the integer $\mathcal{N}$ satisfies $|\mathcal{N}| < 1$.
   (This step usually requires some clever insights from analysis to estimate complicated infinite series or exotic integrals.)
5. **Apply the Fundamental Principle of Number Theory.**   We have produced an integer $|\mathcal{N}|$ between 0 and 1, so we are faced with a mathematical dilemma. If we have not taken any missteps in our previous steps, then the only possible problem is our assumption in Step 1 that α is algebraic.
6. **Big finish.**   With a great sense of accomplishment, we conclude that α is transcendental.

Providing the technical details for Steps 2, 3, and 4—especially Step 3—is at the very heart of transcendental number theory.

1.2 One small rational for Liouville, one giant step for transcendence

The mere existence of transcendental numbers was not established until 1844, when Joseph Liouville explicitly constructed a number and demonstrated that it is transcendental. His ingenious idea was to establish a simple property satisfied by all algebraic numbers. Thus any number that failed to satisfy this property would necessarily be transcendental.

The intuitive idea of Liouville's method

Liouville made the elegant discovery that an algebraic number cannot be approximated particularly well by rational numbers having denominators of relatively

Continued

modest size. Specifically, if α is an algebraic number and p/q is a rational number *extremely* close α, then q must be *extremely* large relative to the closeness of p/q to α. It follows that if a number $\mathcal{L}$ *can* be approximated extremely well by an infinite sequence of rationals, each having a relatively small denominator, then $\mathcal{L}$ cannot be algebraic and thus must be a transcendental number.

THEOREM 1.1 (LIOUVILLE'S THEOREM)   *Let α be an irrational algebraic number of degree d. Then there exists a positive constant depending only on α, $c = c(\alpha)$, such that for every rational number p/q, the inequality*

$$\frac{c}{q^d} \le \left| \alpha - \frac{p}{q} \right| \tag{1.1}$$

is satisfied.

We first verify that Liouville's Theorem does, in fact, capture the spirit we proposed. Suppose that α is an irrational algebraic number of degree d, and that for some $\varepsilon > 0$ there is a rational number p/q satisfying the inequality

$$\left| \alpha - \frac{p}{q} \right| < \varepsilon.$$

Then by inequality (1.1), we have that

$$\frac{c}{q^d} < \varepsilon, \quad \text{or equivalently,} \quad \left(\frac{c}{\varepsilon} \right)^{1/d} < q.$$

Hence if ε is *incredibly small*, then q must be *impressively large*.

Thus, in view of Liouville's Theorem, if we find an α that has an infinite number of amazingly good rational approximations, approximations so good that they violate (1.1) for any choice of d and c, then α must be transcendental.

1.3 Our first transcendental number

Before proving Liouville's Theorem, we illustrate its importance by considering the very first number shown to be transcendental.

COROLLARY 1.2   *The number*

$$\mathcal{L} = \sum_{n=1}^{\infty} 10^{-n!} = 0.110001000000000000000000010000\ldots$$

is transcendental.

Proof.   As always, we begin by assuming that $\mathcal{L}$ is algebraic of degree d. To apply Theorem 1.1, we must first establish that $\mathcal{L}$ is irrational, that is, that $d > 1$. Since the length of the runs of consecutive zeros in the decimal expansion of $\mathcal{L}$ grows without bound, the decimal expansion cannot be eventually periodic. Therefore $\mathcal{L}$ cannot be rational. Thus, assuming that $\mathcal{L}$ is algebraic, it must be of degree at least 2.

We have just established that $\mathcal{L}$ must have degree $d \geq 2$, and so we may apply Liouville's Theorem to conclude that there exists a constant $c > 0$ satisfying the inequality

$$\frac{c}{q^d} \leq \left| \mathcal{L} - \frac{p}{q} \right|, \tag{1.2}$$

for all rational numbers p/q.

We now construct a sequence of amazingly good rational approximations to $\mathcal{L}$ that contradict inequality (1.2). We obtain these approximations by truncating the decimal expansion for $\mathcal{L}$ immediately before each long run of zeros. For a positive integer N, let $r_N = \sum_{n=1}^{N} 10^{-n!}$. Since r_N has a terminating decimal expansion, r_N is a rational number, p_N/q_N, where the integers p_N and q_N are defined by

$$p_N = 10^{N!} \sum_{n=1}^{N} 10^{-n!} \quad \text{and} \quad q_N = 10^{N!}$$

Thus we have

$$\left| \mathcal{L} - \frac{p_N}{q_N} \right| = \sum_{n=1}^{\infty} 10^{-n!} - \sum_{n=1}^{N} 10^{-n!} = \sum_{n=N+1}^{\infty} 10^{-n!}. \tag{1.3}$$

We note that

$$\sum_{n=N+1}^{\infty} 10^{-n!} = 10^{-(N+1)!} + 10^{-(N+2)!} + 10^{-(N+3)!} + \cdots,$$

and so we see that the exponents of 10^{-1} in the infinite series are $(N+1)!$, $(N+2)!$, $(N+3)!, \ldots$, namely consecutive factorials starting with $(N+1)!$. Certainly we could increase the sum in (1.3) by adding additional terms to the series. In particular, we could insert all terms of the form 10^{-m} where m is an integer greater than $(N+1)!$ and not a factorial. That is, we could include *all* integer powers of 10^{-1} starting with the power $(N+1)!$ rather than only those powers that are factorials.

Therefore we can bound the series from above by a geometric series that we can easily sum:

$$\sum_{n=N+1}^{\infty} 10^{-n!} < \sum_{n=(N+1)!}^{\infty} 10^{-n} = \frac{10^{-(N+1)!}}{1 - 10^{-1}} = \frac{10}{9} 10^{-(N+1)!}.$$

This inequality together with (1.3) yields

$$\left| \mathcal{L} - \frac{p_N}{q_N} \right| < \frac{10}{9} 10^{-(N+1)!},$$

which, after applying inequality (1.2) and recalling that $q_N = 10^{N!}$, reveals that

$$c10^{-dN!} < \frac{10}{9} 10^{-(N+1)!}.$$

Thus for all N we have

$$0 < \frac{9}{10} c < 10^{dN! - (N+1)!},$$

or equivalently,

$$0 < 9 < \left(\frac{10}{c} \right) 10^{dN! - (N+1)!}. \tag{1.4}$$

However, for $N \geq d$, the exponent $dN! - (N+1)!$ is negative, and thus as N approaches infinity, the right-hand side of inequality (1.4) gets arbitrarily close to zero. Thus for sufficiently large N, inequality (1.4) reveals that $0 < 9 < 1$, which contradicts the Fundamental Principle of Number Theory (we are assuming, without proof, that 9 is a nonzero integer). Hence our assumption that $\mathcal{L}$ is algebraic is false. Therefore $\mathcal{L}$ is a transcendental number. ∎

Some readers might view Liouville's transcendental number as one we might not accidentally stumble over on the real line, since its decimal expansion contains those incredible, ever-growing, runs of zeros followed by the lone digit 1. While this observation is certainly valid, Liouville's number $\mathcal{L}$ does establish the existence of transcendental numbers. The observation also brings up an interesting point, namely, while Cantor, in 1872, proved that almost all numbers are transcendental—if we pick a real number at random, the probability that it is transcendental is 1—it is nearly impossible to show that a particular "random" real number is transcendental.

The numbers that we are able to prove transcendental are *special* numbers that exhibit *special* properties. In the case of the number 0.1100010000000000000000000010 000 . . . , for example, its decimal expansion is special in that it easily gives rise to incredible rational approximations that allow us to establish that the number must be transcendental. So while essentially all numbers are transcendental, we are able to prove the transcendence only of very elite ones.

1.4 The proof of Liouville's Theorem

Before we turn our attention to the proof of Theorem 1.1, we recall some basic results from algebra.

Algebraic Excursion: Irreducible polynomials evaluated at rational values

We write $\mathbb{Q}[z]$ for the collection of all polynomials having *rational* coefficients. That is,

$$\mathbb{Q}[z] = \{a_d z^d + a_{d-1} z^{d-1} + \cdots + a_1 z + a_0 : d \geq 0 \quad \text{and} \quad a_i \in \mathbb{Q}$$
$$\text{for all } i = 1, 2, \ldots, d\}.$$

In fact, $\mathbb{Q}[z]$ forms a ring under the usual polynomial addition and multiplication and shares many important features with the ring of integers $\mathbb{Z}$. In particular, the *division algorithm* holds, that is, we can perform long division with polynomials to produce quotient and remainder polynomials. We state this result as follows.

THE DIVISION ALGORITHM OVER $\mathbb{Q}[z]$. *Let $f(z)$ and $g(z)$ be two polynomials in $\mathbb{Q}[z]$ with $g(z)$ not the zero polynomial. Then there exist polynomials $q(z)$ and $r(z)$ in $\mathbb{Q}[z]$ satisfying*

$$f(z) = g(z)q(z) + r(z) \quad \text{with} \quad \deg(r) < \deg(g).$$

A polynomial $f(z) \in \mathbb{Q}[z]$ is said to be *reducible* or *reducible over* $\mathbb{Q}$ if it factors in a nontrivial way, that is, if there exist polynomials $p_1(z)$ and $p_2(z) \in \mathbb{Q}[z]$, each of degree less than the degree of $f(z)$, for which $f(z) = p_1(z)p_2(z)$. So, for example, $z^2 - z - 6$ is reducible, since $z^2 - z - 6 = (z - 3)(z + 2)$, while $z^2 + 1$ is not reducible (over $\mathbb{Q}$). A polynomial that is not reducible is said to be *irreducible*.

Challenge 1.1 Use the division algorithm over $\mathbb{Q}[z]$ to prove that p/q is a rational number satisfying $f(p/q) = 0$ if and only if $f(z)$ can be factored as $f(z) = (qz - p)g(z)$ for some polynomial $g(z) \in \mathbb{Q}[z]$. Conclude that if $f(z) \in \mathbb{Q}[z]$ is an irreducible polynomial of degree greater than 1, then for all rational numbers $p/q, f(p/q) \neq 0$.

As we now turn our attention to the proof of Liouville's result, we are faced with a basic question: What information do we have to even *attempt* a proof of Liouville's Theorem? All we have at our disposal is the algebraic number α. But that algebraic number comes equipped with its associated minimal polynomial. Happily, a polynomial, when viewed as a function, is differentiable, and thus we can utilize the power of basic calculus. As we will see, the minimal polynomial together with some very elementary inequalities and a clever application of the Mean Value Theorem will produce the desired result, which we restate here.

LIOUVILLE'S THEOREM   *Let α be an irrational algebraic number of degree d. Then there exists a positive constant depending only on α, $c = c(\alpha)$, such that for every* *rational number p/q, the inequality*

$$\frac{c}{q^d} \le \left| \alpha - \frac{p}{q} \right|$$

(1.5)

is satisfied.

Proof of Liouville's Theorem.   We first consider the case in which α is a real number. Our objective is to demonstrate the existence of a constant c that depends only on α for which inequality (1.5) is satisfied for *all* rational numbers p/q. To accomplish this task we let p/q be an arbitrary rational number, where we always understand that $q > 0$. We seek a lower bound for

$$\left| \alpha - \frac{p}{q} \right|$$

in terms of q and α. To establish this lower bound, we examine two cases separately: The case in which p/q is not very close to α and the case in which it is.

The Uninteresting Case.   Suppose that $1 < \left| \alpha - \frac{p}{q} \right|$. In this case, the rational p/q is relatively far away from α. In particular, since $q \ge 1$, we have that

$$\frac{1}{q^d} \le 1 < \left| \alpha - \frac{p}{q} \right|,$$

and therefore in this case inequality (1.5) will hold for *any* constant c satisfying $c \le 1$. For example, we could take $c = 1$. We are now left with the task of producing a constant c when p/q is actually close to α.

The Interesting Case.   Suppose that $\left| \alpha - \frac{p}{q} \right| \le 1$. Thus we are assuming that p/q is close to α, and we wish to show that it cannot be *too* close. This scenario is clearly at the heart of the matter.

Let $f(x)$ be the minimal polynomial for α; that is, $f(x)$ has integer coefficients, is irreducible, and $f(\alpha) = 0$. Moreover, since α has degree d, we know that $f(x)$ may be expressed as

$$f(x) = a_d x^d + a_{d-1} x^{d-1} + \cdots + a_1 x + a_0,$$

where $a_d, a_{d-1}, \ldots, a_0 \in \mathbb{Z}$ and $a_d > 0$.

We now attempt to produce an inequality whose lower bound is in sympathy with that of inequality (1.5). If we consider the only quantity we have at our disposal, namely, $f(p/q)$, then

$$f\left(\frac{p}{q}\right) = a_d \frac{p^d}{q^d} + a_{d-1} \frac{p^{d-1}}{q^{d-1}} + \cdots + a_1 \frac{p}{q} + a_0,$$

which can be expressed with a common denominator as

$$f\left(\frac{p}{q}\right) = \frac{a_d p^d + a_{d-1} p^{d-1} q + \cdots + a_1 p q^{d-1} + a_0 q^d}{q^d}.$$

Certainly, the numerator looks extremely complicated, and for a good reason—it is. In fact, the only immediate observation we can make is that the numerator is an integer. A slightly more subtle observation is that, because α is irrational, the numerator is a *nonzero* integer. This observation is based on the result from Challenge 1.1: Since $f(x)$ is irreducible of degree greater than 1, we know that $f(p/q) \neq 0$. If we let N denote the numerator, then N is a *nonzero* integer.

We now make the previous expression for $f(p/q)$ look less complicated, but perhaps more mysterious, by writing it as

$$f\left(\frac{p}{q}\right) = \frac{N}{q^d}.$$

Since q is a positive integer, upon taking absolute values, the above identity reveals that

$$\frac{|N|}{q^d} = \left| f\left(\frac{p}{q}\right) \right|. \tag{1.6}$$

We now employ the Fundamental Principle of Number Theory and conclude that since $N \neq 0$, $1 \leq |N|$. This observation allows us to weaken equality (1.6) by replacing the mysterious integer $|N|$ by the more familiar 1 and thus obtaining

$$\frac{1}{q^d} \leq \left| f\left(\frac{p}{q}\right) \right|. \tag{1.7}$$

Believe it or not, we have made progress. We have produced an inequality whose lower bound is similar to that of the inequality we seek. Unfortunately, the right-hand side of (1.7) does not contain the desired expression $\left| \alpha - \frac{p}{q} \right|$. How do we bring that difference $\alpha - \frac{p}{q}$ into the picture? One way to make a difference is to subtract zero.

Since $f(\alpha) = 0$, we can trivially rewrite inequality (1.7) as

$$\frac{1}{q^d} \leq \left| f(\alpha) - f\left(\frac{p}{q}\right) \right|, \tag{1.8}$$

which slightly resembles our sought-after inequality (1.5). Sadly, we see the polynomial f still appearing in (1.8); so we need to relate the quantity $f(\alpha) - f(p/q)$ and the quantity $\alpha - \frac{p}{q}$. Mercifully, we are saved by the Mean Value Theorem, which we can employ, since we are assuming that $\alpha \in \mathbb{R}$.

If we view $y = f(x)$ as a differentiable function, then the Mean Value Theorem allows us to compare those two differences. Specifically, given α and p/q, the Mean Value Theorem asserts that there must exist a real number φ between α and p/q such that

$$f(\alpha) - f\left(\frac{p}{q}\right) = f'(\varphi)\left(\alpha - \frac{p}{q}\right). \tag{1.9}$$

From this identity we can rewrite (1.8) as

$$\frac{1}{q^d} \leq |f'(\varphi)| \left| \alpha - \frac{p}{q} \right|. \tag{1.10}$$

Inequality (1.10) may appear to be perfect. In fact, we may be tempted to take the constant c in (1.5) to be the reciprocal of $|f'(\varphi)|$ (we note that $f'(\varphi) \neq 0$ follows from identity (1.9) by recalling that $f(\alpha) = 0$ and $f(p/q) \neq 0$). However, we must avoid this natural temptation, since we require the constant c to depend only upon α, and unfortunately, the real number φ depends on *both* α and the rational p/q.

How do we avoid this small but serious technical difficulty? We exploit the fact that we are considering the interesting case $\left| \alpha - \frac{p}{q} \right| \leq 1$. Therefore, since φ is between α and p/q, we have that $|\alpha - \varphi| \leq 1$. That is, $\alpha - 1 \leq \varphi \leq \alpha + 1$. Hence we can produce an upper bound for $|f'(\varphi)|$ that depends only on α by observing that

$$|f'(\varphi)| \leq \max\{|f'(\theta)| : \theta \in [\alpha - 1, \alpha + 1]\}.$$

Since $f'(x)$ is a continuous function, it will attain its maximum value on the closed interval $[\alpha - 1, \alpha + 1]$, and we denote this maximum value by M. The critical feature of M is that it depends only on α and not on the particular choice of p/q. Additionally, $M > 0$, since $f'(\varphi) \neq 0$. These observations together with inequality (1.10) yield

$$\frac{1}{q^d} \leq |f'(\varphi)| \left| \alpha - \frac{p}{q} \right| \leq M \left| \alpha - \frac{p}{q} \right|,$$

and thus we have

$$\frac{M^{-1}}{q^d} \leq \left| \alpha - \frac{p}{q} \right|.$$

Finally, combining the previous inequality with our remarks from the uninteresting case, we conclude that for *any* rational number p/q,

$$\frac{c}{q^d} \leq \left| \alpha - \frac{p}{q} \right|,$$

where the constant $c = c(\alpha)$ is defined by

$$c = \min\{1, M^{-1}\},$$

which completes the proof of Liouville's Theorem in the case when $\alpha \in \mathbb{R}$. ∎

Challenge 1.2 Complete the proof by establishing Liouville's Theorem in the case in which $\alpha \in \mathbb{C}$ is not real. (Hint: In this simple case, we can take the constant $c = |\mathrm{Im}(\alpha)|$.)

1.5 Liouville numbers

Liouville's Theorem concerns algebraic numbers, but its contrapositive provides us with an explicit transcendence result.

THEOREM 1.3 (LIOUVILLE'S THEOREM RESTATED) *Let α be a real number. Suppose that for each positive real number c and each positive integer d, there exists a rational number p/q satisfying the inequality*

$$\left| \alpha - \frac{p}{q} \right| < \frac{c}{q^d}.$$

Then α is transcendental.

Challenge 1.3 *Show that for a real algebraic number α, Theorem 1.3 is equivalent to Theorem 1.1.*

The statement of Theorem 1.3 is slightly awkward, since it involves *all* positive real numbers c and all positive integers d. We can avoid this awkwardness by restating Liouville's result yet again as:

THEOREM 1.4 (LIOUVILLE'S THEOREM RESTATED YET AGAIN) *Let α be a real number. Suppose there exists an infinite sequence of rational numbers p_n/q_n satisfying the inequality*

$$\left| \alpha - \frac{p_n}{q_n} \right| < \frac{1}{q_n^n}.$$

Then α is transcendental.

Challenge 1.4 *Prove that Theorems 1.3 and 1.4 are equivalent.*

In honor of Liouville, any real number that has amazingly good rational approximations is called a Liouville number. Specifically, we say that a number $\mathcal{L}$ is a *Liouville number* if there exists an infinite sequence of rational numbers p_n/q_n satisfying

$$\left| \mathcal{L} - \frac{p_n}{q_n} \right| < \frac{1}{q_n^n}. \tag{1.11}$$

We can now restate Theorem 1.4 as

THEOREM 1.5 *All Liouville numbers are transcendental.*

If we monkey with the number in Corollary 1.2 by considering any number of the form

$$a_1 10^{-1!} + a_2 10^{-2!} + a_3 10^{-3!} + \cdots, \tag{1.12}$$

where each coefficient a_k is either 0 or 1, and where infinitely many of the coefficients are 1's, then aping the proof of Corollary 1.2 allows us to swing to the conclusion that each of the numbers of the form given in (1.12) is a Liouville number.

Challenge 1.5 *Go bananas and verify that numbers of the form* (1.12) *are Liouville numbers.*

Using the previous observation, we could adopt a Cantor-esque diagonalization argument to conclude that there are uncountably many Liouville numbers. Despite their abundance, as we will discover in Chapter 6, the collection of all Liouville numbers has *measure zero*. That is, if we select a real number at random, the probability that the selected number is a Liouville number is 0. So, even though there are uncountably many Liouville numbers, they are difficult to hunt down.

Given the fact that the set of Liouville numbers is, in some sense, an invisibly thin set, it is perhaps surprising to discover the following curious fact.

THEOREM 1.6 *Every real number can be expressed as a sum of two Liouville numbers.*

The idea of the proof. Let α be a real number. If α is a rational number, then it can be expressed as a nonterminating decimal with infinitely many nonzero digits (if the decimal terminates, say as 0.452, then we rewrite it as $0.45199999\ldots$). Thus we see that whether α is rational or irrational, it has a decimal expansion with infinitely many nonzero digits.

We now write the decimal expansion described above for α as

$$\alpha = I.d_1 d_2 d_3 d_4 \ldots,$$

where I denotes an integer and d_n denotes the nth decimal digit. Let's now define two Liouville numbers $\mathcal{L}_1$ and $\mathcal{L}_2$ by

$$\mathcal{L}_1 = I.\ 0 \quad d_2 \ \ d_3 \ 0 \ \ 0 \ \ 0 \ \ 0 \ \ 0 \ \ 0 \ \ d_{10} \ \ d_{11} \ \cdots \ d_{33} \ 0 \quad 0 \quad \cdots,$$

$$\mathcal{L}_2 = 0.\ d_1 \ 0 \ \ 0 \ \ d_4 \ \ d_5 \ \ d_6 \ \ d_7 \ \ d_8 \ \ d_9 \ 0 \quad 0 \quad \cdots \ 0 \quad d_{34} \ \ d_{35} \ \cdots,$$

where the lengths of the runs of zeros are, alternately, at least $1!, 2!, 3!, 4!, \ldots$. We allow the runs to be longer, if necessary, to ensure that the digit d_l to be followed by the run of zeros is itself nonzero. If we do have to prolong the run by say 12 additional digits, then we also need to extend the next run and all subsequent runs of zeros from $n!$ to $(n + 12)!$.

Just as we observed while studying the number

$$\sum_{n=1}^{\infty} 10^{-n!} = 0.11000100000000000000000010000\ldots,$$

truncating each of the two decimal expansions before the blocks of zeros produces a sequence of rational approximations that satisfy the approximation condition necessary for $\mathcal{L}_1$ and $\mathcal{L}_2$ to be Liouville numbers. Plainly, $\alpha = \mathcal{L}_1 + \mathcal{L}_2$, which completes the idea of the proof. ∎

The lesson above is clear: Longer and longer runs of consecutive zeros in the decimal expansion for a real number lead to very good rational approximations to that real number. If the lengths of the blocks increase rapidly enough, we are able to use Liouville's Theorem to prove that the real number is transcendental. But transcendental numbers need not have incredibly long runs of zeros. In fact most—almost all—transcendental numbers have the proportion of zeros occurring in their decimal expansions equal to what we would expect on average—namely about 1 in 10.

1.6 The number 0.12345678910111213141516 ...

The numbers we have considered so far are transcendental because their decimal expansions have infinitely many runs of zeros whose lengths grow so quickly that simple truncation of the decimal expansion before each run of zeros leads to amazingly good rational approximations. However, for most numbers, a collection of best rational approximations is not so easily detected from their decimal expansions. Moreover, for most numbers α, even the best rational approximations are not close enough to allow us to conclude that α is transcendental.

As an illustration of the difficulty of finding suitable rational approximations in general, we consider *Mahler's number* (also known as *Champernowne's number*):

$$\mathcal{M} = 0.12345678910111213141516171819202 1 \ldots , \qquad (1.13)$$

where the decimal digits are formed by writing the consecutive natural numbers in juxtaposition. Kurt Mahler, in the 1930s, was the first to show that $\mathcal{M}$ is transcendental. The transcendence of $\mathcal{M}$ is a rather delicate issue. To illustrate the difficulty, we first try as before to use rational approximations created by long runs of zeros. That is, we truncate $\mathcal{M}$ just after the 1 that appears whenever we reach a power of 10.

If we attempt this truncation procedure, we will see that the number of decimal digits *before* each run of zeros far exceeds the length of that run. For example, to get a run of just one zero, we must truncate after 10 digits: 0.1234567891. To get a run of two zeros, we have to wade through an additional whopping run of 179 more digits: 1112131415 ... 9798991. In general, if we want to come across a run of k zeros, we have to travel on the order of $k10^k$ digits from the previous run of $k - 1$ zeros. This observation leads to the conclusion that the truncation method we used before will not generate rational approximations having relatively small denominators that are sufficiently close to $\mathcal{M}$ to contradict Liouville's Theorem.

Using a more clever construction to build rational approximations to $\mathcal{M}$, we can find approximations all having relatively small denominators that allow us to use Liouville's Theorem to prove the following partial result.

THEOREM 1.7   *The number* $\mathcal{M} = 0.1234567891011121314151617181920 21 \ldots$ *is either a transcendental number or an algebraic number of degree at least 5.*

So while this theorem does not guarantee that $\mathcal{M}$ is transcendental, it does imply that $\mathcal{M}$ is not a quadratic, cubic, or even a quartic algebraic number. We first build an intuitive understanding of why this result is true, and then discuss an improvement to Liouville's Theorem that ensures $\mathcal{M}$'s transcendence.

An intuitive sketch of the proof.   In order to apply Liouville's Theorem, we must establish that $\mathcal{M}$ is irrational, which follows from the observation that $\mathcal{M}$ contains arbitrarily long runs of consecutive zeros and therefore cannot have a periodic decimal expansion.

Our goal now is to construct a sequence of excellent rational approximations to $\mathcal{M}$ that will allow us to apply Liouville's result to conclude that *if* $\mathcal{M}$ is algebraic of degree d, then $d \geq 5$. In particular, it will be enough to show that for some $\delta > 4$, there exist infinitely many rational numbers p_n/q_n satisfying

$$\left| \mathcal{M} - \frac{p_n}{q_n} \right| < \frac{1}{q_n^{\delta}}.$$

Challenge 1.6   *Suppose that for some $\delta > 4$, there are infinitely many rational solutions to*

$$\left| \mathcal{M} - \frac{p}{q} \right| < \frac{1}{q^{\delta}}.$$

Using this supposition and Liouville's Theorem, show that $\mathcal{M}$ cannot be an algebraic number of degree less than 5.

Challenge 1.6 implies that to prove Theorem 1.7, it is enough to find an infinite sequence of rational numbers satisfying the previous inequality. To obtain good rational approximations to $\mathcal{M}$, we will not work directly with $\mathcal{M}$ but instead with the rational numbers

$$\mathcal{M}_1 = 0.123456789,$$

$$\mathcal{M}_2 = 0.10111213 \ldots 979899,$$

$$\mathcal{M}_3 = 0.100101102 \ldots 997998999,$$

$$\vdots$$

and so forth—all which come from strings of consecutive digits within the decimal expansion of $\mathcal{M}$. These rational numbers are good approximations to the fractional parts of $\mathcal{M}$, $10^9 \mathcal{M}$, $10^{189} \mathcal{M}$, and so forth, respectively. So upon dividing $\mathcal{M}_n$ by the appropriate power of 10 and adding the appropriate integer part, we are led to a good approximation to $\mathcal{M}$. Unfortunately, these approximations are not sufficiently close to $\mathcal{M}$ for us to apply Liouville's Theorem to establish Theorem 1.7. To better appreciate this phenomenon and discover how to remedy the difficulty, let us first carefully consider $\mathcal{M}_1$.

Since $\mathcal{M}$ and $\mathcal{M}_1$ agree to 9 decimal places, we know that

$$|\mathcal{M} - \mathcal{M}_1| < \frac{1}{10^9}.$$

If we simply express $\mathcal{M}_1$ as the fraction $123456789/10^9$ and denote it by p/q, then the previous inequality becomes

$$\left| \mathcal{M} - \frac{p}{q} \right| < \frac{1}{q},$$

and we see that the upper bound is not of the desired form $1/q^\delta$ for some $\delta > 4$. In other words, we see that $\mathcal{M}_1$'s denominator, $q = 10^9$, is too large given the relatively modest quality of approximation to $\mathcal{M}$. The exceptionally clever new idea to generate an improved rational approximation to $\mathcal{M}$ is to *first* approximate $\mathcal{M}_1$ by a rational number having a very small denominator that is *so* close to $\mathcal{M}_1$ that it provides an equally good approximation to $\mathcal{M}$ as $\mathcal{M}_1$ did, but with a much, much smaller denominator as compared with 10^9. To find this incredible approximation, we must look at $\mathcal{M}_1$ in a different way.

There are many ways to approximate $\mathcal{M}_1$ by rational numbers; for example, $\mathcal{M}_1$ agrees with the fraction

$$\frac{1}{9} + \frac{1}{90} + \cdots + \frac{1}{9 \times 10^8} = 0.123456789999999999\ldots$$

to nine decimal places, but its denominator 9×10^8 is only slightly smaller than the denominator 10^9 of $\mathcal{M}_1$. A better rational approximation to $\mathcal{M}_1$ can be obtained by modifying our first attempt and noticing that each decimal digit of $\mathcal{M}_1$ is obtained by adding 1 to the previous digit. Thus we see that

$$10\mathcal{M}_1 - \mathcal{M}_1 = 1.111111101,$$

which agrees with the decimal $1.111\ldots = 10/9$ for seven digits. Therefore $9\mathcal{M}_1$ is relatively well approximated by $10/9$. Specifically,

$$\left| 9\mathcal{M}_1 - \frac{10}{9} \right| < 0.000000010111\ldots.$$

Dividing through by 9 yields

$$\left| \mathcal{M}_1 - \frac{10}{81} \right| < \frac{0.000000010111\ldots}{9}.$$

Of course, $\mathcal{M}$ and $\mathcal{M}_1$ agree for their first ten digits, so the previous inequality continues to hold when $\mathcal{M}_1$ is replaced by $\mathcal{M}$. Hence we need only express the awkward upper bound of $(0.000000010111...)/9$ in a form that will allow us to compare it with the denominator of the rational approximation to $\mathcal{M}$, namely 81.

There are many ways to connect the upper bound with the denominator. The natural upper bound of

$$\frac{0.000000010111...}{9} < \frac{1}{9 \times 10^7},$$

together with the observation that

$$\frac{1}{9 \times 10^7} < \frac{1}{81^4},$$

leads to the inequality

$$\left| \mathcal{M} - \frac{10}{81} \right| < \frac{1}{81^4}.$$

If the exponent of 4 were to occur for all future rational approximations, then Liouville's Theorem would, at best, allow us to conclude only that if $\mathcal{M}$ were to be algebraic, then its degree would be at least 4. The stronger result that the degree of $\mathcal{M}$ must be at least 5 requires the slightly sharper inequality

$$\frac{0.000000010111...}{9} < \frac{11}{9 \times 10^9},$$

together with the easily verifiable inequality $\frac{11}{9 \times 10^9} < \frac{1}{81^{4.5}}$. If we write $p_1/q_1 = 10/81$, then we obtain

$$\left| \mathcal{M} - \frac{p_1}{q_1} \right| < \frac{1}{q_1^{4.5}}.$$

Finding the first good rational approximation to $\mathcal{M}$ required several clever ideas. However, armed with those ideas, finding the other approximations is somewhat more straightforward. Adopting our previous method in order now to approximate $\mathcal{M}_2$, we observe that

$$100\mathcal{M}_2 - \mathcal{M}_2 = 10.010101...010001,$$

which agrees with $991/99$ to 177 decimal places. This observation implies that $99\mathcal{M}_2$ is well approximated by $991/99$, and consequently $\mathcal{M}_2$ is equally well approximated by $991/99^2$.

But $991/99^2$, in turn, gives rise to a good rational approximation to $\mathcal{M}$, albeit one with a slightly larger denominator. To find this rational approximation, we simply notice that $10^9 \mathcal{M}$ and $123456789 + \mathcal{M}_2$ agree for the first 180 decimal digits. This observation, together with the excellent approximation to $\mathcal{M}_2$, implies that we have

$$\left| 10^9 \mathcal{M} - \left(123456789 + \frac{991}{99^2} \right) \right| < \frac{1}{10^{176}}.$$

Dividing through by 10^9 we obtain

$$\left| \mathcal{M} - \frac{p}{10^9 99^2} \right| < \frac{1}{10^{185}},$$

for some large integer p. Remarkably, we have an amazing approximation to $\mathcal{M}$ by a rational number with a considerably smaller denominator! In fact, if we write $p_2/q_2 = p/10^9 99^2$, then it turns out that

$$\left| \mathcal{M} - \frac{p_2}{q_2} \right| < \frac{1}{q_2^{13}} < \frac{1}{q_2^{4.5}}.$$

The strategy is now beginning to come into focus, and this pattern, in fact, turns out to continue. Specifically, by successively approximating $\mathcal{M}$ with rational numbers with denominators $9^2, 10^9 99^2, \ldots, 10^{n_k}(10^k - 1)^2$, where n_k denotes the number of digits occupied by the k-digit integers written consecutively, starting with 10^{k-1}, we can obtain infinitely many rational numbers p_n/q_n satisfying

$$\left| \mathcal{M} - \frac{p_n}{q_n} \right| < \frac{1}{q_n^{4.5}}, \tag{1.14}$$

which in view of Challenge 1.6 completes our sketch of the proof of Theorem 1.7. ∎

1.7 Roth's Theorem: The ultimate Liouville result

How can we establish the transcendence of Mahler's number $\mathcal{M}$? An obvious idea would be to try to improve Liouville's 1844 result and then use that stronger theorem together with inequality (1.14) to deduce that α cannot be algebraic. In fact, this is the strategy Mahler used for the specific number $\mathcal{M}$. In actuality, however, a much more general improvement can be found. Over the ensuing 100 years following Liouville's result, mathematicians worked on improving Liouville's inequality. Several mathematicians, including Axel Thue, Carl Siegel, and Freeman Dyson made important improvements to Liouville's Theorem, but in 1955 Klaus Roth provided the best possible improvement.

THEOREM 1.8 (ROTH'S THEOREM)   *Let α be an algebraic number of degree $d \geq 2$ and let $\varepsilon > 0$. Then there exists a constant $c(\alpha, \varepsilon) > 0$ such that for all p/q,*

$$\frac{c(\alpha, \varepsilon)}{q^{2+\varepsilon}} < \left| \alpha - \frac{p}{q} \right|.$$

Roth's result had such far-reaching ramifications and was so groundbreaking that in 1958 he was awarded the Fields Medal for his seminal work. The proof of Roth's Theorem is both ingenious and difficult, but including it here would take us too far afield from the study of classical transcendental number theory.

However, we can apply Roth's Theorem together with (1.14) to quickly conclude that $0.123456789101112\ldots$ is transcendental.

Suppose $\mathcal{M} = 0.123456789101112\ldots$ were algebraic of degree $d \geq 2$. Then letting $\varepsilon = 0.5$, we can apply Roth's Theorem and conclude that there exists a constant $c > 0$ such that for all p/q,

$$\frac{c}{q^{2.5}} < \left| \mathcal{M} - \frac{p}{q} \right|.$$

Applying this inequality with the rational numbers p_n/q_n constructed in our previous argument and recalling inequality (1.14), we see that for all n,

$$\frac{c}{q_n^{2.5}} < \left| \mathcal{M} - \frac{p_n}{q_n} \right| < \frac{1}{q_n^{4.5}},$$

which implies that for all n,

$$0 < c < q_n^{-2}. \tag{1.15}$$

But as $q_n \to \infty$, we see that inequality (1.15) cannot hold for all sufficiently large n. Thus $\mathcal{M}$ is transcendental.

1.8 Life after Liouville

As we have already remarked, most transcendental numbers are *not* Liouville numbers. In fact, Mahler's number, $\mathcal{M}$, is not Liouville. Are there any known Liouville numbers that we have seen in some other context? For example, is either of the ever-popular numbers e or π a Liouville number? In the case of e, it is possible to show that it satisfies the following Roth-type inequality: Given any $\varepsilon > 0$, there exists a constant $c(e, \varepsilon) > 0$ such that for all p/q,

$$\frac{c(e, \varepsilon)}{q^{2+\varepsilon}} < \left| e - \frac{p}{q} \right|. \tag{1.16}$$

The case of π is much more complicated and much less understood. For example, in 1993, Masayoshi Hata proved that given any $\varepsilon > 0$, there exists a constant $c = c(\pi, \varepsilon)$ such that for all p/q,

$$\frac{c(\pi, \varepsilon)}{q^{8.02+\varepsilon}} < \left| \pi - \frac{p}{q} \right|.$$

Challenge 1.7 *Use the two previous inequalities to show that neither e nor π is a Liouville number.*

What *can* be said about the numbers e and π? Are they transcendental? Since they are not Liouville numbers, new ideas are required in order to answer this fundamental question. As we suggested earlier in this chapter, the only numbers that we can prove are transcendental are those that are "special." So do e and π have some distinguishing features that might allow us to develop insights into why they are "special"? The answer in each case is, "Absolutely." The number e enjoys an extremely simple representation as an infinite series,

$$e = \sum_{n=0}^{\infty} \frac{1}{n!},$$

while π is connected to e through the famous identity $e^{\pi i} + 1 = 0$.

In the next two chapters, we develop more sophisticated methods that allow us to utilize these unique features of e and π in order to establish that they, in fact, are transcendental numbers. However, as we close this chapter we return to the question: Are there specific "naturally occurring" numbers that are known to be Liouville numbers? The answer is a resounding "No." Thus, while the well-understood structure of Liouville numbers provided us with an entrance into the enigmatic world of transcendental numbers, those numbers themselves remain veiled in a shroud of mystery.

Number 2

2.718281828459045235360287471 3 ...

The Powerful Power Series for e:
Polynomial vanishing and the transcendence of *e*

2.1 Fourier's proof of Euler's slick result

Here we will investigate several features of one of the most famous and important numbers in mathematics, namely, Leonhard Euler's "*e*." Our journey through this chapter sets the stage for much of what follows in our future explorations. To foreshadow the fundamental strategies to come, we open with Joseph Fourier's 1815 clever proof of Euler's result that *e* is irrational.

THEOREM 2.1 *The number e is irrational.*

The intuitive idea behind Fourier's approach

Fourier's strategy was to assume that *e* is rational, say $e = \frac{r}{s}$, and then use the alleged denominator s to construct another rational number t/u that is amazingly close to r/s. Thus the difference $\left|\frac{r}{s} - \frac{t}{u}\right|$ is a *positive* rational number. Fourier then showed that this positive number is incredibly small; in fact, if d is the least common multiple of s and u, then $\left|\frac{r}{s} - \frac{t}{u}\right| < \frac{1}{d}$. Thus clearing denominators by multiplying through by d, we discover that the awkward-looking *integer* $\left|d\frac{r}{s} - d\frac{t}{u}\right|$ satisfies

$$0 < \left|d\frac{r}{s} - d\frac{t}{u}\right| < 1,$$

which contradicts the Fundamental Principle of Number Theory. Hence *e* is irrational.

Proof. As we suggested at the end of the previous chapter, the most important property the number *e* possesses that allows us to classify it as "special" is its representation as an extremely simple infinite series,

$$e = \sum_{n=0}^{\infty} \frac{1}{n!}.$$

Let us now assume that e is rational, say $e = \frac{r}{s}$, where $s \geq 1$. Using s, we construct an excellent rational approximation to r/s. In particular, we consider the rational number formed by truncating the infinite series for e at $n = s$:

$$\sum_{n=0}^{s} \frac{1}{n!}.$$

It immediately follows that $\frac{r}{s} - \sum_{n=0}^{s} \frac{1}{n!}$ is positive. We can clear denominators and thus produce a positive integer by multiplying both sides by $s!$. In doing so, we see that

$$s!\left(\frac{r}{s} - \sum_{n=0}^{s}\frac{1}{n!}\right) = s!\left(e - \sum_{n=0}^{s}\frac{1}{n!}\right) = s!\left(\sum_{n=0}^{\infty}\frac{1}{n!} - \sum_{n=0}^{s}\frac{1}{n!}\right)$$

$$= s!\left(\frac{1}{(s+1)!} + \frac{1}{(s+2)!} + \frac{1}{(s+3)!} + \cdots\right)$$

$$= \frac{1}{s+1} + \frac{1}{(s+2)(s+1)} + \frac{1}{(s+3)(s+2)(s+1)} + \cdots$$

$$(2.1)$$

is a positive integer. However, since $s \geq 1$, we can bound the positive integer in (2.1) from above by a geometric series:

$$\frac{1}{s+1} + \frac{1}{(s+2)(s+1)} + \frac{1}{(s+3)(s+2)(s+1)} + \cdots < \frac{1}{2} + \frac{1}{2^2} + \frac{1}{2^3} + \cdots = 1.$$

Thus we have constructed an integer between 0 and 1, which is a direct violation of the Fundamental Principle of Number Theory. Thus we conclude that e is irrational. ∎

The key step in the previous argument was the construction of the integer in (2.1) by finding a spectacular rational approximation to the assumed-rational number e and then clearing all denominators by multiplying through by $s!$. In fact, this basic theme can be developed into a proof of the transcendence of e. In order to appreciate the subtleties involved in extending Fourier's basic idea, we first consider a proof of the irrationality of $e^{a/b}$ for any nonzero rational number a/b. While the fundamental strategy used in demonstrating that e is irrational will remain intact in the more general argument, producing a spectacular rational approximation will require a considerable amount of ingenuity. Once we have developed the ideas central to the proof of the irrationality of $e^{a/b}$, we will be well prepared to establish the transcendence of e.

THEOREM 2.2 *For any nonzero rational number a/b, the number $e^{a/b}$ is irrational.*

2.2 A first attempt at a proof

We begin by observing that establishing Theorem 2.2 is equivalent to proving that e^a is irrational for positive integers a.

Challenge 2.1 *Prove that if e^m is irrational for all integers $m \geq 1$, then for any nonzero rational number a/b, $e^{a/b}$ is irrational.*

Thus by the challenge, we need to prove that e^a is irrational only for positive integers a. The strategy of our argument is straightforward: We adopt the basic plan of attack used in the proof of the irrationality of e. Unfortunately, as we will quickly discover, the most obvious extension of those ideas fails to actually lead to a proof. However, pursuing that obvious, albeit ill-fated, attempt will illustrate the need for a more elaborate adaptation of the argument and also provide some insight into the subtle refinements to come.

We embark on our star-crossed attempt by viewing e^a as a value of the function e^z expressed as the well-known power series

$$e^z = \sum_{n=0}^{\infty} \frac{z^n}{n!}.$$

Let us suppose that e^a equals the rational number r/s. For any index N, we can approximate r/s by $\sum_{n=0}^{N-1} \frac{a^n}{n!}$ and thus see that their difference

$$\frac{r}{s} - \sum_{n=0}^{N-1} \frac{a^n}{n!} = \sum_{n=0}^{\infty} \frac{a^n}{n!} - \sum_{n=0}^{N-1} \frac{a^n}{n!} = \sum_{n=N}^{\infty} \frac{a^n}{n!}$$

is a positive *rational* number. We now wish to estimate this difference.

Challenge 2.2 *Make a change in variables in the indices of the series above to verify the identity*

$$\frac{r}{s} - \sum_{n=0}^{N-1} \frac{a^n}{n!} = \frac{a^N}{N!} \sum_{n=0}^{\infty} \frac{N!a^n}{(N+n)!}. \tag{2.2}$$

The reason for the inclusion of the seemingly superfluous factor $N!$ in both the numerator and denominator of the series in (2.2) is that it allows us to produce a simple and clean upper bound. In particular, since $\frac{(N+n)!}{N!n!}$ is the binomial coefficient $\binom{N+n}{n}$, it is a *positive integer*. Therefore, $1 \le \frac{(N+n)!}{N!n!}$, and hence we have

$$\frac{N!}{(N+n)!} \le \frac{1}{n!}. \tag{2.3}$$

In view of this inequality, identity (2.2), and the power series expansion for e^a, we conclude that

$$0 < \frac{r}{s} - \sum_{n=0}^{N-1} \frac{a^n}{n!} \le \frac{a^N}{N!} e^a.$$

Multiplying the previous inequality by $s(N-1)!$ clears all the denominators of $\frac{r}{s} - \sum_{n=0}^{N-1} \frac{a^n}{n!}$ and yields

$$0 < s(N-1)! \left(\frac{r}{s} - \sum_{n=0}^{N-1} \frac{a^n}{n!} \right) \le s e^a \left(\frac{a^N}{N} \right), \tag{2.4}$$

where the awkward-appearing quantity $s(N-1)! \left(\frac{r}{s} - \sum_{n=0}^{N-1} \frac{a^n}{n!} \right)$ is an *integer*. We recall that the index N is a free parameter. In the special case $a = 1$, we see that for all sufficiently large N, the upper bound in (2.4) is less than 1, and thus we contradict the Fundamental Principle of Number Theory. Hence this argument allows us to conclude—yet again—that e is irrational. In view of Theorem 2.1, however, this conclusion is nothing new.

In the more interesting case $a \ne 1$, if we could select an N such that the upper bound in (2.4) is less than 1, then we would again arrive at a contradiction and would have the irrationality of e^a in the palms of our hands. Unfortunately, by applying inequality (2.4), the irrationality slips through our fingers, since there is no value of N for which the upper bound in (2.4) is less than 1 for $a \ne 1$. Thus, as we cautioned at the opening, this approach fails to yield the desired result.

We now look ahead and foreshadow an improved version of the crucial inequality (2.4). Suppose that we could construct an integer $\mathcal{I}$ satisfying an inequality of the basic shape

$$0 < \mathcal{I} \le (\text{some constant}) \times \left(\frac{a^N}{(N-1)!} \right) \tag{2.5}$$

(note the appearance of the factorial in the denominator!). Then as N approaches infinity, the upper bound in (2.5) would approach 0, and hence for all sufficiently large choices of N, this new upper bound would indeed be less than 1, and we would have our much sought-after contradiction. This observation is a clue as to how to modify our failed attempt. We desire a rational approximation that is so close to the assumed-rational number e^a that their difference, after clearing denominators, gives rise to a positive integer less than 1. Basically, we require an *improved* rational approximation to e^a.

The intuitive idea for the refinement of the argument

In our first attempt, we obtained a rational approximation to e^a by truncating the power series after N terms to obtain a polynomial, and then evaluating that polynomial at a. That is, we wrote

$$e^z = \sum_{n=0}^{N-1} \frac{z^n}{n!} + \sum_{n=N}^{\infty} \frac{z^n}{n!}$$

and took the first sum to be the approximating polynomial. In particular, if we let $\mathcal{P}(z) = \sum_{n=0}^{N-1} \frac{z^n}{n!}$, then the natural rational approximation we considered was $\mathcal{P}(a)$.

Unfortunately, simply truncating the power series for e^z does not lead to a sufficiently good rational approximation. The fundamental problem with the truncation strategy is that it leads to a polynomial $\mathcal{P}(z)$ that approximates the function e^z reasonably well for *all* z. In fact, we only require an approximation at the particular value $z = a$; but that particular approximation should be an incredibly good one.

Our new point of attack is to find a polynomial that is an amazingly good approximation to e^z at the point $z = a$, but that is not necessarily any better than the previous truncation attempt for other values of z. The basic idea is to split the polynomial $\mathcal{P}(z)$ into two terms and write

$$e^z = \left(\sum_{n=0}^{N-p} \frac{z^n}{n!} + \sum_{n=N-p+1}^{N-1} \frac{z^n}{n!} \right) + \sum_{n=N}^{\infty} \frac{z^n}{n!}. \tag{2.6}$$

If the second polynomial term, $\sum_{n=N-p+1}^{N-1} \frac{z^n}{n!}$, in the previous expression were to vanish at $z = a$, then the first polynomial $\sum_{n=0}^{N-p} \frac{z^n}{n!}$ would give rise to an *amazing* rational approximation to e^z at $z = a$, which, in turn, would allow us to deduce an inequality of the form (2.5). Unfortunately, it is abundantly clear that the middle term $\sum_{n=N-p+1}^{N-1} \frac{z^n}{n!}$ will never vanish at $z = a$.

Since N is a free variable, we can decompose e^z into three terms as in (2.6) for different values of N, say for example, $N_1, N_2, \ldots, N_L$. In this case we would have L different "middle term" polynomials:

$$\sum_{n=N_1-p+1}^{N_1-1} \frac{z^n}{n!}, \sum_{n=N_2-p+1}^{N_2-1} \frac{z^n}{n!}, \ldots, \sum_{n=N_L-p+1}^{N_L-1} \frac{z^n}{n!}.$$

Of course, none of those polynomials vanish at $z = a$. However, perhaps we could string them all together as a linear combination so as to create a *new* polynomial that *would* vanish at our desired point.

Continued

To illustrate this possibility, let us consider the polynomials z^3, z^2, z, together with the constant polynomial 1. Certainly none of these vanish at $z = a$. However, if we consider the linear combination of these polynomials

$$f(z) = z^3 - (1+a)z^2 + (1+a)z - (a)1,$$

then we immediately see that

$$f(a) = a^3 - a^3 - a^2 + a^2 + a - a = 0,$$

and hence we have combined our original polynomials to construct a new polynomial that *does* vanish at $z = a$.

Inspired by the previous illustration, we wonder whether it is possible to find indices $N_1, N_2, \ldots, N_L$ and integer coefficients $k_{N_1}, k_{N_2}, \ldots, k_{N_L}$, not all zero, such that if we were to decompose e^z into three terms as in (2.6) for each $N_1, N_2, \ldots, N_L$ and consider the linear combination

$$\sum_{\ell=1}^{L} \left(k_{N_\ell} e^z \right) = \sum_{\ell=1}^{L} \left(k_{N_\ell} \sum_{n=0}^{N_\ell - p} \frac{z^n}{n!} \right) + \sum_{\ell=1}^{L} \left(k_{N_\ell} \sum_{n=N_\ell - p + 1}^{N_\ell - 1} \frac{z^n}{n!} \right)$$

$$+ \sum_{\ell=1}^{L} \left(k_{N_\ell} \sum_{n=N_\ell}^{\infty} \frac{z^n}{n!} \right). \tag{2.7}$$

then the combined "middle term" polynomial $\sum_{\ell=1}^{L} \left(k_{N_\ell} \sum_{n=N_\ell - p + 1}^{N_\ell - 1} \frac{z^n}{n!} \right)$ would vanish at $z = a$. This strategy is precisely the approach that eventually leads to success. Our challenge at hand is now clear: Discover how to construct that linear combination.

2.3 The classic vanishing polynomial trick

We wish to find a polynomial of the form

$$\sum_{\ell=1}^{L} \left(k_{N_\ell} \sum_{n=N_\ell - p + 1}^{N_\ell - 1} \frac{z^n}{n!} \right) \tag{2.8}$$

that vanishes at $z = a$. Such a polynomial will give rise to an amazing rational approximation to κe^a, for some nonzero integer κ; which, in turn, will allow us to construct an integer less than 1. However, the polynomial in (2.8) must possess some additional structure in order to allow us to conclude that our integer is also positive and therefore contradicts the Fundamental Principle of Number Theory. How do we build a polynomial having the shape of (2.8)? The answer is that we can start with just about any polynomial we wish. To illustrate this vague claim, let us consider the generic polynomial

$$f(z) = c_6 z^6 + c_5 z^5 + c_4 z^4 + c_3 z^3 = \sum_{N=3}^{6} c_N z^N$$

and notice that if we sum its first three derivatives, $f^{(1)}(z) + f^{(2)}(z) + f^{(3)}(z)$, then we have

$$
\begin{aligned}
f^{(1)}(z) + f^{(2)}(z) + f^{(3)}(z) = \quad & 6c_6 z^5 + \ 5c_5 z^4 + \ 4c_4 z^3 + 3c_3 z^2 \\
& + \ 30c_6 z^4 + 20c_5 z^3 + 12c_4 z^2 + 6c_3 z \\
& + 120c_6 z^3 + 60c_5 z^2 + 24c_4 z \ + 6c_3.
\end{aligned}
$$

If we now factor out the factoral $N!$ from those terms possessing the coefficient c_N, then we are faced with an expression that has an uncanny resemblance to (2.8):

$$
\begin{aligned}
\sum_{n=1}^{3} f^{(n)}(z) &= \sum_{N=3}^{6} N c_N z^{N-1} + \sum_{N=3}^{6} N(N-1)c_N z^{N-2} \\
&\quad + \sum_{N=3}^{6} N(N-1)(N-2)c_N z^{N-3} \\
&= \sum_{N=3}^{6} N! c_N \left(\frac{z^{N-1}}{(N-1)!} + \frac{z^{N-2}}{(N-2)!} + \frac{z^{N-3}}{(N-3)!} \right) \\
&= \sum_{N=3}^{6} \left(N! c_N \sum_{n=N-3}^{N-1} \frac{z^n}{n!} \right).
\end{aligned}
$$

Hence we discover that when we sum the appropriate derivatives of the polynomial $f(z)$, we magically arrive at an expression of the form (2.8). This observation can be generalized as follows.

Challenge 2.3 *For integers j and k satisfying $1 \le j \le k$, let $f(z)$ be the polynomial defined by $f(z) = \sum_{n=j}^{k} c_n z^n$. Show that*

$$\sum_{n=1}^{j} f^{(n)}(z) = \sum_{N=j}^{k} \left(N! c_N \sum_{n=N-j}^{N-1} \frac{z^n}{n!} \right).$$

Thus conclude that for any polynomial $f(z)$ having a factor of z^j, for some $j \ge 1$, the sum of its first j derivatives can be expressed in the form (2.8).

The only way a polynomial with integer coefficients can vanish at $z = a$ is for it to have $(z - a)$ as a factor. So one scheme to create a polynomial of the form (2.8) that also vanishes at $z = a$ is to begin with an auxiliary polynomial $f(z)$ that has both a factor of z^j and a factor of $(z - a)^m$. The factor z^j allows us to apply the result from Challenge 2.3, and for a sufficiently large choice of m, the polynomials $f^{(n)}(z)$, for $n = 1, 2, \ldots, j$, will all vanish at $z = a$. Thus we require that the exponent m be greater than the exponent j. These remarks lead us to conclude that a natural choice for $f(z)$ is a polynomial of the form $z^j(z - a)^{j+1}$, for some integer $j \geq 1$.

Challenge 2.4 *Let* $f(z) = z^j(z - a)^{j+1}$, *for some integer* $j \geq 1$, *and write it as* $f(z) = \sum_{n=j}^{2j+1} c_n z^n$. *Show*

$$\sum_{n=1}^{j} f^{(n)}(a) = 0.$$

After studying Challenges 2.3 and 2.4 together with identity (2.6), we find that we should take $j = p - 1$ and select our indices appearing in (2.7) to be $N_1 = p - 1$, $N_2 = p, N_3 = p + 1, \ldots, N_L = 2p - 1$. Thus we are led to consider the polynomial

$$f(z) = z^{p-1}(z - a)^p.$$

2.4 The first part of the proof of Theorem 2.2—The elusive estimate

As we remarked earlier, it is enough to prove that e^a is irrational for positive integers a. We now assume that e^a is a rational number, say $e^a = \frac{r}{s}$. Thus we have that

$$\frac{r}{s} = e^a = \sum_{n=0}^{\infty} \frac{a^n}{n!} = \sum_{n=0}^{N-p} \frac{a^n}{n!} + \sum_{n=N-p+1}^{N-1} \frac{a^n}{n!} + \sum_{n=N}^{\infty} \frac{a^n}{n!},$$

for any integers N and p. Next we write the polynomial $f(z) = z^{p-1}(z - a)^p$ as $f(z) = \sum_{n=p-1}^{2p-1} c_n z^n \in \mathbb{Z}[z]$. Thus, given Challenge 2.3, if we consider the linear combination

$$\sum_{N=p-1}^{2p-1} N! c_N \frac{r}{s} = \sum_{N=p-1}^{2p-1} N! c_N e^a = \sum_{N=p-1}^{2p-1} \left(N! c_N \sum_{n=0}^{N-p} \frac{a^n}{n!} \right)$$

$$+ \sum_{N=p-1}^{2p-1} \left(N! c_N \sum_{n=N-p+1}^{N-1} \frac{a^n}{n!} \right) + \sum_{N=p-1}^{2p-1} \left(N! c_N \sum_{n=N}^{\infty} \frac{a^n}{n!} \right),$$

then we conclude that the middle term appearing on the right-hand side is a sum of derivatives of $f(z)$ evaluated at a. Specifically, in view of Challenges 2.3 and 2.4, that middle term $\sum_{N=p-1}^{2p-1} \left(N! c_N \sum_{n=N-p+1}^{N-1} \frac{a^n}{n!} \right)$ equals 0, and hence the previous identity can be expressed simply as

$$\frac{r}{s} \sum_{N=p-1}^{2p-1} N! c_N = \sum_{N=p-1}^{2p-1} \left(N! c_N \sum_{n=0}^{N-p} \frac{a^n}{n!} \right) + \sum_{N=p-1}^{2p-1} \left(N! c_N \sum_{n=N}^{\infty} \frac{a^n}{n!} \right). \tag{2.9}$$

We now define the polynomial approximation $\mathcal{P}_p(z)$ and the tail of the series $\mathcal{T}_p(z)$ by

$$\mathcal{P}_p(z) = \sum_{N=p-1}^{2p-1} \left(N! c_N \sum_{n=0}^{N-p} \frac{z^n}{n!} \right) \quad \text{and} \quad \mathcal{T}_p(z) = \sum_{N=p-1}^{2p-1} \left(N! c_N \sum_{n=N}^{\infty} \frac{z^n}{n!} \right),$$

where we remark that for $N = p - 1$, the inner sum in $\mathcal{P}_p(z)$ is empty and thus equals 0. Hence we observe that each coefficient of $\mathcal{P}_p(z)$ is divisible by p. We can now rewrite (2.9) as

$$\frac{r}{s} \sum_{N=p-1}^{2p-1} N! c_N = \mathcal{P}_p(a) + \mathcal{T}_p(a),$$

which immediately implies

$$\left| \frac{r}{s} \sum_{N=p-1}^{2p-1} N! c_N - \mathcal{P}_p(a) \right| = |\mathcal{T}_p(a)|.$$

In order to produce an upper bound for the quantity $|\mathcal{T}_p(a)|$, we note that after the change of variables $m = n - N$, we have

$$\mathcal{T}_p(a) = \sum_{N=p-1}^{2p-1} \left(c_N \sum_{m=0}^{\infty} \frac{N!}{(m+N)!} a^{m+N} \right).$$

By inequality (2.3), we recall that $\frac{N!}{(m+N)!} \leq \frac{1}{m!}$, which together with the triangle inequality, the power series expansion for e^a, and the assumption that $a \geq 1$ reveals that

$$|\mathcal{T}_p(a)| \leq \sum_{N=p-1}^{2p-1} |c_N| a^N \sum_{m=0}^{\infty} \frac{a^m}{m!}$$

$$\leq \sum_{N=p-1}^{2p-1} |c_N| a^{2p-1} \sum_{m=0}^{\infty} \frac{a^m}{m!}$$

$$= e^a a^{2p-1} \sum_{N=p-1}^{2p-1} |c_N|. \tag{2.10}$$

Challenge 2.5 *Recall that $f(z) = z^{p-1}(z-a)^p = \sum_{N=p-1}^{2p-1} c_N z^N$, where a is a positive integer. Apply the Binomial Theorem to show that*

$$f(z) = \sum_{\ell=0}^{p} \binom{p}{\ell} (-a)^{p-\ell} z^{\ell+p-1},$$

and then take $z = -1$ to conclude that

$$\sum_{N=p-1}^{2p-1} |c_N| = (1+a)^p.$$

In view of the bound in (2.10) and Challenge 2.5 we see that

$$\left| \mathcal{T}_p(a) \right| \le e^a a^{2p-1} (1+a)^p.$$

Thus if we define the constants $K_1 = \frac{1}{a} e^a$ and $K_2 = a^2(1+a)$, then we obtain the now-not-so elusive estimate we desire:

$$\left| \frac{r}{s} \sum_{N=p-1}^{2p-1} N! c_N - \mathcal{P}_p(a) \right| = \left| \mathcal{T}_p(a) \right| \le K_1 (K_2)^p. \tag{2.11}$$

2.5 The dramatic conclusion of the proof Theorem 2.2—Arithmetic conquers all

We are finally in position to construct the infamous integer that will contradict the Fundamental Principle of Number Theory. That integer is inspired by the quantity bounded by inequality (2.11). If we multiply inequality (2.11) by s, we then see that

$$\left| r \sum_{N=p-1}^{2p-1} N! c_N - s \mathcal{P}_p(a) \right| \le s K_1 (K_2)^p. \tag{2.12}$$

Challenge 2.6 *Prove that $r \sum_{N=p-1}^{2p-1} N! c_N - s \mathcal{P}_p(a)$ is an integer.*

In fact, we can actually deduce some divisibility properties for the integer in Challenge 2.6 that will allow us to divide both sides of (2.12) by an appropriate integer in order to obtain an integer whose absolute value is less than 1. We will then show that this integer is nonzero.

Challenge 2.7 *Prove that $(p-1)!$ is a factor of both the integer $r \sum_{N=p-1}^{2p-1} N! c_N$ and the integer $s \mathcal{P}_p(a)$, and therefore is a factor of their difference.*

Thus if we divide inequality (2.12) by $(p-1)!$, then we have

$$\left| \frac{r}{(p-1)!} \sum_{N=p-1}^{2p-1} N! c_N - \frac{s}{(p-1)!} \mathcal{P}_p(a) \right| \le s K_1 \frac{K_2^p}{(p-1)!}, \tag{2.13}$$

where the unwieldy quantity

$$\frac{r}{(p-1)!} \sum_{N=p-1}^{2p-1} N! c_N - \frac{s}{(p-1)!} \mathscr{P}_p(a)$$

is an *integer*. It is certainly worth taking a moment to catch our breath and appreciate how far we have journeyed. In particular, notice how closely the previous inequality resembles the upper bound of (2.5), which up until this moment has been only a fantasy.

Our mission now is clear: We need to show that the unwieldy integer in (2.13) is, in fact, *nonzero*. Fortunately, we have a degree of freedom at our disposal that will assist us in our mission—the parameter p. We now will select p to be any *prime* number satisfying $p > \max\{a, r\}$, so we are certain that p will not divide either a or r.

Challenge 2.8   *Prove that for our choice of p given above,*

$$\frac{r}{(p-1)!} \sum_{N=p-1}^{2p-1} N! c_N \not\equiv 0 \bmod p,$$

while

$$\frac{s}{(p-1)!} \mathscr{P}_p(a) \equiv 0 \bmod p.$$

(Hint: We remark that $c_{p-1} \neq 0$, since in view of the definition of $f(z)$, $|c_{p-1}| = |a^p| \neq 0$.)

From Challenge 2.8, we conclude that the integer

$$\frac{r}{(p-1)!} \sum_{N=p-1}^{2p-1} N! c_N - \frac{s}{(p-1)!} \mathscr{P}_p(a)$$

is not congruent to 0 modulo p, and thus must be a *nonzero* integer. Putting this observation together with inequality (2.13), we discover that our unwieldy integer satisfies

$$0 < \left| \frac{r}{(p-1)!} \sum_{N=p-1}^{2p-1} N! c_N - \frac{s}{(p-1)!} \mathscr{P}_p(a) \right| \leq s K_1 \frac{K_2^p}{(p-1)!}.$$

If we now let the prime number p approach infinity, we see that our upper bound will eventually be less than 1, and thus our unwieldy integer clashes head-on with the Fundamental Principle of Number Theory. This contradiction implies that our assumption that e^a is rational is false. Thus we have established the irrationality of e^a and hence by Challenge 2.1, the irrationality of $e^{a/b}$ for nonzero rational numbers a/b.

2.6 The transcendence of e

The previous argument was certainly elaborate and delicate. Some exhausted readers may exclaim, "All that effort just to show the *irrationality* of $e^{a/b}$!" Happily, those readers will now become reinvigorated as we discover that the circle of ideas we have just developed in the previous argument can be quickly adapted and applied to establish the *transcendence* of e.

THEOREM 2.3 *The number e is transcendental.*

Proof. We begin by assuming that e is algebraic. Thus there exist integers $r_0, r_1, \ldots, r_d$, with $r_d \neq 0$, such that

$$r_0 + r_1 e + r_2 e^2 + \cdots + r_d e^d = 0 \qquad (2.14)$$

In our demonstration of the irrationality of e^a, we assumed that $r - se^a = 0$ and then found a polynomial $\mathscr{P}_p(z)$ such that $\mathscr{P}_p(a)$ is an amazing approximation to e^a. Thus if we wish to follow the same line of attack in the present context, we must construct a polynomial $\mathscr{P}_p(z)$ such that $\mathscr{P}_p(1), \mathscr{P}_p(2), \ldots, \mathscr{P}_p(d)$ provide amazing rational approximations to $e, e^2, \ldots, e^d$, respectively. Inspired by our previous work, we immediately consider

$$f(z) = z^{p-1}(z-1)^p(z-2)^p \cdots (z-d)^p,$$

which we write as $f(z) = \sum_{n=p-1}^{(d+1)p-1} c_n z^n$. Applying Challenge 2.3, we find that

$$\sum_{n=1}^{p-1} f^{(n)}(z) = \sum_{N=p-1}^{(d+1)p-1} \left(N! c_N \sum_{n=N-p+1}^{N-1} \frac{z^n}{n!} \right).$$

Challenge 2.9 *Given $f(z)$ as defined above, show that*

$$\sum_{n=1}^{p-1} f^{(n)}(t) = 0,$$

for $t = 1, 2, \ldots, d$.

Just as in our earlier argument, here we now use the coefficients of the polynomial $f(z)$ to produce the following particularly advantageous linear combination

$$\sum_{N=p-1}^{(d+1)p-1} N! c_N e^z = \sum_{N=p-1}^{(d+1)p-1} \left(N! c_N \sum_{n=0}^{N-p} \frac{z^n}{n!} \right)$$

$$+ \sum_{N=p-1}^{(d+1)p-1} \left(N! c_N \sum_{n=N-p+1}^{N-1} \frac{z^n}{n!} \right) + \sum_{N=p-1}^{(d+1)p-1} \left(N! c_N \sum_{n=N}^{\infty} \frac{z^n}{n!} \right).$$

By Challenge 2.9, we see that for $t = 1, 2, \ldots, d$, the middle sum vanishes, and so we have

$$e^t \sum_{N=p-1}^{(d+1)p-1} N! c_N = \sum_{N=p-1}^{(d+1)p-1} \left(N! c_N \sum_{n=0}^{N-p} \frac{t^n}{n!} \right) + \sum_{N=p-1}^{(d+1)p-1} \left(N! c_N \sum_{n=N}^{\infty} \frac{t^n}{n!} \right),$$

which, as before, we write as a polynomial term plus a tail term:

$$e^t \sum_{N=p-1}^{(d+1)p-1} N! c_N = \mathcal{P}_p(t) + \mathcal{T}_p(t). \tag{2.15}$$

Arguing as we did in (2.10), we conclude that for $t = 1, 2, \ldots, d$,

$$|\mathcal{T}_p(t)| \le e^t t^{(d+1)p-1} \sum_{N=p-1}^{(d+1)p-1} |c_N|. \tag{2.16}$$

Our (or, more accurately, your) next challenge is to provide an upper bound for the sum in (2.16).

Challenge 2.10 *Given that $(z - t)^p = \sum_{n=0}^{p} \binom{p}{n} (-t)^{p-n} z^n$, show that*

$$\max_{n=0,1,\ldots,p} \left\{ \left| \binom{p}{n} (-t)^{p-n} \right| \right\} \le t^p \sum_{n=0}^{p} \binom{p}{n} = (2t)^p.$$

Recalling that $z^{p-1}(z - 1)^p(z - 2)^p \cdots (z - d)^p = \sum_{n=p-1}^{(d+1)p-1} c_n z^n$, use the previous inequality to conclude that

$$|c_n| \le \prod_{t=1}^{d} (2t)^p \le ((2d)^d)^p. \tag{2.17}$$

Combining inequalities (2.16) and (2.17), together with the observation that the number of coefficients c_n is $dp + 1$, yields

$$|\mathcal{T}_p(t)| \le e^t t^{(d+1)p-1} (dp + 1)((2d)^d)^p \le e^t t^{(d+1)p-1} d^p ((2d)^d)^p,$$

which, for $1 \le t \le d$, implies

$$|\mathcal{T}_p(t)| \le e^d d^{(d+2)p-1} ((2d)^d)^p = K_1 (K_2)^p, \tag{2.18}$$

where the constants K_1 and K_2 are defined by $K_1 = e^d/d$ and $K_2 = d^2(2d^2)^d$. In view of identities (2.14) and (2.15), together with the observation that $\mathcal{T}_p(0) = 0$, we have that

$$r_0 \mathcal{P}_p(0) + r_1 \mathcal{P}_p(1) + r_2 \mathcal{P}_p(2) + \cdots + r_d \mathcal{P}_p(d)$$
$$= -r_0 \mathcal{T}_p(0) - r_1 \mathcal{T}_p(1) - \cdots - r_d \mathcal{T}_p(d)$$
$$= -r_1 \mathcal{T}_p(1) - r_2 \mathcal{T}_p(2) - \cdots - r_d \mathcal{T}_p(d).$$

Dividing the previous equality by $(p-1)!$ and then applying inequality (2.18) yields

$$\left| \frac{r_0}{(p-1)!} \mathcal{P}_p(0) + \sum_{t=1}^{d} \frac{r_t}{(p-1)!} \mathcal{P}_p(t) \right| = \left| \sum_{t=1}^{d} \frac{r_t}{(p-1)!} \mathcal{T}_p(t) \right|$$

$$\leq K_1 \left(\sum_{t=1}^{d} |r_t| \right) \frac{(K_2)^p}{(p-1)!}. \qquad (2.19)$$

Challenge 2.11 *Adopting the ideas used in Challenge 2.8, prove that for all sufficiently large prime numbers p,*

$$\frac{r_0}{(p-1)!} \mathcal{P}_p(0) + \sum_{t=1}^{d} \frac{r_t}{(p-1)!} \mathcal{P}_p(t)$$

is a nonzero integer.

So for all sufficiently large prime numbers p, inequality (2.19) violates the Fundamental Principle of Number Theory, and thus we have arrived at a contradiction. Hence e is not algebraic and therefore is, in fact, transcendental. ∎

2.7 Foreshadowing algebraic exponents—The irrationality of $e^{\sqrt{n}}$ and π

In order to inspire the themes we will develop in the next chapter, where we establish the transcendence of e^α for nonzero algebraic numbers α, we close our discussion here by considering numbers of the form $e^{\sqrt{a/b}}$, where a/b is a nonzero rational number, and discovering how to modify the arguments of this chapter in order to demonstrate the irrationality of $e^{\sqrt{a/b}}$.

THEOREM 2.4 *Let a/b be a nonzero rational number. Then $e^{\sqrt{a/b}}$ is irrational.*

Before considering the proof of this theorem, we pause momentarily to acknowledge and appreciate an immediate, but enormous, consequence.

COROLLARY 2.5 *The number π is irrational.*

Proof. Suppose that π is a rational number, say $\pi = \frac{c}{d}$. Then by Theorem 2.4 we see that $e^{\sqrt{-c^2/d^2}}$ is irrational. However, in view of one of the most famous identities in mathematics, we have

$$e^{\sqrt{-c^2/d^2}} = e^{(\sqrt{-1})(c/d)} = e^{i\pi} = -1.$$

Thus we are forced to conclude that -1 is an irrational number, which happens to be utterly false. Hence π is indeed irrational. ∎

Proof of Theorem 2.4.   Let a/b be a nonzero rational number and let $\alpha = \sqrt{a/b}$. We wish to prove that e^{α} is irrational, so we assume that e^{α} is rational, say $e^{\alpha} = \frac{r}{s}$. Thus we have that

$$r - se^{\alpha} = 0. \tag{2.20}$$

As in our previous arguments, we wish to replace e^{α} by a polynomial approximation $\mathcal{P}(\alpha)$, where $\mathcal{P}(z) \in \mathbb{Z}[z]$, and use it to construct an integer violating the Fundamental Principle of Number Theory. The immediate difficulty with this approach is that if $\mathcal{P}(z)$ is a polynomial with integral coefficients, then $\mathcal{P}(\alpha)$ is an algebraic number, but not necessarily an integer or even a rational number.

To make this crucial point concrete, let us consider the polynomial $\mathcal{P}(z) = z^3 - 4z^2 + 5z + 3$ and notice that $\mathcal{P}(\sqrt{2}) = -5 + 7\sqrt{2}$, which is certainly not a rational number and thus would not lead us, in any immediate manner, to an integer that would contradict the Fundamental Principle of Number Theory. However, let us notice that if we evaluate that same polynomial at the conjugate of $\sqrt{2}$, namely $-\sqrt{2}$, then we have $\mathcal{P}(-\sqrt{2}) = -5 - 7\sqrt{2}$. While that value is also irrational, we see an interesting phenomenon:

$$\mathcal{P}(\sqrt{2}) + \mathcal{P}(-\sqrt{2}) = -10,$$

that is, the sum of these values yields an integer. This simple observation inspires us to bring the conjugate of α into the approximation picture in the hope of producing our impossible integer. Indeed, the specific result we require is given by the following challenge.

Challenge 2.12 *Suppose that $\mathcal{P}(z) \in \mathbb{Z}[z]$ has degree d. Then show that $\mathcal{P}(\sqrt{a/b}) + \mathcal{P}(-\sqrt{a/b})$ is a rational number and can be written having a denominator equal to b^d.*

The symmetry introduced by considering both α and its conjugate is the critical new step that allows us to move forward. Thus, rather than considering the now unbalanced-looking quantity in (2.20), we consider the more symmetrically appealing identity $(r - se^{\alpha})(r - se^{-\alpha}) = 0$, which gives rise to

$$(s^2 + r^2) - rs(e^{\alpha} + e^{-\alpha}) = 0. \tag{2.21}$$

Next we construct a polynomial, $\mathcal{P}_p(z) \in \mathbb{Z}[z]$, that simultaneously provides a good approximation to *both* e^{α} and $e^{-\alpha}$. Toward this end, we proceed precisely as in our previous arguments by defining $f(z)$ to be

$$f(z) = b^p z^{p-1} (z - \alpha)^p (z + \alpha)^p = z^{p-1} (bz^2 - a)^p \in \mathbb{Z}[z], \tag{2.22}$$

which, as before, we write as $f(z) = \sum_{n=p-1}^{3p-1} c_n z^n$.

We now proceed exactly as we did in the proof of the transcendence of e. Specifically, we apply the polynomial approximation formed by the appropriate linear combinations to the identity in (2.21) to conclude that

$$\left| \frac{s^2 + r^2}{(p-1)!} \mathscr{P}_p(0) - \frac{rs}{(p-1)!} (\mathscr{P}_p(\alpha) + \mathscr{P}_p(-\alpha)) \right| = \left| \frac{rs}{(p-1)!} (\mathscr{T}_p(\alpha) + \mathscr{T}_p(-\alpha)) \right|,$$

(2.23)

where the polynomial $\mathscr{P}_p(z)$ and the tail $\mathscr{T}_p(z)$ are as they were defined in the proof of the transcendence of e.

We are now ready to utilize the quantity

$$\frac{s^2 + r^2}{(p-1)!} \mathscr{P}_p(0) - \frac{rs}{(p-1)!} (\mathscr{P}_p(\alpha) + \mathscr{P}_p(-\alpha))$$

to construct our nonzero integer. As we have seen in our previous argument, the first term is an integer, but now the second term involves the irrational number α. However, here is where we exploit the symmetry we introduced through the use of the conjugate of α. Specifically, we first notice that by definition of $\mathscr{P}_p(z)$, $\frac{1}{(p-1)!} \mathscr{P}_p(z)$ has integer coefficients. Thus, in view of Challenge 2.12, we could clear denominators and conclude that

$$\frac{b^{3p-1}(s^2 + r^2)}{(p-1)!} \mathscr{P}_p(0) - \frac{b^{3p-1} rs}{(p-1)!} (\mathscr{P}_p(\alpha) + \mathscr{P}_p(-\alpha))$$

is an *integer*. The fact that this integer is nonzero follows from the identical argument given in the proof of the transcendence of e. Thus identity (2.23) can be rewritten as

$$\left| \frac{b^{3p-1}(s^2 + r^2)}{(p-1)!} \mathscr{P}_p(0) - \frac{b^{3p-1} rs}{(p-1)!} (\mathscr{P}_p(\alpha) + \mathscr{P}_p(-\alpha)) \right|$$

$$= \left| \frac{b^{3p-1} rs}{(p-1)!} (\mathscr{T}_p(\alpha) + \mathscr{T}_p(-\alpha)) \right|.$$

(2.24)

Challenge 2.13 *Using (2.10) as a guide, find the analogue to the upper bound of (2.11) in this context. Then apply (2.24) to produce an inequality similar to (2.13). Finally, apply this new inequality to show that the integer in (2.24) can be made less than 1 for all sufficiently large primes p.*

Thus we have constructed a positive integer less than 1. This contradiction leads us to the conclusion that $e^{\sqrt{a/b}}$ is irrational and brings us to the end of our proof. ∎

The important new idea introduced in the proof of Theorem 2.4 was the balanced application of *both* zeros of the polynomial $bz^2 - a$ in the construction of the function $f(z)$ defined in (2.22). The symmetry occurring in $f(z)$ led to the critical fact that $f(z) \in \mathbb{Z}[z]$. The deep idea of considering all the conjugates of an algebraic number α in the construction of the auxiliary polynomial allows us to extend the themes we have developed in this chapter to prove the spectacular result that for any nonzero algebraic number α, the number e^{α} is transcendental. We carry out this program and explore some of the result's far-reaching and beautiful consequences in the next chapter.

4.1132503787829275171735818151 ...

Conjugation and Symmetry as a Means Towards Transcendence:
The Lindemann–Weierstrass Theorem and the transcendence of $e^{\sqrt{2}}$

3.1 Algebraic exponents through the looking glass—A wonderland of transcendence

In this chapter we consider numbers of the form e^{α}, where α is a nonzero algebraic number. As we indicated to at the close of the previous chapter, here we will prove the following result due to Charles Hermite and Ferdinand Lindemann.

THEOREM 3.1 *The number* e^{α} *is transcendental for any nonzero algebraic number* α.

One immediate but spectacular corollary to this result is:

COROLLARY 3.2 *The number* π *is transcendental.*

Proof. Suppose that π were algebraic. Since the product of two algebraic numbers is algebraic, we see that $i\pi$ is a nonzero algebraic number. Hence in view of Theorem 3.1, we are forced to conclude that $e^{i\pi}$ is transcendental. However, we know that $e^{i\pi} = -1$, a value that can be shown to be highly algebraic. This contradiction reveals that π is, in fact, a transcendental number. ∎

Instead of beginning with a proof of Theorem 3.1, we first explore the result very naïvely and see where those naïve explorations lead. For a nonzero algebraic number α, Theorem 3.1 is equivalent to the assertion that the number e^{α} is *not* algebraic. That is, $e^{\alpha} \neq \beta$ for any algebraic number β. Since the quantities e^{α} and β remain unequal even if we multiply them by any nonzero algebraic number, we see that Theorem 3.1 implies that for all nonzero algebraic numbers β_0, β_1, we have

$$\beta_0 + \beta_1 e^{\alpha} \neq 0,$$

which can be rewritten in a more balanced form as

$$\beta_0 e^0 + \beta_1 e^{\alpha} \neq 0.$$

Thus a more abstruse, although equivalent, version of Theorem 3.1 is the following.

THEOREM 3.3 *Suppose that α is a nonzero algebraic number. If β_0, β_1 are algebraic numbers, not both zero, then*

$$\beta_0 e^0 + \beta_1 e^\alpha \neq 0.$$

Theorem 3.3 implies that there is no linear combination of e^0 and e^α with algebraic coefficients that produces 0 unless both of the coefficients are 0. In other words, the theorem asserts that e^0 and e^α are *linearly independent over the algebraic numbers*. Thus we could recast Theorem 3.1 in the language of linear algebra by stating that for the distinct algebraic numbers 0 and α, the quantities e^0 and e^α are linearly independent over the algebraic numbers. When phrased in this manner, a natural question arises. Suppose we are given several distinct algebraic numbers, say $\alpha_0, \alpha_1, \ldots, \alpha_M$. Will the values $e^{\alpha_0}, e^{\alpha_1}, \ldots, e^{\alpha_M}$ be linearly independent over the algebraic numbers? This natural question leads to a beautiful generalization of the Hermite–Lindemann Theorem due to Lindemann and Karl Weierstrass, which we now state.

THEOREM 3.4 *Let $\alpha_0, \alpha_1, \ldots, \alpha_M$ be $M + 1$ distinct algebraic numbers. Then*

$$e^{\alpha_0}, e^{\alpha_1}, \ldots, e^{\alpha_M}$$

are linearly independent over the algebraic numbers. That is, if $\beta_0, \beta_1, \ldots, \beta_M$ are nonzero algebraic numbers, then

$$\sum_{m=0}^{M} \beta_m e^{\alpha_m} \neq 0.$$

As we have just seen, Theorem 3.1 is an immediate corollary of Theorem 3.4. But this generalization does far more than just impress fans of linear algebra. As we will see, knowing that numbers are linearly independent over algebraic numbers leads to extremely powerful and important results about transcendental numbers. To illustrate this point, consider the following beautiful consequences.

Challenge 3.1 *Deduce the following two impressive transcendence results from Theorem 3.4.*

COROLLARY 3.5 *If α is a nonzero real algebraic number, then $\sin \alpha$ is transcendental. Thus, if r/s is a rational number, then $\sin(r/s)$ is algebraic if and only if $\frac{r}{s} = 0$. (Hint: Verify the identity*

$$\sin \alpha = \frac{e^{i\alpha} - e^{-i\alpha}}{2i},$$

and then observe that $\{-i\alpha, 0, i\alpha\}$ is a set of distinct algebraic numbers.)

COROLLARY 3.6 *If α is a real algebraic number, $\alpha \neq 0, 1$, then $\log \alpha$ is transcendental.*

So from just one theorem we instantly obtain the transcendence of an enormous variety of numbers such as $e^{\sqrt{2}}$ (the chapter title number), $\sin(1/4)$, $\log(\sqrt{2} +$

$\sqrt[3]{7}$),... the list goes on and on. In fact, another immediate consequence of Theorem 3.4 is the following general result.

COROLLARY 3.7   *Let $\alpha_0, \alpha_1, \ldots, \alpha_M$ be $M + 1$ distinct nonzero algebraic numbers. Then for nonzero algebraic numbers $\beta_0, \beta_1, \ldots, \beta_M$, the number*

$$\sum_{m=0}^{M} \beta_m e^{\alpha_m}$$

is transcendental.

Challenge 3.2   *Deduce Corollary 3.7 from Theorem 3.4.*

We embark upon our journey toward a proof of Theorem 3.4 by first considering a simple version and then applying the insights we develop as a springboard to catapult us all the way to a proof of the general result. We caution that the trajectory caused by that catapulting action is dramatic, and thus we need to coast aloft a bit in order to avoid a rough landing. We then close this chapter with some commentary on how to extend the notion of linear independence and why such an extension has important ramifications in the theory of transcendence.

3.2   Heading towards Hermite—A partial result casts some foreshadowing

The proof of Theorem 3.4 is actually an adaptation of the ideas employed in the proofs of Chapter 2. Thus we will apply the strategies we have already developed and then work to modify those arguments when new complications arise. In order to develop an appreciation for the technical difficulties we must overcome, we first explore a special case of the Hermite–Lindemann result.

In the previous chapter we discovered that $e^{\sqrt{2}}$ is irrational. But is it algebraic? Before we attempt to answer that question, we ask a much more modest one: Can $e^{\sqrt{2}}$ equal the square root of an integer? If we look at one decimal digit of accuracy, then we notice the amazing digit coincidence

$$e^{\sqrt{2}} \approx 4.1 \quad \text{and} \quad \sqrt{17} \approx 4.1.$$

Based on this compelling numerical evidence, it is possible that these two numbers are equal, and if so, then it would follow that $e^{\sqrt{2}}$ is a quadratic irrational. A look at one more digit of accuracy reveals that $e^{\sqrt{2}} \approx 4.11$, while $\sqrt{17} \approx 4.12$; thus we make the not-so-shocking realization that these two numbers are, in fact, not the same. But the question of whether $e^{\sqrt{2}}$ is quadratic or not remains open, at least until the next result.

THEOREM 3.8 *Let a/b be a nonzero rational number. Then $e^{\sqrt{a/b}}$ is neither a rational number nor a quadratic irrational. That is, $e^{\sqrt{a/b}}$ is not equal to an expression of the form*

$$\frac{r + \sqrt{q}}{s},$$

for any integers $r, s,$ and q.

$\mathcal{T}he\ intuitive\ idea\ behind\ our\ approach$

Theorem 3.8 generalizes Theorem 2.4, and thus it seems reasonable to follow a line of reasoning that reflects the one that established Theorem 2.4. There is one new fundamental complication to overcome, but after that lone hurdle the proof parallels the proofs of Theorems 2.3 and 2.4. To provide a thread of intuition into the fabric of the argument, the new wrinkle that will appear, and how it can be ironed out, we now outline a proof—mirroring the one for Theorem 2.4—of an assertion that we already believe: $e^{\sqrt{2}}$ is indeed *not* equal to $\sqrt{17}$.

If we assume that $e^{\sqrt{2}} = \sqrt{17}$, then clearly $\sqrt{17} - e^{\sqrt{2}} = 0$. Just as in the proof of Theorem 2.4, we are faced with an annoyingly irrational exponent. As before, we circumvent this obstacle by considering a more symmetric expression formed by introducing the conjugate of the exponent. In particular, we consider

$$\left(\sqrt{17} - e^{\sqrt{2}} \right) \left(\sqrt{17} - e^{-\sqrt{2}} \right) = 0,$$

which is equivalent to the identity

$$18 - \sqrt{17} \left(e^{\sqrt{2}} + e^{-\sqrt{2}} \right) = 0. \tag{3.1}$$

As we discovered in the proof of Theorem 2.4, we could apply Challenge 2.12 to see that the symmetry of the exponents implies that our polynomial approximation to $e^{\sqrt{2}} + e^{-\sqrt{2}}$ will produce a rational number. Unfortunately, there is a fundamental and critical difference between (3.1) and (2.21). Specifically, in (2.21), the coefficients of the expression are all *integers*, while in (3.1) one coefficient is irrational. Thus identity (3.1) cannot be utilized, in its current state, to produce the integer that will contradict the Fundamental Principle of Number Theory. We need to modify the expression in (3.1) so that the coefficients are all integers.

A natural and healthy temptation is to apply the "conjugate philosophy" we have already embraced—namely, create objects that are symmetric with respect to the conjugates of an algebraic number and thereby reduce their arithmetic complexity. Hence in order to make (3.1) appear more symmetric with respect to the conjugates of the *coefficients*, we multiply (3.1) by a similar expression in

Continued

which the coefficient $\sqrt{17}$ is replaced by its conjugate. Hence we are led to the expression

$$\left(18 - \sqrt{17}\left(e^{\sqrt{2}} + e^{-\sqrt{2}}\right)\right)\left(18 + \sqrt{17}\left(e^{\sqrt{2}} + e^{-\sqrt{2}}\right)\right) = 0,$$

which can be rewritten as

$$290 - 17\left(e^{2\sqrt{2}} + e^{-2\sqrt{2}}\right) = 0.$$

Happily, we have built an expression having *integer* coefficients, and while the exponents have been altered slightly, we have retained the symmetry with respect to the conjugates. Thus the previous identity perfectly mirrors identity (2.21). The argument employed to prove Theorem 2.4 can now be directly applied to reach a contradiction and thus allow us to make the startling rediscovery that $e^{\sqrt{2}} \neq \sqrt{17}$.

The new complication that requires our attention arises from the realization that, as we saw above, after we make our expression symmetric with respect to the conjugates of the algebraic exponents, we are left with an identity involving algebraic *irrational* coefficients. As we have just discovered in the previous simple illustration, producing a symmetric expression with respect to the conjugates of those coefficients is the key step in constructing a new identity with integer coefficients.

Preproof Precaution. In our development of the proof below, we will come face to face with a number of impressively long and complicated identities. They may intimidate some readers, and for good reason: They are intimidating. But if these identities are taken slowly, they can be verified and thereby tamed. However, taming algebraically intricate expressions is *not* our goal. Thus we recommend that the reader forgo the messy algebra required to verify the long identities in the proof ahead and merely observe the identities' general structure.

The important feature on which to focus is the basic *shape* of the coefficients and the exponents. In particular, as the identities grow in length, notice that the exponents begin to display some attractive symmetry with respect to their conjugates, and the coefficients are transformed into integers. It is this algebraically attractive general structure that is critical and should be appreciated—even by those readers who choose not to invest the time or energy to verify the otherwise obscenely long identities looming ahead. *You have been duly warned.*

Proof of Theorem 3.8. As always, we proceed by contradiction and assume that there exist integers $r, s,$ and q such that

$$e^{\sqrt{a/b}} = \frac{r + \sqrt{q}}{s}.$$

Thus we have

$$(r + \sqrt{q}) - se^{\sqrt{a/b}} = 0, \tag{3.2}$$

which, as in the proof of Theorem 2.4, suffers from the feature that there is an irrational value in the exponent of e. Therefore we again introduce some symmetry by including the conjugate of $\sqrt{a/b}$, namely $-\sqrt{a/b}$, so that we can later apply Challenge 2.12. Hence we quickly move to consider the expression

$$\left((r + \sqrt{q}) - se^{\sqrt{a/b}}\right)\left((r + \sqrt{q}) - se^{-\sqrt{a/b}}\right) = 0, \qquad (3.3)$$

which is equivalent to

$$(r + \sqrt{q})^2 + s^2 - (r + \sqrt{q})s\left(e^{\sqrt{a/b}} + e^{-\sqrt{a/b}}\right) = 0. \qquad (3.4)$$

It is worth noting how closely (3.4) resembles identity (2.21)—in fact they are identical if we take $q = 0$. Unfortunately, since q is not necessarily 0, the coefficients occurring in (3.4) are not necessarily rational, and thus we cannot use this identity in its current state to produce an integer as we did in the proof of Theorem 2.4.

One way to modify identity (3.4) so that it contains only integer coefficients is to inject additional symmetry by also considering the corresponding version of the left-hand side of (3.4) with the coefficients $r + \sqrt{q}$ replaced by their conjugate $r - \sqrt{q}$. In other words, we consider the more symmetric albeit more complicated identity

$$\left((r + \sqrt{q})^2 + s^2 - (r + \sqrt{q})s\left(e^{\sqrt{a/b}} + e^{-\sqrt{a/b}}\right)\right)$$
$$\times \left((r - \sqrt{q})^2 + s^2 - (r - \sqrt{q})s\left(e^{\sqrt{a/b}} + e^{-\sqrt{a/b}}\right)\right) = 0, \qquad (3.5)$$

which, after expanding, yields the unwieldy

$$\left((r + \sqrt{q})^2 + s^2\right)\left((r - \sqrt{q})^2 + s^2\right) - \left((r + \sqrt{q})s\left((r - \sqrt{q})^2 + s^2\right)\right.$$
$$+ (r - \sqrt{q})s\left((r + \sqrt{q})^2 + s^2\right)\left.\right)\left(e^{\sqrt{a/b}} + e^{-\sqrt{a/b}}\right)$$
$$+ (r + \sqrt{q})(r - \sqrt{q})s^2\left(e^{\sqrt{a/b}} + e^{-\sqrt{a/b}}\right)^2 = 0, \qquad (3.6)$$

which, in turn, after even greater algebraic efforts, gives rise to the expression

$$\left\{\left((r^2 - q)^2 + s^4 + s^2(2r^2 + 2q)\right)\right\} - \left\{(2r^3 - 2rq + 2rs^2)s\left(e^{\sqrt{a/b}} + e^{-\sqrt{a/b}}\right)\right\}$$
$$+ \left\{(r^2 - q)s^2\left(e^{\sqrt{a/b}} + e^{-\sqrt{a/b}}\right)^2\right\} = 0, \qquad (3.7)$$

where we include the superfluous braces to call our attention to the attractive conjugate symmetry of the exponents within those mighty brackets.

The symmetry introduced by the conjugate factor allowed us to realize our goal of producing an expression with *integer* coefficients. We do pay a small price for such an important feature: We are faced with the term $\left(e^{\sqrt{a/b}} + e^{-\sqrt{a/b}}\right)^2$. However, after squaring this expression we see that (3.7) becomes

$$\left((r^2 - q)^2 + s^4 + s^2(2r^2 + 2q) + 2(r^2 - q)s^2\right)$$

$$- (2r^3 - 2rq + 2rs^2)s \left(e^{\sqrt{a/b}} + e^{-\sqrt{a/b}}\right)$$

$$+ (r^2 - q)s^2 \left(e^{2\sqrt{a/b}} + e^{-2\sqrt{a/b}}\right) = 0,$$

which can be simplified slightly to

$$\left\{\left((r^2 - q)^2 + s^4 + 4r^2s^2\right)\right\} - \left\{(2r^3 - 2rq + 2rs^2)s \left(e^{\sqrt{a/b}} + e^{-\sqrt{a/b}}\right)\right\}$$

$$+ \left\{(r^2 - q)s^2 \left(e^{2\sqrt{a/b}} + e^{-2\sqrt{a/b}}\right)\right\} = 0. \tag{3.8}$$

We note that identity (3.8) is the analogue of (2.21) in this context. Thus we wish to simultaneously approximate $e^{\sqrt{a/b}}, e^{-\sqrt{a/b}}, e^{2\sqrt{a/b}}$, and $e^{-2\sqrt{a/b}}$. Inspired by our proofs of the transcendence of e and the irrationality of $e^{\sqrt{a/b}}$, we are led to consider the auxiliary polynomial

$$f(z) = b^{2p} z^{p-1} \left(z - \sqrt{\frac{a}{b}}\right)^p \left(z + \sqrt{\frac{a}{b}}\right)^p \left(z - 2\sqrt{\frac{a}{b}}\right)^p \left(z + 2\sqrt{\frac{a}{b}}\right)^p$$

$$= z^{p-1}(bz^2 - a)^p (bz^2 - 4a)^p \in \mathbb{Z}[z], \tag{3.9}$$

which we write as $f(x) = \sum_{n=p-1}^{5p-1} c_n z^n$. It follows from the arguments of Challenges 2.3 and 2.4 that

$$\sum_{n=1}^{p-1} f^{(n)}(z) = \sum_{N=p-1}^{5p-1} \left(N! c_N \sum_{n=N-p+1}^{N-1} \frac{z^n}{n!}\right) \tag{3.10}$$

and

$$\sum_{n=1}^{p-1} f^{(n)}(\tau) = 0, \tag{3.11}$$

for $\tau \in \{\sqrt{a/b}, -\sqrt{a/b}, 2\sqrt{a/b}, -2\sqrt{a/b}\}$. Arguing as in the proofs of Chapter 2, the two previous identities imply that for $\tau \in \{\sqrt{a/b}, -\sqrt{a/b}, 2\sqrt{a/b}, -2\sqrt{a/b}\}$,

$$e^\tau \sum_{N=p-1}^{5p-1} N! c_N = \sum_{N=p-1}^{5p-1} \left(N! c_N \sum_{n=0}^{\infty} \frac{\tau^n}{n!}\right)$$

$$= \sum_{N=p-1}^{5p-1} \left(N! c_N \sum_{n=0}^{N-p} \frac{\tau^n}{n!}\right) + \sum_{N=p-1}^{5p-1} \left(N! c_N \sum_{n=N}^{\infty} \frac{\tau^n}{n!}\right),$$

which, as before, we write as the polynomial term plus the tail term:

$$e^{\tau} \sum_{N=p-1}^{5p-1} N! c_N = \mathcal{P}_p(\tau) + \mathcal{T}_p(\tau), \tag{3.12}$$

where we again remark that for $N = p - 1$, the inner sum in $\mathcal{P}_p(z)$ is empty and thus equals 0. Thus applying Challenge 2.7 allows us to make the important arithmetical observation that $\frac{1}{p!}\mathcal{P}_p(z)$ is a polynomial with *integer* coefficients. Multiplying identity (3.8) by $N! c_N / (p - 1)!$, then summing both sides from $N = p - 1$ to $5p - 1$, and applying (3.12) leads us to the following exotic identity for 0:

$$\frac{(r^2 - q)^2 + s^4 + 4r^2 s^2}{(p-1)!} \left(\sum_{N=p-1}^{5p-1} N! c_N \right) - \frac{(2r^3 - 2rq + 2rs^2)s}{(p-1)!}$$

$$\times \left(\mathcal{P}_p(\sqrt{a/b}) + \mathcal{T}_p(\sqrt{a/b}) + \mathcal{P}_p(-\sqrt{a/b}) + \mathcal{T}_p(-\sqrt{a/b}) \right) + \frac{(r^2 - q)s^2}{(p-1)!}$$

$$\times \left(\mathcal{P}_p(2\sqrt{a/b}) + \mathcal{T}_p(2\sqrt{a/b}) + \mathcal{P}_p(-2\sqrt{a/b}) + \mathcal{T}_p(-2\sqrt{a/b}) \right) = 0. \tag{3.13}$$

In view of Challenge 2.12 we see that

$$\mathcal{P}_p(\sqrt{a/b}) + \mathcal{P}_p(-\sqrt{a/b}) = \frac{j}{b^{5p-1}} \text{ and } \mathcal{P}_p(2\sqrt{a/b}) + \mathcal{P}_p(-2\sqrt{a/b}) = \frac{k}{b^{5p-1}},$$

for some integers j and k. Given that $\frac{1}{p!}\mathcal{P}_p(z) \in \mathbb{Z}[z]$, it follows that $p!$ is a factor of both j and k. Thus multiplying (3.13) by b^{5p-1} leads to the following complicated identity:

$$\frac{b^{5p-1}\left((r^2 - q)^2 + s^4 + 4r^2 s^2\right)\left(\sum_{N=p-1}^{5p-1} N! c_N\right)}{(p-1)!}$$

$$- \frac{js(2r^3 - 2rq + 2rs^2)}{(p-1)!} + \frac{ks^2(r^2 - q)}{(p-1)!}$$

$$= \frac{b^{5p-1}s(2r^3 - 2rq + 2rs^2)}{(p-1)!}\left(\mathcal{T}_p(\sqrt{a/b}) + \mathcal{T}_p(-\sqrt{a/b})\right)$$

$$- \frac{b^{5p-1}s^2(r^2 - q)}{(p-1)!}\left(\mathcal{T}_p(2\sqrt{a/b}) + \mathcal{T}_p(-2\sqrt{a/b})\right). \tag{3.14}$$

We will now show that the complicated quantity above is our much sought after integer.

Since $p!$ divides both j and k, we immediately have that the quantity

$$-\frac{js(2r^3 - 2rq + 2rs^2)}{(p-1)!} + \frac{ks^2(r^2 - q)}{(p-1)!}$$

is an integer having a factor of p. In addition, by writing $\sum_{N=p-1}^{5p-1} N! c_N$ as $(p-1)! c_{p-1} + \sum_{N=p}^{5p-1} N! c_N$, it is easy to see that

$$\frac{b^{5p-1}\left((r^2 - q)^2 + s^4 + 4r^2 s^2\right)\left(\sum_{N=p-1}^{5p-1} N! c_N\right)}{(p-1)!}$$

$$= b^{5p-1}\left((r^2 - q)^2 + s^4 + 4r^2 s^2\right) c_{p-1}$$

$$+ \frac{b^{5p-1}\left((r^2 - q)^2 + s^4 + 4r^2 s^2\right)\left(\sum_{N=p}^{5p-1} N! c_N\right)}{(p-1)!}$$

is also an integer and furthermore that p divides the second term on the right-hand side of this equality.

We now let p be a prime satisfying

$$p > \max\left\{b,\ (r^2 - q)^2 + s^4 + 4r^2 s^2,\ |c_{p-1}|\right\}.$$

We note that given the definition of $f(z)$, it follows that $c_{p-1} = 4^p a^{2p}$ and thus is nonzero. Similarly, $b \neq 0$, and thus we see that the integer $b^{5p-1}((r^2 - q)^2 + s^4 + 4r^2 s^2) c_{p-1}$ is nonzero and therefore, by our choice of our large prime, not divisible by p. If we let $\mathcal{N}$ denote the left-hand side of the messy identity from (3.14), then our previous remarks reveal that $\mathcal{N}$ is an integer that can be expressed as the sum of four integers, exactly one of which is *not* divisible by p. Hence we conclude that the integer $\mathcal{N}$ itself is not divisible by p and therefore is *nonzero*.

We remark that it is easy to verify that for $\tau \in \left\{\sqrt{a/b}, -\sqrt{a/b}, 2\sqrt{a/b}, -2\sqrt{a/b}\right\}$,

$$|\mathcal{T}_p(\tau)| \leq K_1 (K_2)^p,$$

which is merely the analogue of inequality (2.18), where the constants K_1 and K_2 are defined by $K_1 = \sqrt{\frac{b}{4a}}\, e^{2\sqrt{a/b}}$ and $K_2 = \frac{4a}{b}\left(\frac{8a}{b}\right)^{2\sqrt{a/b}}$. Thus an application of the triangle inequality to identity (3.14) together with the previous bound yields

$$0 < |\mathcal{N}| \leq \left(|s(2r^3 - 2rq + 2rs^2)| + |s^2(r^2 - q)|\right)\frac{2K_1(b^5 K_2)^p}{(p-1)!}.$$

We now see that for all sufficiently large primes p, the right-hand side of the above inequality is less than 1. Thus we have an constructed an integer $|\mathcal{N}|$ strictly between 0 and 1, contradicting the Fundamental Principle of Number Theory. Therefore our original assumption was false; hence $e^{\sqrt{a/b}}$ is not a quadratic irrational.   ∎

We close this section by making several important observations regarding our previous argument. Once we arrived at identity (3.14), the steps that followed were identical to those in the proofs of Theorems 2.3 and 2.4. There are three key features that (3.8) possesses that allowed us to quickly adopt our previous strategies to complete the proof. The first important feature is that all of the coefficients are integers, and the second is the existence of a nonzero integer term free of any e^{α} factors. The last critical feature that (3.8) exhibits is that if α_1 and α_2 are conjugates, then the coefficients of e^{α_1} and e^{α_2} are equal. This feature enables us to factor out that common coefficient, leaving us with a sum of exponential terms having symmetric exponents.

In the next section, we isolate these three key features and prove a special case of the Lindemann–Weierstrass Theorem. Just as we discovered in our previous analysis, in Section 3.4 we will again see that the general case follows from a seemingly specialized version.

3.3 A surprisingly non-special, special case of the Lindemann–Weierstrass Theorem

Inspired by the key features of identity (3.8) that allowed us to apply the ideas developed in Chapter 2 to complete the proof of Theorem 3.8, here we consider the following seemingly special case of the Lindemann–Weierstrass Theorem in which those features appear as hypotheses in the theorem.

THEOREM 3.9 *Let $\alpha_1, \alpha_2, \ldots, \alpha_M$ be distinct nonzero algebraic numbers with the property that if $\alpha_m \in \{\alpha_1, \alpha_2, \ldots, \alpha_M\}$, then each conjugate of α_m is also an element of the set $\{\alpha_1, \alpha_2, \ldots, \alpha_M\}$. Suppose that $\beta_0, \beta_1, \ldots, \beta_M$ are nonzero integers satisfying the condition that if α_i and α_j are conjugates, then $\beta_i = \beta_j$. Then*

$$\beta_0 + \sum_{m=1}^{M} \beta_m e^{\alpha_m} \neq 0.$$

As we hinted at in the previous section, the strategy of the proof is identical to that of Theorem 2.4. The only new requirement is an appropriate generalization of Challenge 2.12. In particular, we will merely need to verify that given a complete set of conjugates $\alpha_1, \alpha_2, \ldots, \alpha_L$ and an arbitrary polynomial $\mathcal{P}(z) \in \mathbb{Z}[z]$, the sum $\mathcal{P}(\alpha_1) + \mathcal{P}(\alpha_2) + \cdots + \mathcal{P}(\alpha_L)$ is a *rational* number with an easily understood denominator. We will now see that the word "merely" in the previous sentence is a bit misplaced, since such a result requires a considerable amount of important algebraic theory. Thus, before we consider the proof of Theorem 3.9, we first embark on a significant detour into the algebraic issues that will be central in this and future arguments.

Algebraic Excursion: Symmetric functions and conjugates

Let $F(x_1, x_2, \ldots, x_L)$ be a function in L variables. We say that F is a *symmetric function* if any rearrangement of the variables does not change the function.

Continued

That is, a function of two variables $F(x_1, x_2)$ is symmetric precisely when $F(x_1, x_2) = F(x_2, x_1)$. A function of three variables $F(x_1, x_2, x_3)$ is symmetric if and only if

$$F(x_1, x_2, x_3) = F(x_1, x_3, x_2) = F(x_2, x_1, x_3) = F(x_2, x_3, x_1)$$
$$= F(x_3, x_1, x_2) = F(x_3, x_2, x_1).$$

In other words, a function in L variables is symmetric if the function remains unchanged after any of the $L!$ permutations of the variables. For example, the following functions are symmetric

$$F(x_1, x_2, x_3) = 4x_1^2 + 4x_2^2 + 4x_3^2 - e^{x_1 x_2 x_3}$$

$$F(x_1, x_2, x_3, x_4) = (x_1 + x_2 + x_3 + x_4)^5 \sin(x_1 x_2 x_3 + x_1 x_2 x_4 + x_1 x_3 x_4 + x_2 x_3 x_4).$$

Challenge 3.3 *Let $F(x_1, x_2, \ldots, x_L)$ and $G(x_1, x_2, \ldots, x_L)$ be two symmetric functions. Show that the sum and product*

$$F(x_1, x_2, \ldots, x_L) + G(x_1, x_2, \ldots, x_L) \quad \text{and} \quad F(x_1, x_2, \ldots, x_L) G(x_1, x_2, \ldots, x_L)$$

are also symmetric functions.

We will be especially interested in symmetric *polynomials*, for example, $P(x_1, x_2) = 3x_1^3 x_2^3 - x_1^2 - x_2^2$. A simple procedure for generating symmetric polynomials in $x_1, x_2, \ldots, x_L$ is to view $x_1, x_2, \ldots, x_L$ as the zeros of a polynomial in z and then consider the *coefficients* of that polynomial:

$$F(z) = (z - x_1)(z - x_2) \cdots (z - x_L) = z^L - \sigma_1 z^{L-1} + \sigma_2 z^{L-2} - \cdots + (-1)^L \sigma_L,$$

where

$$\sigma_1(x_1, x_2, \ldots, x_L) = x_1 + x_2 + \cdots + x_L,$$
$$\sigma_2(x_1, x_2, \ldots, x_L) = x_1 x_2 + x_1 x_3 + \cdots + x_2 x_3 + x_2 x_4 + \cdots + x_{L-1} x_L,$$

$$\vdots$$

$$\sigma_l(x_1, x_2, \ldots, x_L) = \text{the sum of all products of } l \text{ different } x_n\text{'s,}$$

$$\vdots$$

$$\sigma_L(x_1, x_2, \ldots, x_L) = x_1 x_2 \cdots x_L.$$

The symmetric polynomials $\sigma_1, \sigma_2, \ldots, \sigma_L$ are referred to as the *elementary symmetric functions in* $x_1, x_2, \ldots, x_L$. We note that by convention, we suppress the dependency on the variables and just write σ_l rather than the usual function notation $\sigma_l(x_1, x_2, \ldots, x_L)$.

Continued

A natural question that begs to be asked is why the polynomials $\sigma_1, \sigma_2, \ldots, \sigma_L$ are referred to as *elementary* symmetric functions? The answer is illustrated by considering the example $P(x_1, x_2) = 3x_1^3 x_2^3 - x_1^2 - x_2^2$ and noticing that we can express $P(x_1, x_2)$ in terms of the elementary symmetric functions in two variables, $\sigma_1 = x_1 + x_2$ and $\sigma_2 = x_1 x_2$, as

$$P(x_1, x_2) = 3(x_1 x_2)^3 - ((x_1 + x_2)^2 - 2x_1 x_2) = 3\sigma_2^3 - \sigma_1^2 - 2\sigma_2 .$$

Thus the symmetric polynomial $P(x_1, x_2)$ can be decomposed into a polynomial in the elementary symmetric functions. This example illustrates a general principle, namely that *any* symmetric polynomial can be decomposed into a polynomial expression in the elementary symmetric functions. Thus the elementary symmetric functions are the building blocks for all symmetric polynomials. We state this important observation as the following result.

THEOREM 3.10 *Let $P(x_1, x_2, \ldots, x_L)$ be a symmetric polynomial with rational coefficients. Then P can be expressed as a polynomial function in the elementary symmetric functions in $x_1, x_2, \ldots, x_L$ with rational coefficients. That is, there exists a polynomial F with rational coefficients such that*

$$P(x_1, x_2, \ldots, x_L) = F(\sigma_1, \sigma_2, \ldots, \sigma_L).$$

In order to prove this fundamental theorem, we recall that a polynomial is a sum of monomial terms. We begin by defining an ordering on those monomials. Suppose that $\mathcal{M} = cx_1^{n_1} x_2^{n_2} \cdots x_L^{n_L}$ and $\mathcal{M}' = c'x_1^{n_1'} x_2^{n_2'} \cdots x_L^{n_L'}$ are two monomials, where we allow for the possibility that all or some of the exponents are equal to zero. We say that the monomial $\mathcal{M}'$ is *less than* the monomial $\mathcal{M}$, denoted by $\mathcal{M}' < \mathcal{M}$, if $n_1' < n_1$; or $n_1' = n_1$ and $n_2' < n_2$; or $n_1' = n_1, n_2' = n_2,$ and $n_3' < n_3$; or $\ldots$; or $n_1' = n_1, n_2' = n_2, n_3' = n_3, \ldots, n_{L-1}' = n_{L-1},$ and $n_L' < n_L$.

In other words, we compare corresponding exponents of $\mathcal{M}'$ and $\mathcal{M}$ starting with the exponents on x_1 and then x_2 and so forth, and we say that $\mathcal{M}' < \mathcal{M}$ if for the first pair of exponents that differ, the exponent of $\mathcal{M}'$ is less than the corresponding exponent of $\mathcal{M}$. To illustrate this notion, we remark that the largest monomial from the polynomial $\sigma_1 = x_1 + x_2 + \cdots + x_L$ is x_1; and the largest monomial from $\sigma_2 = x_1 x_2 + x_1 x_3 + \cdots + x_{L-1} x_L$ is $x_1 x_2$. In order to solidify this ordering scheme, we offer the following two helpful challenges.

Challenge 3.4 *Given the ordering of monomials as defined above, find the greatest monomial term in each of the elementary symmetric functions in $x_1, x_2, \ldots, x_L$: $\sigma_1, \sigma_2, \ldots, \sigma_L$.*

Challenge 3.5 *Prove that the greatest monomial term of the polynomial $\sigma_1^{n_1} \sigma_2^{n_2} \cdots \sigma_L^{n_L}$ is*

$$x_1^{n_1+n_2+\cdots+n_L} x_2^{n_2+n_3+\cdots+n_L} \cdots x_L^{n_L} .$$

Continued

Given $\mathcal{M} = cx_1^{n_1}x_2^{n_2}\cdots x_L^{n_L}$, we define the *degree of the monomial*, $\deg(\mathcal{M})$, by $\deg(\mathcal{M}) = n_1 + n_2 + \cdots + n_L$. We define the *degree* of a polynomial to be the maximum degree of its monomials. If all the monomials of a polynomial have the same degree, then the polynomial is said to be *homogeneous*. We are now ready to take an important step towards the proof of Theorem 3.10.

Challenge 3.6   *Suppose that $P(x_1, x_2, \ldots, x_L)$ is a homogeneous, symmetric polynomial written so that all monomials having identical exponents have been combined. Thus each monomial has a unique ordered list of exponents—for if not, then we could further combine terms. Let $\mathcal{M} = cx_1^{n_1}x_2^{n_2}\cdots x_L^{n_L}$ be the greatest monomial term of P. Prove that $n_L \leq n_{L-1} \leq \cdots \leq n_2 \leq n_1$. Use this observation together with Challenge 3.5 to show that*

$$\mathcal{M} - c\sigma_1^{n_1-n_2}\sigma_2^{n_2-n_3}\cdots\sigma_{L-1}^{n_{L-1}-n_L}\sigma_L^{n_L}$$

is a polynomial whose largest monomial is strictly less than $\mathcal{M}$. Thus conclude that

$$P(x_1, x_2, \ldots, x_L) - c\sigma_1^{n_1-n_2}\sigma_2^{n_2-n_3}\cdots\sigma_{L-1}^{n_{L-1}-n_L}\sigma_L^{n_L}$$

either is the identically zero polynomial or is a homogeneous, symmetric polynomial with its greatest monomial term strictly less than $\mathcal{M}$.

Of course in mathematics, once we have a good idea, we try to apply it as often as possible. The next, critical challenge illustrates the power of this important maxim.

Challenge 3.7   *Prove the following lemma by repeated applications of Challenge 3.6.*

LEMMA 3.11   *Let $P(x_1, x_2, \ldots, x_L)$ be a homogeneous, symmetric polynomial with rational coefficients. Then $P(x_1, x_2, \ldots, x_L)$ can be expressed as a polynomial F in the elementary symmetric functions $\sigma_1, \sigma_2, \ldots, \sigma_L$ with rational coefficients. Moreover, $\deg(F) \leq \deg(P)$.*

We are now ready for the big finish, which, after the next challenge, we will discover is slightly bigger than we even anticipated.

Challenge 3.8   *Using Lemma 3.11, give a proof of Theorem 3.10. In addition, show that given P and F as defined in the theorem, it follows that $\deg(F) \leq \deg(P)$.*

We remark that our proof of Theorem 3.10 actually provides an algorithm for computing the polynomial F. It is clear by the algorithm that the coefficients of F are integer linear combinations of the coefficients of P. Thus we have actually proved more:

Continued

THEOREM 3.12 *Let $P(x_1, x_2, \ldots, x_L)$ be a symmetric polynomial. Then P can be expressed as a polynomial function of the elementary symmetric functions in $x_1, x_2, \ldots, x_L$. That is, there exists a polynomial F such that*

$$P(x_1, x_2, \ldots, x_L) = F(\sigma_1, \sigma_2, \ldots, \sigma_L).$$

Moreover, $\deg(F) \leq \deg(P)$, and each coefficient of F is an integer linear combination of coefficients of P.

We now bring the conjugates of algebraic numbers onto the scene and, generalizing what we saw in the proof of Theorem 3.8, discover how beautifully they fit together when combined in a symmetric manner. Let $G(z)$ be an integral polynomial, which may or may not be irreducible. We write $G(z) = \sum_{l=0}^{L} a_l z^l$ and factor $G(z)$ as

$$G(z) = a_L(z - \alpha_1)(z - \alpha_2) \cdots (z - \alpha_L).$$

If we factor $G(z)$ into irreducible factors, then we discover that the set of algebraic numbers $\{\alpha_1, \alpha_2, \ldots, \alpha_L\}$ is either a complete set of conjugates or several collections of complete sets of conjugates. We immediately conclude that the elementary symmetric functions in the α_l's are *rational* numbers. In particular, by noting that

$$
\begin{aligned}
G(z) &= a_L\left(z^L + \frac{a_{L-1}}{a_L}z^{L-1} + \cdots + \frac{a_1}{a_L}z + \frac{a_0}{a_L}\right) \\
&= a_L(z - \alpha_1)(z - \alpha_2) \cdots (z - \alpha_L) \\
&= a_L\Big(z^L - (\alpha_1 + \alpha_2 + \cdots + \alpha_L)z^{L-1} \\
&\qquad + (\alpha_1\alpha_2 + \alpha_1\alpha_3 + \cdots + \alpha_{L-1}\alpha_L)z^{L-2} + \cdots \\
&\qquad + (-1)^L(\alpha_1\alpha_2 \cdots \alpha_L)\Big),
\end{aligned}
$$

we discover that

$$\sigma_1 = \alpha_1 + \alpha_2 + \cdots + \alpha_L = -\frac{a_{L-1}}{a_L},$$

$$\sigma_2 = \alpha_1\alpha_2 + \alpha_1\alpha_3 + \cdots + \alpha_{L-1}\alpha_L = \frac{a_{L-2}}{a_L},$$

$$\vdots$$

$$\sigma_L = \alpha_1\alpha_2 \cdots \alpha_L = (-1)^L\frac{a_0}{a_L}.$$

(3.15)

Continued

In view of the identities in (3.15) and Theorem 3.12 we have the following useful result.

THEOREM 3.13 *Let $S = \{\alpha_1, \alpha_2, \ldots, \alpha_L\}$ be a set of algebraic numbers with the property that if $\alpha \in S$, then all of the conjugates of α are also in S. Moreover, if α appears in the set S m times, then so does each of its conjugates. Let $P(x_1, x_2, \ldots, x_L)$ be a symmetric polynomial. Then $P(\alpha_1, \alpha_2, \ldots, \alpha_L)$ is a rational linear combination of the coefficients of P.*

We close this excursion by finally giving a generalization of Challenge 2.12. In this direction we begin with a very simple challenge.

Challenge 3.9 Let $\mathcal{P}(z) \in \mathbb{Z}[z]$. Show that the polynomial

$$P(x_1, x_2, \ldots, x_L) = \mathcal{P}(x_1) + \mathcal{P}(x_2) + \cdots + \mathcal{P}(x_L)$$

is a symmetric polynomial with integer coefficients.

Our generalization of the result in Challenge 2.12 can now be given by

LEMMA 3.14 *Let $G(z) \in \mathbb{Z}[z]$ be given by $G(z) = \sum_{l=0}^{L} a_l z^l$ and let $\alpha_1, \alpha_2, \ldots, \alpha_L$ denote the zeros of $G(z)$. Let $\mathcal{P}(z)$ be any polynomial with integer coefficients. Then the quantity*

$$\mathcal{P}(\alpha_1) + \mathcal{P}(\alpha_2) + \cdots + \mathcal{P}(\alpha_L) \tag{3.16}$$

is a rational number with denominator equal to $a_L^{\deg(\mathcal{P})}$.

Proof. From Challenge 3.9 we know that (3.16) is a symmetric function in the α_l's. Thus by Theorem 3.12 we know that there exists a polynomial F with integer coefficients of degree at most the degree of $\mathcal{P}$ such that

$$\mathcal{P}(\alpha_1) + \mathcal{P}(\alpha_2) + \cdots + \mathcal{P}(\alpha_L) = F(\sigma_1, \sigma_2, \ldots, \sigma_L).$$

By the identities of (3.15), we see that each of the elementary symmetric functions in the α_l's is a rational number having denominator a_L. Therefore when we evaluate F at these rational numbers, we produce a rational number with denominator $a_L^{\deg(F)}$. Given that $\deg(F) \leq \deg(\mathcal{P})$, we can multiply the numerator and denominator of this rational number by a suitable power of a_L and thus write $\mathcal{P}(\alpha_1) + \mathcal{P}(\alpha_2) + \cdots + \mathcal{P}(\alpha_L)$ as a rational number having denominator equal to $a_L^{\deg(\mathcal{P})}$, which completes our proof and our extended algebraic excursion. ∎

3.4 A conjugate appetizer—The delicate but surprisingly non-special, special case

We are now in position to prove the special case of the Lindemann–Weierstrass Theorem, which we state again:

THEOREM 3.9   *Let $\alpha_1, \alpha_2, \ldots, \alpha_M$ be distinct nonzero algebraic numbers with the property that if $\alpha_m \in \{\alpha_1, \alpha_2, \ldots, \alpha_M\}$, then each conjugate of α_m is also an element of the set $\{\alpha_1, \alpha_2, \ldots, \alpha_M\}$. Suppose that $\beta_0, \beta_1, \ldots, \beta_M$ are nonzero integers satisfying the condition that if α_i and α_j are conjugates, then $\beta_i = \beta_j$. Then*

$$\beta_0 + \sum_{m=1}^{M} \beta_m e^{\alpha_m} \neq 0.$$

Proof of Theorem 3.9.   As always, we proceed by contradiction. In this case, we assume that there exist nonzero integers $\beta_0, \beta_1, \ldots, \beta_M$ satisfying the hypotheses for which

$$\beta_0 + \sum_{m=1}^{M} \beta_m e^{\alpha_m} = 0. \tag{3.17}$$

Our next step is to reorder the α_m's so that their conjugates are grouped together.

Since the set $\{\alpha_1, \alpha_2, \ldots, \alpha_M\}$ consists of complete collections of conjugates, we can rename and group the α_m's by

$$\alpha_{11}, \alpha_{12}, \ldots, \alpha_{1M_1}, \quad \alpha_{21}, \alpha_{22}, \ldots, \alpha_{2M_2}, \quad \ldots, \quad \alpha_{L1}, \alpha_{L2}, \ldots, \alpha_{LM_L},$$

where $\{\alpha_{l1}, \alpha_{l2}, \ldots, \alpha_{lM_l}\}$ is a complete collection of conjugates. By the hypothesis on the β_m's, we can reorder and rename the coefficients in sympathy with our renaming of the α_m's. In particular, we now let β_l denote the coefficients $\beta_{l1} = \beta_{l2} = \cdots = \beta_{lM_l}$. Thus we can rewrite (3.17) as

$$\beta_0 + \sum_{l=1}^{L} \beta_l \left(\sum_{m=1}^{M_l} e^{\alpha_{lm}} \right) = 0. \tag{3.18}$$

If we write

$$f_l(z) = (z - \alpha_{l1})(z - \alpha_{l2}) \cdots (z - \alpha_{lM_l}),$$

then since $\{\alpha_{l1}, \alpha_{l2}, \ldots, \alpha_{lM_l}\}$ is a complete set of conjugates, we know that $f_l(z)$ is a polynomial with rational coefficients. Hence there exists a positive integer d_l such that the polynomial $d_l f_l(z)$ has *integer* coefficients.

We now define our auxiliary polynomial $f(z)$ by

$$f(z) = (d_1 d_2 \cdots d_L)^p z^{p-1} f_1(z)^p f_2(z)^p \cdots f_L(z)^p, \tag{3.19}$$

which we write as

$$f(z) = \sum_{n=p-1}^{(M+1)p-1} c_n z^n.$$

By our previous observations we see that $f(z) \in \mathbb{Z}[z]$; that is, the coefficients c_n are integers. Clearly, we have that the integer c_{p-1} satisfies

$$c_{p-1} = \pm(d_1 d_2 \cdots d_L \alpha_1 \alpha_2 \cdots \alpha_M)^p,$$

which in view of the hypothesis that all the α_m's are nonzero leads to the trivial observation that

$$c_{p-1} \neq 0. \tag{3.20}$$

This simple observation will be crucial in our ultimate contradiction to the Fundamental Principle of Number theory.

We remark that the function $f(z)$ above is a generalization of the function defined in (3.9). Again, as we have seen in Challenges 2.3 and 2.4, we may conclude that

$$\sum_{n=1}^{p-1} f^{(n)}(z) = \sum_{N=p-1}^{(M+1)p-1} \left(N! c_N \sum_{n=N-p+1}^{N-1} \frac{z^n}{n!} \right), \tag{3.21}$$

and

$$\sum_{n=1}^{p-1} f^{(n)}(\alpha) = 0, \tag{3.22}$$

for any $\alpha \in \{\alpha_1, \alpha_2, \ldots, \alpha_M\}$.

For each $\alpha \in \{\alpha_1, \alpha_2, \ldots, \alpha_M\} = \{\alpha_{11}, \ldots, \alpha_{1M_1}, \ldots, \alpha_{L1}, \ldots, \alpha_{LM_L}\}$, the power series for e^α allows us to write

$$e^\alpha \sum_{N=p-1}^{(M+1)p-1} N! c_N = \sum_{N=p-1}^{(M+1)p-1} \left(N! c_N \sum_{n=0}^{\infty} \frac{\alpha^n}{n!} \right)$$

$$= \sum_{N=p-1}^{(M+1)p-1} \left(N! c_N \sum_{n=0}^{N-p} \frac{\alpha^n}{n!} \right) + \sum_{N=p-1}^{(M+1)p-1} \left(N! c_N \sum_{n=N-p+1}^{N-1} \frac{\alpha^n}{n!} \right)$$

$$+ \sum_{N=p-1}^{(M+1)p-1} \left(N! c_N \sum_{n=N}^{\infty} \frac{\alpha^n}{n!} \right).$$

The middle summand on the right-hand side is precisely $\sum_{n=1}^{p-1} f^{(n)}(\alpha)$, so in view of (3.22) we have

$$e^\alpha \sum_{N=p-1}^{(M+1)p-1} N! c_N = \sum_{N=p-1}^{(M+1)p-1} \left(N! c_N \sum_{n=0}^{N-p} \frac{\alpha^n}{n!} \right) + \sum_{N=p-1}^{(M+1)p-1} \left(N! c_N \sum_{n=N}^{\infty} \frac{\alpha^n}{n!} \right),$$

which yet again we rewrite in terms of a polynomial and a tail as

$$e^{\alpha} \sum_{N=p-1}^{(M+1)p-1} N!c_N = \mathcal{P}_p(\alpha) + \mathcal{T}_p(\alpha). \tag{3.23}$$

We now return to our original assumption (3.17): $\beta_0 + \sum_{l=1}^{L} \beta_l \left(\sum_{m=1}^{M_l} e^{\alpha_{lm}} \right) = 0$. If we multiply both sides of this identity by $N!c_N/(p-1)!$, and sum both sides from $N = p-1$ to $N = (M+1)p - 1$, then in view of (3.18) we deduce that

$$\frac{\beta_0}{(p-1)!} \sum_{N=p-1}^{(M+1)p-1} N!c_N + \frac{1}{(p-1)!} \sum_{l=1}^{L} \beta_l \left(\sum_{m=1}^{M_l} \mathcal{P}_p(\alpha_{lm}) + \mathcal{T}_p(\alpha_{lm}) \right) = 0. \tag{3.24}$$

Rewriting this equation produces

$$\beta_0 \sum_{N=p-1}^{(M+1)p-1} \frac{N!c_N}{(p-1)!} + \sum_{l=1}^{L} \beta_l \left(\sum_{m=1}^{M_l} \frac{\mathcal{P}_p(\alpha_{lm})}{(p-1)!} \right) = -\sum_{l=1}^{L} \beta_l \left(\sum_{m=1}^{M_l} \frac{\mathcal{T}_p(\alpha_{lm})}{(p-1)!} \right). \tag{3.25}$$

Once again recalling that for $N = p-1$, the inner sum in $\mathcal{P}_p(z)$ is empty and thus equals 0, we can apply Challenge 2.7 to conclude that $\frac{1}{p!} \mathcal{P}_p(z) \in \mathbb{Z}[z]$. Combining this observation with Lemma 3.14, we conclude that for each l there exists an integer a_l having a factor of p such that the quantity

$$\sum_{m=1}^{M_l} \frac{\mathcal{P}_p(\alpha_{lm})}{(p-1)!}$$

is a rational number of the form a_l/d_l^{Mp-1}. Hence (3.25) reduces to

$$\beta_0 \sum_{N=p-1}^{(M+1)p-1} \frac{N!c_N}{(p-1)!} + \sum_{l=1}^{L} \frac{a_l \beta_l}{d_l^{Mp-1}} = -\sum_{l=1}^{L} \beta_l \left(\sum_{m=1}^{M_l} \frac{\mathcal{T}_p(\alpha_{lm})}{(p-1)!} \right). \tag{3.26}$$

Multiplying both sides of (3.26) by the integer D^{Mp}, where $D = d_1 d_2 \cdots d_L$, yields

$$\beta_0 D^{Mp} \sum_{N=p-1}^{(M+1)p-1} \frac{N!c_N}{(p-1)!} + \sum_{l=1}^{L} a_l \beta_l d_l (D/d_l)^{Mp} = -\sum_{l=1}^{L} \beta_l D^{Mp} \left(\sum_{m=1}^{M_l} \frac{\mathcal{T}_p(\alpha_{lm})}{(p-1)!} \right). \tag{3.27}$$

Since D/d_l is an integer, we note that the quantity

$$\mathcal{N} = \beta_0 D^{Mp} \sum_{N=p-1}^{(M+1)p-1} \frac{N!c_N}{(p-1)!} + \sum_{l=1}^{L} a_l \beta_l d_l (D/d_l)^{Mp}$$

is also an integer. To verify that $\mathcal{N}$ is nonzero, we recall our trivial observation from (3.20) that $c_{p-1} \neq 0$, and rewrite $\mathcal{N}$ as

$$\mathcal{N} = \beta_0 D^{Mp} c_{p-1} + \left(\beta_0 D^{Mp} \sum_{N=p}^{(M+1)p-1} \frac{N! c_N}{(p-1)!} + \sum_{l=1}^{L} a_l \beta_l d_l (D/d_l)^{Mp} \right). \quad (3.28)$$

Next we observe that since p divides a_l for each l, then p divides the integer contained in the parentheses. On the other hand, we remark that by our hypothesis, β_0 is a nonzero integer. Thus if p is a prime number such that $p > \max\{|\beta_0|, D, |c_{p-1}|\}$, then $\beta_0 D^{Mp} c_{p-1}$ is a nonzero integer *not* divisible by p. Since the prime p does not divide $\beta_0 D^{Mp} c_{p-1}$ but does divide the parenthetical integer in (3.28), we conclude that p does not divide the integer $\mathcal{N}$ and thus $\mathcal{N}$ is a *nonzero* integer.

We now turn our attention to showing that $|\mathcal{N}|$ violates the Fundamental Principle of Number Theory. By (3.27) and the triangle inequality, we conclude that the integer $\mathcal{N}$ satisfies

$$0 < |\mathcal{N}| \leq \sum_{l=1}^{L} \left(\sum_{m=1}^{M_l} \frac{|\beta_l D^{Mp} \mathcal{T}_p(\alpha_{lm})|}{(p-1)!} \right). \quad (3.29)$$

Challenge 3.10 *Modify the argument that produced inequality (2.18) to deduce that there exist two constants K_1 and K_2 that do not depend on p such that for any $\alpha \in \{\alpha_1, \alpha_2, \ldots, \alpha_M\}$,*

$$|\mathcal{T}_p(\alpha)| \leq K_1 (K_2)^p.$$

If we let $B = \max\{|\beta_1|, |\beta_2|, \ldots, |\beta_L|\}$, then, in view of the previous challenge and (3.29), we see that

$$0 < |\mathcal{N}| \leq BMK_1 \frac{(D^M K_2)^p}{(p-1)!}.$$

Given that BMK_1 and $D^M K_2$ are constants independent of p, we conclude that for all sufficiently large primes p, the integer $\mathcal{N}$ satisfies

$$0 < |\mathcal{N}| < 1,$$

which, at long last, is a contradiction to the Fundamental Principle of Number Theory, and thus our assumption (3.17) is false, and we therefore have that

$$\beta_0 + \sum_{m=1}^{M} \beta_m e^{\alpha_m} \neq 0,$$

which completes our proof.   ∎

3.5 The main dish—Serving up a spaghetti of symmetry

We are now ready to delve into the proof of the original, conjugate-neutral Lindemann–Weierstrass Theorem, which, in order to whet our appetites, we state again.

THEOREM 3.4 *Let $\alpha_0, \alpha_1, \ldots, \alpha_M$ be $M + 1$ distinct algebraic numbers. Then*

$$e^{\alpha_0}, e^{\alpha_1}, \ldots, e^{\alpha_M}$$

are linearly independent over the algebraic numbers. That is, if $\beta_0, \beta_1, \ldots, \beta_M$ are nonzero algebraic numbers, then

$$\sum_{m=0}^{M} \beta_m e^{\alpha_m} \neq 0.$$

To prove this theorem, it will be enough to show that it follows from the special case we already proved, namely Theorem 3.9. That special case was special because of all those helpful hypotheses that the full-strength version lacks. However, in our discussion of the intuitive idea behind Theorem 3.8, we have already touched on a method for massaging the general case into the special case. In particular, we can clean up the exponents by symmetrizing with respect to their conjugates. This step leads to an expression in which all the conjugate exponents possess the same coefficients. We then clean up the algebraic coefficients once again by the symmetry introduced by their conjugates to create rational coefficients. We can then simply clear denominators so we are left with an expression having integer coefficients and thus precisely the hypotheses of Theorem 3.9.

As the previous paragraph may suggest, the argument ahead is notationally treacherous. Before embarking upon that perilous journey, we encourage the reader to move to Section 3.6 and safely enjoy the big picture together with some important consequences of Theorem 3.4.

The intuitive idea for transforming the general case into the special case

Our proof of Theorem 3.4 involves three major steps.

1. Clean up the exponents so that terms with conjugate exponents have equal coefficients.
2. Clean up the coefficients so that they are all integers.
3. Ensure that one of the exponents equals zero—so we have a lone β_0 term.

In moving forward we must be cautious of one delicate, nontrivial point: We must be sure that while carrying each of the above steps, we do not accidently create

Continued

some dramatic conspiracy among the coefficients causing all of them to become zero. Ensuring that we avoid a coefficient massacre is the one last remaining detail we must address.

An illustration of Step 1. To illustrate the first step, let us suppose that $\alpha_0 = 0$, $\alpha_1 = \sqrt{2}$, and $\alpha_2 = \sqrt{3}$ and

$$\beta_0 + \beta_1 e^{\sqrt{2}} + \beta_2 e^{\sqrt{3}} = 0. \tag{3.30}$$

Inspired by the intuitive overview from Section 3.2, we multiply (3.30) by analogous copies of that sum with all possible conjugates in place of the exponents $\sqrt{2}$ and $\sqrt{3}$:

$$\left(\beta_0 + \beta_1 e^{\sqrt{2}} + \beta_2 e^{\sqrt{3}}\right)\left(\beta_0 + \beta_1 e^{-\sqrt{2}} + \beta_2 e^{\sqrt{3}}\right)$$
$$\times \left(\beta_0 + \beta_1 e^{\sqrt{2}} + \beta_2 e^{-\sqrt{3}}\right)\left(\beta_0 + \beta_1 e^{-\sqrt{2}} + \beta_2 e^{-\sqrt{3}}\right) = 0. \tag{3.31}$$

Expanding the previous expression involves straightforward but tedious algebra. That headache-inducing exercise is easy to do incorrectly—we know this from personal experience. However, once it is expanded and our headache subsides, we are left with an expression of the form

$$b_0 + b_1\left(e^{\sqrt{2}} + e^{-\sqrt{2}}\right) + b_2\left(e^{-\sqrt{3}} + e^{\sqrt{3}}\right) + b_3\left(e^{2\sqrt{2}} + e^{-2\sqrt{2}}\right)$$
$$+ b_4\left(e^{-2\sqrt{3}} + e^{2\sqrt{3}}\right) + b_5\left(e^{\sqrt{2}-\sqrt{3}} + e^{-\sqrt{2}-\sqrt{3}} + e^{\sqrt{2}+\sqrt{3}} + e^{-\sqrt{2}+\sqrt{3}}\right)$$
$$+ b_6\left(e^{2\sqrt{2}-\sqrt{3}} + e^{-2\sqrt{2}-\sqrt{3}} + e^{2\sqrt{2}+\sqrt{3}} + e^{-2\sqrt{2}+\sqrt{3}}\right)$$
$$+ b_7\left(e^{\sqrt{2}-2\sqrt{3}} + e^{-\sqrt{2}-2\sqrt{3}} + e^{\sqrt{2}+2\sqrt{3}} + e^{-\sqrt{2}+2\sqrt{3}}\right) = 0,$$

where each of the coefficients $b_0, b_1, \ldots, b_7$ is a polynomial in three variables with integer coefficients evaluated at $(\beta_0, \beta_1, \beta_2)$. For example, we note that

$$b_0 = \beta_0^4 + \beta_1^4 + \beta_2^4 + 2\beta_1^2\beta_2^2 + 4\beta_0^2\beta_1^2 + 4\beta_0^2\beta_2^2,$$

while $b_6 = \beta_0\beta_1^2\beta_2$ and $b_7 = \beta_0\beta_1\beta_2^2$. Thus while it is clear that b_6 and b_7 are nonzero, we note that it is entirely possible that $b_0 = 0$ since some β_m's may be complex. Indeed, any, or all, of the coefficients $b_0, b_1, \ldots, b_5$ *might* vanish. We promise to revisit this concerning coefficient-vanishing issue.

While some may be in awe of our algebraic skills and perhaps others may have verified our assertions about the precise values of b_0, b_6, and b_7, we hope that everyone is able to appreciate the one critical feature regarding that expansive identity: If two exponents γ_1 and γ_2 are conjugates, then the coefficient of e^{γ_1}

Continued

is equal to the coefficient of e^{y_2}. To illustrate this observation more explicitly, let us examine the exponents $\sqrt{2} - 2\sqrt{3}, -\sqrt{2} - 2\sqrt{3}, \sqrt{2} + 2\sqrt{3}, -\sqrt{2} + 2\sqrt{3}$ by considering the polynomial having all those numbers as zeros, that is,

$$p(z) = \left(z - (\sqrt{2} - 2\sqrt{3})\right)\left(z - (-\sqrt{2} - 2\sqrt{3})\right)$$
$$\times \left(z - (\sqrt{2} + 2\sqrt{3})\right)\left(z - (-\sqrt{2} + 2\sqrt{3})\right).$$

Notice that both the first two factors and the last two factors are symmetric in $\sqrt{2}$ and $-\sqrt{2}$. If we multiply the first two terms and the last two terms together, we see that

$$p(z) = (z^2 + 4\sqrt{3}z + 10)(z^2 - 4\sqrt{3}z + 10);$$

in particular, we notice that by multiplying similar factors containing the conjugates $\sqrt{2}$ and $-\sqrt{2}$ together, we generate elementary symmetric functions in $\sqrt{2}$ and $-\sqrt{2}$, and thus the $\sqrt{2}$ coefficients dissappear. The two remaining factors are symmetric in $\sqrt{3}$ and $-\sqrt{3}$. If we multiply those two factors together, we discover that

$$p(z) = z^4 - 28z^2 + 100,$$

which, not surprisingly, has integer coefficients. We will prove below that this observation implies that the conjugate of each element of the collection

$$\left\{\sqrt{2} - 2\sqrt{3}, -\sqrt{2} - 2\sqrt{3}, \sqrt{2} + 2\sqrt{3}, -\sqrt{2} + 2\sqrt{3}\right\}$$

must be contained in that collection.

Thus we see that multiplying (3.30) by similiar expressions formed by all possible replacements of exponents with their corresponding conjugates, we arrive at an expression in which terms with conjugate exponents have the same coefficients. Hence we have realized Step 1 above. We need only manipulate that expression a bit further to produce *integer* coefficients. As we observed earlier, this last step is accomplished by repeating the above process with respect to the coefficients rather than the exponents.

An illustration of Step 2. We illustrate the process of clearing algebraic coefficients with a very simple expression in which the exponents have already been grouped with their conjugate terms:

$$3 + \sqrt{5}\left(e^{\sqrt{6}} + e^{-\sqrt{6}}\right) + \frac{\sqrt{7}}{2}\left(e^{\sqrt{11}} + e^{-\sqrt{11}}\right) = 0.$$

Continued

Multiplying by all possible versions of the above obtained by replacing the coefficients with their conjugates gives rise to the product

$$
\left(3 + \sqrt{5}\left(e^{\sqrt{6}} + e^{-\sqrt{6}}\right) + \frac{\sqrt{7}}{2}\left(e^{\sqrt{11}} + e^{-\sqrt{11}}\right)\right)
$$

$$
\times \left(3 - \sqrt{5}\left(e^{\sqrt{6}} + e^{-\sqrt{6}}\right) + \frac{\sqrt{7}}{2}\left(e^{\sqrt{11}} + e^{-\sqrt{11}}\right)\right)
$$

$$
\times \left(3 + \sqrt{5}\left(e^{\sqrt{6}} + e^{-\sqrt{6}}\right) - \frac{\sqrt{7}}{2}\left(e^{\sqrt{11}} + e^{-\sqrt{11}}\right)\right)
$$

$$
\times \left(3 - \sqrt{5}\left(e^{\sqrt{6}} + e^{-\sqrt{6}}\right) - \frac{\sqrt{7}}{2}\left(e^{\sqrt{11}} + e^{-\sqrt{11}}\right)\right) = 0.
$$

If we let $E_1 = e^{\sqrt{6}} + e^{-\sqrt{6}}$ and $E_2 = e^{\sqrt{11}} + e^{-\sqrt{11}}$, then the above becomes the slightly more manageable

$$
\left(3 + \sqrt{5}E_1 + \frac{\sqrt{7}}{2}E_2\right)\left(3 - \sqrt{5}E_1 + \frac{\sqrt{7}}{2}E_2\right)
$$

$$
\times \left(3 + \sqrt{5}E_1 - \frac{\sqrt{7}}{2}E_2\right)\left(3 - \sqrt{5}E_1 - \frac{\sqrt{7}}{2}E_2\right) = 0.
$$

By the symmetry of the $\sqrt{5}$ and $-\sqrt{5}$ terms in the first and last pairs of factors, we know that if those two pairs are combined, the $\sqrt{5}$ terms will vanish. Multiplying the first two terms and the last two terms confirms our hunch and produces

$$
\left(9 - 5E_1^2 + \frac{7}{4}E_2^2 + 3\sqrt{7}E_2\right)\left(9 - 5E_1^2 + \frac{7}{4}E_2^2 - 3\sqrt{7}E_2\right) = 0.
$$

Again the symmetry among the conjugates $3\sqrt{7}$ and $-3\sqrt{7}$ indicates that the final product will be free of those pesky $\sqrt{7}$ coefficients. In fact, that product can be expanded to reveal

$$
\left(9 - 5E_1^2 + \frac{7}{4}E_2^2\right)^2 - 63E_2^2 = 0,
$$

which if we multiply through by 16 produces an expression with *integer* coefficients as we desired for Step 2.

This coefficient-cleaning appears to come at a significant cost—for now we no longer see sums of only E_1 and E_2 terms, but here we have sums of

Continued

$E_1^2, E_2^2, E_1^2 E_2^2, E_1^4$, and E_2^4. Thus we now need to multiply all those exponential terms, which will introduce *new* algebraic exponents. After performing all that multiplication, two crucial questions remain: Have we preserved the first feature that we worked so hard to achieve; namely, do terms that have conjugate exponents possess the same coefficients? And are we guaranteed that we avoided some amazing conspiracy where by *all* the coefficients magically combine to equal 0? The answer to both questions is "yes" (we leave the verification of these assertions in the specific example above as an exercise to the restless reader). Therefore what remains in completing the move from the general theorem to the not-so-special case of Theorem 3.9 is verifying the two previous assertions in general and realizing Step 3.

Before turning to the proof of the theorem, we address the two assertions alluded to at the end of our intuitive overview. We begin by establishing a result that guarantees we never turn all the coefficients to 0, and we then demonstrate that once we have conjugate exponents possessing equal coefficients, they will continue to have equal coefficients as we modify our identity. The following challenge will allow us to guarantee that we will never wipe out an entire colony of coefficients.

Challenge 3.11 *Let* $\rho_1, \rho_2, \ldots, \rho_L, \tau_1, \tau_2, \ldots, \tau_M$ *be distinct complex numbers and let* $r_1, r_2, \ldots, r_L, t_1, t_2, \ldots, t_M$ *be nonzero complex numbers. If the product*

$$\left(\sum_{l=1}^{L} r_l e^{\rho_l} \right) \left(\sum_{m=1}^{M} t_m e^{\tau_m} \right) \tag{3.32}$$

is expanded and like exponential terms are combined, then show we are left with an expression

$$\sum_{n=1}^{N} s_n e^{\lambda_n},$$

for some positive integer N, *complex numbers* s_n, *and distinct exponents* λ_n *of the form* $\lambda_n = \rho_l + \tau_m$. *Then prove that at least one of the coefficients* s_n *is nonzero. (Hint: Let* $\{\rho_{l_1}, \rho_{l_2}, \ldots, \rho_{l_K}\} \subseteq \{\rho_1, \rho_2, \ldots, \rho_L\}$ *be the subset of all numbers having maximal real parts, and show there is a unique element* $\rho_s \in \{\rho_{l_1}, \rho_{l_2}, \ldots, \rho_{l_K}\}$ *with maximal imaginary part. Let* τ_t *be the analogous unique maximal element from* $\{\tau_1, \tau_2, \ldots, \tau_M\}$. *Now show that when the product in* (3.32) *is expanded, the coefficient of* $e^{\rho_s + \tau_t}$ *is nonzero.)*

We now codify an idea that has been floating around this entire section. We say that a finite set of algebraic numbers $\mathcal{S}$ is *conjugate complete* if for every $\alpha \in \mathcal{S}$, $\mathcal{S}$ contains all the conjugates of α, and moreover, if α appears in $\mathcal{S}$ m times, then so does each of its conjugates. So, for example, $\{\sqrt{2}, \sqrt{3}, -\sqrt{2}, -\sqrt{3}\}$ is a conjugate complete set, while $\{\sqrt{2}, \sqrt{3}, \sqrt{3}, -\sqrt{2}, -\sqrt{3}\,\}$ is not, since it is missing an additional $-\sqrt{3}$ element.

Challenge 3.12 *Prove that a set* $\{\alpha_1, \alpha_2, \ldots, \alpha_L\}$ *is conjugate complete if and only if the polynomial*

$$(x - \alpha_1)(x - \alpha_2) \cdots (x - \alpha_L)$$

has rational coefficients.

LEMMA 3.15 *Suppose that* $\{\gamma_1, \gamma_2, \ldots, \gamma_J\}$ *is a set of arbitrary algebraic numbers and* $f(x_1, x_2, \ldots, x_J) = a_1 x_1 + a_2 x_2 + \cdots + a_J x_J$ *where* $a_j \in \mathbb{Z}$ *for all* j. *Then the set*

$$\{f(\tau_1, \tau_2, \ldots, \tau_J) : \tau_j \text{ is a conjugate of } \gamma_j \text{ for all } j = 1, 2, \ldots, J\}$$

is conjugate complete.

Proof. Let us consider the associated polynomial

$$p(z) = \prod_{\substack{\tau_j \text{ a conjugate of } \gamma_j \\ \text{for } j=1,2,\ldots,J}} \left(z - f(\tau_1, \tau_2, \ldots, \tau_J) \right).$$

If we view $p(z)$ as

$$p(z) = \prod_{\substack{\tau_j \text{ a conjugate of } \gamma_j \\ \text{for } j=2,3,\ldots,J}} \left(\prod_{\tau_1 \text{ a conjugate of } \gamma_1} \left(z - f(\tau_1, \tau_2, \ldots, \tau_J) \right) \right),$$

then we see that the coefficients of the polynomial within the large parentheses are symmetric polynomial functions in the conjugates of γ_1. Thus it follows from Theorem 3.12 that the conjugates of γ_1 combine to create rational numbers, and we are thus left with coefficients in terms of the coefficients of $\tau_2, \ldots, \tau_J$. However, the coefficients of the polynomial $p(z)$ are also symmetric in the conjugates of the other γ_j's, and thus the same reasoning can be applied for each set of conjugates for each γ_j to conclude that the coefficients of $p(z)$ are rational. The lemma now follows from Challenge 3.12. ∎

A finite sum of exponential terms $e^{\alpha_1} + e^{\alpha_2} + \cdots + e^{\alpha_L}$ is a *conjugate complete exponential sum* if the collection of exponents $\{\alpha_1, \alpha_2, \ldots, \alpha_L\}$ is conjugate complete. We are now in a position to establish the last assertion from our intuitive discussion.

LEMMA 3.16 *The product of two conjugate complete exponential sums is a conjugate complete exponential sum.*

Proof. Suppose that $e^{\alpha_1} + e^{\alpha_2} + \cdots + e^{\alpha_L}$ and $e^{\gamma_1} + e^{\gamma_2} + \cdots + e^{\gamma_J}$ are two conjugate complete exponential sums. That is, the sets $\{\alpha_1, \alpha_2, \ldots, \alpha_L\}$ and $\{\gamma_1, \gamma_2, \ldots, \gamma_J\}$ are each conjugate complete. If we consider the product

$$\left(e^{\alpha_1} + e^{\alpha_2} + \cdots + e^{\alpha_L} \right) \left(e^{\gamma_1} + e^{\gamma_2} + \cdots + e^{\gamma_J} \right),$$

then we see that it equals an exponential sum of the form

$$\sum_{\substack{1 \le l \le L \\ 1 \le j \le J}} e^{\alpha_l + \gamma_j}.$$

Applying Lemma 3.15 with $f(x_1, x_2) = x_1 + x_2$, it follows that the set $\{\alpha_l + \gamma_j : 1 \le l \le L \text{ and } 1 \le j \le J\}$ is conjugate complete. Thus the product of two conjugate complete exponential sums is another conjugate complete exponential sum. ∎

We are, at last, ready to face the long-awaited proof of the Lindemann–Weierstrass Theorem.

Proof of Theorem 3.4. We proceed by contradiction and assume that there exist algebraic numbers $\beta_0, \beta_1, \ldots, \beta_M$, not all zero, such that

$$\sum_{m=0}^{M} \beta_m e^{\alpha_m} = 0. \tag{3.33}$$

We first clean up the exponents by multiplying the expression in (3.33) by analogous expressions with the collection of exponents replaced by all possible combinations of their corresponding conjugates:

$$\prod_{\substack{\rho_m \text{ a conjugate of } \alpha_m \\ \text{for } m=0,1,2,\ldots,M}} \left(\beta_0 e^{\rho_0} + \beta_1 e^{\rho_1} + \beta_2 e^{\rho_2} + \cdots + \beta_M e^{\rho_M} \right) = 0.$$

In view of the symmetry with respect to the conjugates, after expanding the previous enormous product and factoring out common coefficients it is easy to verify that we are left with an expression of the form

$$\kappa_0 E_0 + \kappa_1 E_1 + \kappa_2 E_2 + \cdots + \kappa_L E_L = 0, \tag{3.34}$$

where the κ_l's are integer linear combinations of products of the β_m's, and the E_l's are conjugate complete exponential sums. Also, by Challenge 3.11 we know that not all of the κ_l's are equal to zero. Thus, by omitting any zero terms and reindexing the subscripts if necessary, we can assume, without loss of generality, that the κ_l's are all nonzero.

Next we transform the coefficients into integers by first multiplying the expression in (3.34) by analogous versions with the collection of coefficients replaced by all possible combinations of their corresponding conjugates:

$$\prod_{\substack{\gamma_l \text{ a conjugate of } \kappa_l \\ \text{for } l=0,1,2,\ldots,L}} \left(\gamma_0 E_0 + \gamma_1 E_1 + \gamma_2 E_2 + \cdots + \gamma_L E_L \right) = 0.$$

Once we expand this impressive product and factor out common coefficients, we are left with an expression of the form

$$\eta_0 \mathcal{E}_0 + \eta_1 \mathcal{E}_1 + \eta_2 \mathcal{E}_2 + \cdots + \eta_K \mathcal{E}_K = 0, \qquad (3.35)$$

where the η_k's are symmetric polynomials in the β_l's and their conjugates with integer coefficients, and where the $\mathcal{E}_k$'s are products of E_l's. Thus by Theorem 3.13 we have that each η_k is a rational number, and by Challenge 3.11 we see that not all of the η_k's can equal zero. Again, without loss of generality, we can assume that we have removed any zero terms and thus each η_k is nonzero. Multiplying (3.35) by the least common multiple of all the denominators of the η_k's will clear all denominators, and hence we can assume we have already performed this bottom-cleaning, so without loss of generality we assume that all the η_k's are nonzero *integers*. By Lemma 3.16 we know that since each E_l is conjugate complete, each of the $\mathcal{E}_k$ is also conjugate complete. Simply by regrouping, if necessary, as illustrated by the following example,

$$\eta \left(e^{2\sqrt{3}} + e^{-5\sqrt{7}} + e^{-2\sqrt{3}} + e^{5\sqrt{7}} \right) = \eta \left(e^{2\sqrt{3}} + e^{-2\sqrt{3}} \right) + \eta \left(e^{-5\sqrt{7}} + e^{5\sqrt{7}} \right),$$

we can further assume, without loss of generality, that all exponents from an exponential sum $\mathcal{E}_k$ are conjugate to each other.

Hence we see that from our original assumption (3.33), we are led to

$$\eta_0 \mathcal{E}_0 + \eta_1 \mathcal{E}_1 + \eta_2 \mathcal{E}_2 + \cdots + \eta_K \mathcal{E}_K = 0, \qquad (3.36)$$

where the η_k's are nonzero integers and the $\mathcal{E}_k$'s are exponential sums with algebraic exponents such that any two exponents from any exponential sum are conjugates. Therefore we are almost perfectly situated to apply our not-so-special Theorem 3.9. However, in that result, we assumed that for some k, $\mathcal{E}_k = e^0 = 1$; that is, we require that one lone integer have no e^α factor. If one of the $\mathcal{E}_k$'s in (3.36) is equal to 1, then we can immediately apply Theorem 3.9 and conclude that (3.33) yields a contradiction.

We now assume that $\mathcal{E}_k \neq 1$ for all k. Thus $\mathcal{E}_0 = e^{\nu_1} + e^{\nu_2} + \cdots + e^{\nu_J}$, where $\nu_1, \nu_2, \ldots, \nu_J$ are nonzero algebraic numbers all of which are conjugates. We now let $\mathcal{E}_0' = e^{-\nu_1} + e^{-\nu_2} + \cdots + e^{-\nu_J}$, and note that $\mathcal{E}_0'$ is a conjugate complete exponential sum. Thus the product $\mathcal{E}_0' \mathcal{E}_0$ is also a conjugate complete exponential sum by Lemma 3.16. This exponential sum contains at least J copies of e^0. Hence we can write the product as

$$\mathcal{E}_0' \mathcal{E}_0 = J + \mathcal{E}_0'',$$

where $\mathcal{E}_0''$ is some conjugate complete exponential sum. Finally, we note that in (3.36), for $k_1 \neq k_2$, each exponent appearing in $\mathcal{E}_{k_1}$ is different from each exponent of $\mathcal{E}_{k_2}$. Thus for $k \neq 0$, we are certain that $\mathcal{E}_0' \mathcal{E}_k$ will not contain any e^0 terms. Hence if we multiply (3.36) through by $\mathcal{E}_0'$, then we are left with

$$J\eta_0 + \eta_0 \mathcal{E}_0'' + \eta_1 \mathcal{E}_1'' + \eta_2 \mathcal{E}_2'' + \cdots + \eta_K \mathcal{E}_K'' = 0, \tag{3.37}$$

where the η_k's are nonzero integers and each $\mathcal{E}_k''$ is a conjugate complete exponential sum having no e^0 term. Since we are certain that $J\eta_0$ is the only term free of any exponential factor, we are guaranteed to have a nonzero integer as one of the terms in the sum (3.37). Thus we can now apply Theorem 3.9 and discover that (3.33) is impossible. Therefore, at long last, we have completed the amazing proof of the Lindemann–Weierstrass Theorem. ∎

3.6 Algebraic Independence—Freeing ourselves of unhealthy dependencies

In this last section we step back from the jungle of technical details and place the Lindemann–Weierstrass result in context within the landscape of transcendence. We begin with two attractive applications, the second of which was foreshadowed by Corollary 3.5.

Given a complex number $\alpha = x + iy$, we recall that $\mathrm{Re}(\alpha) = x$ and $\mathrm{Im}(\alpha) = y$. We now consider the following refinement to Hermite's original result.

THEOREM 3.17 *Let α be a nonreal algebraic number. Then both $\mathrm{Re}(e^\alpha)$ and $\mathrm{Im}(e^\alpha)$ are transcendental numbers.*

Proof. Since α is a nonreal algebraic number, it has the form $\alpha = a + bi$, where a and b are *real* algebraic numbers, and $b \neq 0$. We recall that by definition, $e^\alpha = e^{a+bi} = e^a \cos b + i e^a \sin b$. We now assume that $\mathrm{Re}(e^\alpha) = e^a \cos b$ is algebraic and call this algebraic number β. We note that if $\beta = 0$, then b must be a nonzero rational multiple of the transcendental number π, which is impossible, since b is algebraic. Thus we see that β is a nonzero algebraic number.

Next we observe that

$$e^{a+bi} + e^{a-bi} = e^a\left(e^{bi} + e^{-bi}\right) = e^a(\cos b + i\sin b + \cos(-b) + i\sin(-b))$$

$$= 2e^a \cos b = 2\beta,$$

which yields

$$2\beta e^0 - e^{a+bi} - e^{a-bi} = 0. \tag{3.38}$$

Thus we have a linear combination of e^0, e^{a+bi}, and e^{a-bi} with nonzero algebraic coefficients that equals zero. Since $0, a + bi$, and $a - bi$ are all algebraic numbers, Theorem 3.4 implies that these three numbers must *not* be distinct—for if they were all distinct, then (3.38) would be nonzero. It follows that $b = 0$, contrary to our hypothesis that α is nonreal. Hence $\mathrm{Re}(e^\alpha)$ is a transcendental number.

Challenge 3.13 *Complete the proof by showing that $\mathrm{Im}(e^\alpha)$ is also transcendental.* ∎

As a second direct consequence of the Lindemann–Weierstrass result, we consider the following:

THEOREM 3.18 *Let α be a nonzero algebraic number. Then $\tan \alpha$ is transcendental.*

Proof.   We recall that in Challenge 3.1 we saw that $\sin \alpha = \frac{e^{i\alpha} - e^{-i\alpha}}{2i}$. Similarly, it is easy to verify the identity $\cos \alpha = \frac{e^{i\alpha} + e^{-i\alpha}}{2}$. Thus we have

$$\tan \alpha = \frac{\sin \alpha}{\cos \alpha} = \frac{e^{i\alpha} - e^{-i\alpha}}{i(e^{i\alpha} + e^{-i\alpha})}.$$

If we now assume that $\tan \alpha$ is algebraic, say $\tan \alpha = \beta$, then the previous identity yields

$$(1 - i\beta)e^{i\alpha} - (1 + i\beta)e^{-i\alpha} = 0. \tag{3.39}$$

Given that $\alpha \neq 0$, we see that $i\alpha$ and $-i\alpha$ are *distinct* algebraic numbers and also note that $1 - i\beta$ and $1 + i\beta$ cannot simultaneously equal 0. Hence Theorem 3.4 reveals that

$$(1 - i\beta)e^{i\alpha} - (1 + i\beta)e^{-i\alpha} \neq 0,$$

which contradicts (3.39). Therefore we conclude that $\tan \alpha$ is transcendental.   ∎

The method employed in the previous proof can be applied to easily show the transcendence of numbers such as

$$\frac{\left(e^{\sqrt{5}}\right)^3 + \sqrt[3]{7}\, e^{1-\sqrt{11}}}{6\sqrt{13} + e^{\sqrt{3}+\sqrt[4]{9}}}.$$

However, the method breaks down with expressions such as

$$\frac{4\left(e^{\sqrt{2}}\right)^3 - 4e^{3\sqrt{2}}}{e^{\sqrt{13}} + e^{\sqrt{5}}}, \tag{3.40}$$

since the two exponents in the numerator are both equal to $3\sqrt{2}$ and we have a coefficient coincidence that causes complete cancellation. In fact, the number in (3.40) equals 0 and thus could not be any less transcendental. Similarly, any attempt to prove that the number

$$\frac{22e^{\sqrt{3}} - 22e^{\sqrt{2}}}{7e^{\sqrt{3}} - 7e^{\sqrt{2}}}$$

is transcendental will only lead to disappointment. However, if we have an expression that is devoid of such amazing conspiracies, then we indeed find ourselves face to face with a transcendental number. We illustrate this basic theme with the following general transcendence result.

THEOREM 3.19 *Let* $P(z_1, z_2, \ldots, z_K)$ *and* $Q(z_1, z_2, \ldots, z_L)$ *be two nonzero polynomials with integer coefficients. If* $\gamma_1, \gamma_2, \ldots, \gamma_K$ *and* $\eta_1, \eta_2, \ldots, \eta_L$ *are algebraic numbers, then*

$$\frac{P(e^{\gamma_1}, e^{\gamma_2}, \ldots, e^{\gamma_K})}{Q(e^{\eta_1}, e^{\eta_2}, \ldots, e^{\eta_L})}$$

either is a rational number or is transcendental.

Proof.   Let us suppose that $P(e^{\gamma_1}, e^{\gamma_2}, \ldots, e^{\gamma_K})/Q(e^{\eta_1}, e^{\eta_2}, \ldots, e^{\eta_L})$ is an *irrational* number and further assume that it is an *algebraic* number, say β. Thus we have

$$P(e^{\gamma_1}, e^{\gamma_2}, \ldots, e^{\gamma_K}) - \beta Q(e^{\eta_1}, e^{\eta_2}, \ldots, e^{\eta_L}) = 0. \qquad (3.41)$$

Since β is irrational, we cannot have any coefficient cancellation after factoring out common exponential terms. Thus we again see that Theorem 3.4 contradicts identity (3.41), and so the expression in the Theorem must be transcendental, which completes our short proof.   ∎

The above applications and observations inspire us to take one final look at the statement of the Lindemann–Weierstrass Theorem itself. As we discovered in this chapter through all our "coefficient cleaning" activities, the statement that the numbers $e^{\alpha_0}, e^{\alpha_1}, \ldots, e^{\alpha_M}$ are linearly independent over the algebraic numbers is equivalent to the statement that $e^{\alpha_0}, e^{\alpha_1}, \ldots, e^{\alpha_M}$ are linearly independent over the integers. Thus the Lindemann–Weierstrass Theorem is actually equivalent to the following:

THEOREM 3.20 *Let* $P(z_0, z_1, \ldots, z_M) \in \mathbb{Z}[z_0, z_1, \ldots, z_M]$ *be a nonzero polynomial defined by*

$$P(z_0, z_1, \ldots, z_M) = b_0 z_0 + b_1 z_1 + \cdots + b_M z_M.$$

If $\alpha_0, \alpha_1, \ldots, \alpha_M$ *are distinct algebraic numbers, then*

$$P\left(e^{\alpha_0}, e^{\alpha_1}, \ldots, e^{\alpha_M}\right) \neq 0.$$

When viewed in this fashion, the Lindemann–Weierstrass Theorem is a statement about linear polynomials evaluated at particular values. Thus we are led to a natural question: Does the previous theorem hold for *arbitrary* polynomials $P(z_0, z_1, \ldots, z_M) \in \mathbb{Z}[z_0, z_1, \ldots, z_M]$?

The answer is quickly seen to be "no" by considering the polynomial $P(z_0, z_1) = z_0^3 - z_1$, and observing that

$$P\left(e^{\sqrt{2}}, e^{3\sqrt{2}}\right) = \left(e^{\sqrt{2}}\right)^3 - e^{3\sqrt{2}} = e^{3\sqrt{2}} - e^{3\sqrt{2}} = 0.$$

However, this exponent conspiracy is identical to the one we already witnessed in (3.40). Therefore in this generalized context it is not enough to insist that the algebraic exponents be merely distinct—it appears we now need them to be linearly independent over $\mathbb{Z}$. The generalization below follows immediately from all our previous hard work.

THEOREM 3.21   *Let* $P(z_0, z_1, \ldots, z_M) \in \mathbb{Z}[z_0, z_1, \ldots, z_M]$ *be a nonzero polynomial. If* $\alpha_0, \alpha_1, \ldots, \alpha_M$ *are* $\mathbb{Z}$-*linearly independent algebraic numbers, then*

$$P(e^{\alpha_0}, e^{\alpha_1}, \ldots, e^{\alpha_M}) \neq 0.$$

Proof.   We express the polynomial $P(z_0, z_1, \ldots, z_M)$ as

$$P(z_0, z_1, \ldots, z_M) = \sum_{(d_0, d_1, \ldots, d_M) \in \Lambda} b_{(d_0, d_1, \ldots, d_M)} z_0^{d_0} z_1^{d_1} \cdots z_M^{d_M},$$

where Λ denotes a suitable finite collection of $(M + 1)$-tuples of nonnegative integers, and the coefficients $b_{(d_0, d_1, \ldots, d_M)}$ are nonzero integers. Thus we have

$$P(e^{\alpha_0}, e^{\alpha_1}, \ldots, e^{\alpha_M}) = \sum_{(d_0, d_1, \ldots, d_M) \in \Lambda} b_{(d_0, d_1, \ldots, d_M)} e^{d_0\alpha_0 + d_1\alpha_1 + \cdots + d_M\alpha_M}. \qquad (3.42)$$

We now claim that the exponents

$$\{d_0\alpha_0 + d_1\alpha_1 + \cdots + d_M\alpha_M : (d_0, d_1, \ldots, d_M) \in \Lambda\}$$

are *distinct*. To establish this claim, let us assume that $(d_0, d_1, \ldots, d_M) \neq (d_0', d_1', \ldots, d_M')$ are two elements from Λ such that

$$d_0\alpha_0 + d_1\alpha_1 + \cdots + d_M\alpha_M = d_0'\alpha_0 + d_1'\alpha_1 + \cdots + d_M'\alpha_M.$$

Therefore we have that

$$(d_0 - d_0')\alpha_0 + (d_1 - d_1')\alpha_1 + \cdots + (d_M - d_M')\alpha_M = 0,$$

where the integer coefficients $d_0 - d_0', d_1 - d_1', \ldots, d_M - d_M'$ are not all zero—which contradicts the hypothesis that $\alpha_0, \alpha_1, \ldots, \alpha_M$ are $\mathbb{Z}$-linearly independent. Hence the exponents in (3.42) *are* distinct. Since the exponents are distinct algebraic numbers, we can apply Theorem 3.4 and deduce that

$$P(e^{\alpha_0}, e^{\alpha_1}, \ldots, e^{\alpha_M}) \neq 0,$$

which completes the proof. ∎

This general idea of "polynomial independence" is an extremely important one, and we formalize it here. We say that the numbers $\gamma_1, \gamma_2, \ldots, \gamma_L$ are *algebraically independent* if for every nonzero polynomial $P(z_1, z_2, \ldots, z_L) \in \mathbb{Z}[z_1, z_2, \ldots, z_L]$, we have

$$P(\gamma_1, \gamma_2, \ldots, \gamma_L) \neq 0.$$

Thus we can restate the previous theorem as:

THEOREM 3.22 *If $\alpha_0, \alpha_1, \ldots, \alpha_M$ are $\mathbb{Z}$-linearly independent algebraic numbers, then $e^{\alpha_0}, e^{\alpha_1}, \ldots, e^{\alpha_M}$ are algebraically independent.*

Challenge 3.14 *Prove that the converse of Theorem 3.22 also holds.*

Using the language of algebraic independence, we can now generalize and improve Theorem 3.17.

THEOREM 3.23 *Suppose that α is an algebraic number whose real and imginary parts are both nonzero. Then $\mathrm{Re}(e^{\alpha})$ and $\mathrm{Im}(e^{\alpha})$ are algebraically independent.*

The following simple lemma allows us to deduce Theorem 3.23 from the Lindemann–Weierstrass Theorem.

LEMMA 3.24 *Two nonzero real numbers α_1 and α_2 are algebraically dependent if and only if $\alpha_1 + i\alpha_2$ and $\alpha_1 - i\alpha_2$ are algebraically dependent.*

Proof. If we write $\beta_1 = \alpha_1 + i\alpha_2$ and $\beta_2 = \alpha_1 - i\alpha_2$, then it follows that $\alpha_1 = \frac{\beta_1 + \beta_2}{2}$ and $\alpha_2 = \frac{\beta_1 - \beta_2}{2i}$. We suppose first that there exists a nonzero polynomial $P(X, Y) = \sum_{m,n} a_{mn} X^m Y^n \in \mathbb{Z}[X, Y]$ such that $P(\alpha_1, \alpha_2) = 0$. Thus we have

$$P\left(\frac{\beta_1 + \beta_2}{2}, \frac{\beta_1 - \beta_2}{2i}\right) = \sum_{m,n} a_{mn} \left(\frac{1}{2}\right)^{m+n} (-i)^n (\beta_1 + \beta_2)^m (\beta_1 - \beta_2)^n = 0.$$

The previous expression inspires us to define the polynomials

$$Q(X, Y) = \sum_{m,n} a_{mn} \left(\frac{1}{2}\right)^{m+n} (-i)^n X^m Y^n, \quad \overline{Q}(X, Y) = \sum_{m,n} a_{mn} \left(\frac{1}{2}\right)^{m+n} (i)^n X^m Y^n.$$

Therefore if we let $G(X, Y) = Q(X, Y)\overline{Q}(X, Y)$, then it follows that there exists an integer d such that $dG(X, Y) \in \mathbb{Z}[X, Y]$ and $dG(\beta_1 + \beta_2, \beta_1 - \beta_2) = 0$. If we now define the polynomial $H(X, Y) = dG(X + Y, X - Y) \in \mathbb{Z}[X, Y]$, then we see that $H(\beta_1, \beta_2) = 0$.

Conversely, if there exists a nonzero polynomial $P(X, Y) \in \mathbb{Z}[X, Y]$ such that $P(\beta_1, \beta_2) = 0$, then $P(\alpha_1 + i\alpha_2, \alpha_1 - i\alpha_2) = 0$. Following the same line of reasoning as above, we can find a polynomial $J(X, Y) \in \mathbb{Z}[X, Y]$ satisfying $J(\alpha_1, \alpha_2) = 0$.

Challenge 3.15   *Complete the proof of the lemma by showing that both $H(X, Y)$ and $J(X, Y)$ are nonzero polynomials. (Hint: Look at the leading coefficient of each as a polynomial in X.)*   ∎

We now apply the lemma to prove the theorem.

Proof of Theorem 3.23.   We write $\alpha = a + bi$ and notice that a and b are each nonzero real algebraic numbers. By the previous lemma we see that $e^a \cos b$ and $e^a \sin b$ are algebraically dependent if and only if $e^a \cos b + ie^a \sin b$ and $e^a \cos b - ie^a \sin b$ are algebraically dependent. Of course we have $e^a \cos b + ie^a \sin b = e^{a+bi}$ and $e^a \cos b - ie^a \sin b = e^{a-bi}$. Now by Theorem 3.22 and Challenge 3.14 we see that e^{a+bi} and e^{a-bi} are algebraically dependent if and only if $a + bi$ and $a - bi$ are $\mathbb{Z}$-linearly dependent, a circumstance that occurs if and only if $a = 0$ or $b = 0$, which completes our proof.   ∎

The concept of algebraic independence is one of the central notions in transcendental number theory. In fact, many open questions in transcendence theory involve algebraic independence. For example, while as an immediate consequence of our work in this chapter we see that the numbers e and $e^{\sqrt{2}}$ are algebraically independent, a famous open question remains: Are e and π algebraically independent? That is, no one knows if there exists a nonzero polynomial $P(z_1, z_2) \in \mathbb{Z}[z_1, z_2]$ such that $P(e, \pi) = 0$. The widely believed conjecture is that no such polynomial exists and thus the conjecture is that e and π are algebraically independent. While we will not answer this notoriously difficult open question here, in the next chapter we will bring e and π together by forgoing algebraic exponents and examining the enigmatic number e^π. That investigation will introduce the next major advance in classical transcendental number theory.

23.1406926327792690057290863 67 ...

The Analytic Adventures of e^z:
Siegel's Lemma and the transcendence of e^π

4.1 Giving e^z a complex by taking away its power series

In the previous two chapters we exploited the simplicity of the power series for the function e^z in order to establish the transcendence results of Hermite–Lindemann and then Lindemann–Weierstrass. In this chapter we describe the next stage in the evolution of classical transcendental number theory, which involves viewing e^z as a function of a complex variable z and applying more sophisticated analytic techniques. We illustrate these new themes by establishing the transcendence of e^π.

While this basic shift in perspective liberates transcendental number theory from the profoundly technical confines of approximating the power series for e^z, it also introduces a plethora of challenges for us to overcome as we attempt to construct an integer $\mathcal{N}$ in violation of the Fundamental Principle of Number Theory. As we will see in what follows, this integer will arise from a value of a complex function rather than through a polynomial approximation to a power series. As is our custom, we foreshadow the technical themes to come by first considering the slightly simpler scenario of showing that e^π is irrational.

The intuitive rationale for our new complex point of view

Here we offer the broadest of overviews to highlight the feature of our new approach that allows us to utilize the power of complex analysis.

We say that a function $G : \mathbb{C} \to \mathbb{C}$ is *entire* if it can be expressed as a power series centered at $z = z_0$ as

$$G(z) = \sum_{n=0}^{\infty} \frac{G^{(n)}(z_0)}{n!}(z - z_0)^n,$$

where the series converges for all $z \in \mathbb{C}$.

Continued

Let us assume that e^π is a rational number and that, as in our previous arguments, we can use this assumption to deduce some information about the exponent, which in this case is π. In particular, we can construct an entire function $G(z)$ with the property that $G(\pi)$ is a nonzero *rational* number having denominator d. Thus $dG(\pi)$ is a nonzero *integer*. In order to contradict the Fundamental Principle of Number Theory, it would be enough to show that $|dG(\pi)| < 1$, or equivalently, $|G(\pi)| < d^{-1}$. One way to find an upper bound for $|G(\pi)|$ is to consider the disk of complex numbers z defined by $|z| \leq R$, for some real number $R > \pi$. Then clearly we have

$$|G(\pi)| \leq \max\{|G(z)| : |z| \leq R\},$$

since $|G(\pi)|$ is one of the candidates considered for the maximum value.

Finding the maximum of $|G(z)|$ among complex numbers z within a closed disk may sound daunting, but it is here that we are saved by the incredibly rich structure of complex analysis that real analysis lacks. In paricular, we may apply the so-called Maximum Modulus Principle, which asserts that the maximum absolute value an entire function takes on among points in a closed disk is attained on the *boundary* circle of the disk. (In complex analysis, people refer to the "absolute value" of a complex number as its *modulus*—hence the name Maximum Modulus Principle.) We describe this incredible and surprising result in our overview of some highlights of complex analysis in the appendix, but here we apply it to see that

$$|G(\pi)| \leq \max\{|G(z)| : |z| \leq R\} = \max\{|G(z)| : |z| = R\}.$$

As we will soon discover, in our particular situation, we can find upper bounds for the maximum of $|G(z)|$ along the circle centered at the origin of radius R that lead to a contradiction. Thus our major challenge in pursuing this approach is the very intricate construction of the entire function $G(z)$.

4.2 Throwing in our two bits as a warm-up to the irrationality of e^π

In order to outline some of the major themes involved in the proof of the irrationality of e^π, we begin here by sketching an argument for the far less impressive realization that $e^\pi \neq 23.25$. That is, we will show that e^π is not twenty-three and two bits.

THEOREM 4.1 $e^\pi \neq 23.25$.

A quick and slick proof. One way to convince ourselves of the validity of this theorem is to compute e^π on a calculator and discover that the answer resembles the number from the title of this chapter—23.140blah blah blah—and be done with it. While this method is both short and reasonably persuasive, it cannot be extended to establish the irrationality of e^π. Thus we need to develop a more elaborate line of reasoning.

A sketch of a more insightful proof.   As always, we proceed by contradiction and assume that $e^\pi = \frac{93}{4}$, or expressed differently,

$$4e^\pi - 93 = 0. \tag{4.1}$$

If we attempt to adopt the methods introduced in the previous two chapters, then we would view e^π as the number given by $\sum_{n=0}^{\infty} \frac{\pi^n}{n!}$, and try to construct an excellent polynomial approximation with rational coefficients by considering linear combinations of truncated pieces of the series. However, since we have already established that π is transcendental, we are unable to apply the techniques we developed for e raised to an *algebraic* power, since there is no obvious way to combine the polynomials in a symmetric manner to create a *rational* number that would lead to our integer in violation of the Fundamental Principle of Number Theory. Hence our old strategy no longer appears viable.

Given the previous failed attempt, we need to look in a completely new direction. In particular, we view (4.1) in a different light—specifically, we observe that if we define the polynomial $P(x) = 4x - 93 \in \mathbb{Z}[x]$, then we have that $P(e^\pi) = 0$. However, from an analytic point of view, the structure of a polynomial is far too simple to move us forward. Thus we create a more analytically interesting function through composition: We let $f(z) = P(e^z) = 4e^z - 93$, and notice that $f(\pi) = 0$. Therefore, in some analytical sense, $f(z)$ must be connected with $(z - \pi)$; the simplest function that vanishes at π. This vague remark provides the inspiration for us to consider the power series of $f(z)$ centered at $z = \pi$, that is,

$$f(z) = \sum_{n=0}^{\infty} \frac{f^{(n)}(\pi)}{n!}(z - \pi)^n,$$

which we note converges for all $z \in \mathbb{C}$. Given our assumption that $f(\pi) = f^{(0)}(\pi) = 0$, we can rewrite the previous power series as

$$f(z) = (z - \pi) \sum_{n=1}^{\infty} \frac{f^{(n)}(\pi)}{n!}(z - \pi)^{n-1},$$

and hence the connection between $f(z)$ and $(z - \pi)$ comes into focus.

It now follows that $f(z)/(z - \pi)$ is an entire function, since it has a power series that converges for all $z \in \mathbb{C}$, namely,

$$\frac{f(z)}{z - \pi} = \sum_{n=1}^{\infty} \frac{f^{(n)}(\pi)}{n!}(z - \pi)^{n-1}. \tag{4.2}$$

This observation is crucial, since it empowers us to evaluate $f(z)/(z - \pi)$ at $z = \pi$ by simply evaluating the power series in (4.2). If we define $G(z) = f(z)/(z - \pi)$, then applying the power series expansion, we conclude that

$$G(\pi) = \frac{f'(\pi)}{1!}.$$

Any fan of calculus who recalls our assumption that $e^{\pi} = \frac{93}{4}$ could quickly verify that $f'(\pi) = 4e^{\pi} = 4\left(\frac{93}{4}\right) = 93$. Thus, as hinted in our intuitive outline in the previous section, we have constructed an entire function $G(z)$ such that $G(\pi)$ is a rational number; in particular, $G(\pi) = 93$. We now follow the strategy foreshadowed in that outline.

If $R > \pi$ is a real number, then we have that

$$93 = |G(\pi)| \leq \max_{|z| \leq R} \{|G(z)|\} = \max_{|z| \leq R} \left\{ \left| \frac{f(z)}{z - \pi} \right| \right\}.$$

Next we apply the Maximum Modulus Principle to conclude that

$$93 \leq \max_{|z|=R} \left\{ \left| \frac{f(z)}{z - \pi} \right| \right\} = \max_{|z|=R} \left\{ \left| \frac{4e^z - 93}{z - \pi} \right| \right\}. \tag{4.3}$$

Unfortunately, while it is not at all obvious at this point, for any choice of $R > \pi$, the upper bound in (4.3) would be substantially larger than 93 and hence no contradiction could be realized.

In order to step back from this temporary setback, we modify our polynomial so that a new upper bound analogous to that of (4.3) would be *so* small that we would be faced with an impossible inequality. Certainly, if we increased the size of the denominator $|z - \pi|$, we would produce a smaller upper bound. As we will now see, there are two basic means of increasing the denominator, and both involve introducing a polynomial having more zeros. Our first attempt is to consider a new polynomial that vanishes at $\frac{93}{4}$ with some multiplicity, say $P(x) = (4x - 93)^T$. If we let $f(z) = P(e^z) = (4e^z - 93)^T$, then we see that we have a high degree of vanishing at $z = \pi$. In particular, it is easy to show that $z = \pi$ is a zero of $f(z)$ with multiplicity T, that is, $f(\pi), f'(\pi), f^{(2)}(\pi), \ldots, f^{(T-1)}(\pi)$ are all equal to 0, while $f^{(T)}(\pi) \neq 0$. Thus, in this case, we have that

$$f(z) = (z - \pi)^T \sum_{n=T}^{\infty} \frac{f^{(n)}(\pi)}{n!} (z - \pi)^{n-T}.$$

If we now let $G(z) = f(z)/(z - \pi)^T$, then we see that G is an entire function with

$$G(\pi) = \frac{f^{(T)}(\pi)}{T!},$$

which is nonzero. In fact, we will now discover that $f^{(T)}(\pi)$ is a *rational* number.

Challenge 4.1 *Let $P(x) = c_T x^T + c_{T-1} x^{T-1} + \cdots + c_1 x + c_0$ be a polynomial with integer coefficients and let $f(z) = P(e^z)$. Suppose that there exists a $z_0 \in \mathbb{C}$ such that e^{z_0} is a rational number, say $e^{z_0} = \frac{r}{s}$. Prove that for every $t \geq 0, f^{(t)}(\pi)$ is a rational number having a denominator equal to a power of s not exceeding T.*

Thus in view of the previous observations together with our assumption that $e^\pi = \frac{93}{4}$, we conclude that

$$4^T T! G(\pi) \text{ is a nonzero integer.}$$

Hence, for any real number $R > \pi$, we can apply the Maximum Modulus Principle to deduce that

$$0 < 4^T T! |G(\pi)| \le 4^T T! \max_{|z| \le R} \{|G(z)|\}$$

$$= 4^T T! \max_{|z|=R} \{|G(z)|\} = 4^T T! \max_{|z|=R} \left\{ \frac{|4e^z - 93|^T}{|z - \pi|^T} \right\}.$$

Thus while we have potentially increased the denominator as compared with the bound in (4.3), we have also increased the numerator of our upper bound, and so no obvious progress has been made. Upper bounds for the right-hand side of the previous inequality will be studied in greater detail in the next section; however, here we just note that while the idea of increasing the multiplicity of the zeros is not fruitful in this case, it remains an excellent idea that will, in fact, have much appeal in the more elaborate construction ahead.

The second manner in which we can increase the denominator of our upper bound is to modify the polynomial $P(x)$ so that it has a second zero. Thus we need to use our assumption $e^\pi = \frac{93}{4}$ to produce another such identity. An easy means of producing such an expression is simply to square both sides of the previous equality, which yields

$$e^{2\pi} = \frac{93^2}{4^2}.$$

This observation leads us to consider the polynomial

$$P(x) = (4x - 93)(16x - 93^2),$$

which vanishes at both e^π and $e^{2\pi}$, and more importantly, if we let $f(z) = P(e^z) = (4e^z - 93)(16e^z - 93^2)$, then we see that $f(z)$ vanishes at both π and 2π. If we set

$$G(z) = \frac{f(z)}{(z - \pi)(z - 2\pi)} = \frac{(4e^z - 93)(16e^z - 93^2)}{(z - \pi)(z - 2\pi)},$$

then $G(z)$ can be shown to be an entire function, since it is the product of two entire functions: $(4e^z - 93)/(z - \pi)$ and $(16e^z - 93^2)/(z - 2\pi)$. If we expand $f(z)$ in its power series about the point $z = \pi$, then because $f(z)$ vanishes at π, we would see that

$$f(z) = (z - \pi) \sum_{n=1}^{\infty} \frac{f^{(n)}(\pi)}{n!} (z - \pi)^{n-1}.$$

Thus we have

$$G(\pi) = \frac{f'(\pi)}{\pi - 2\pi} = -\frac{769761}{\pi},$$

which, after multiplying by π produces a nonzero integer. Hence for $R > \pi$, the Maximum Modulus Principle reveals that

$$769761 = \pi|G(\pi)| \le \pi \max_{|z|\le R}\{|G(z)|\} = \pi \max_{|z|=R}\{|G(z)|\}$$

$$= \pi \max_{|z|=R}\left\{\left|\frac{(4e^z - 93)(16e^z - 93^2)}{(z - \pi)(z - 2\pi)}\right|\right\}. \tag{4.4}$$

A sneaky computational trick. If we now let $R = \pi + 0.05$, then some elementary complex analysis and calculus reveal that the maximum in (4.4) is attained for $z = \pi + 0.05 = (\pi + 0.05) + 0i$. If we evaluate the function at $\pi + 0.05$, then we obtain

$$\max_{|z|=R}\left\{\left|\frac{(4e^z - 93)(16e^z - 93^2)}{(z - \pi)(z - 2\pi)}\right|\right\} = 230222.695580\ldots,$$

and therefore we conclude that

$$769761 < 230223\pi,$$

which, at long last, is a contradiction, since in fact, $230223\pi < 769761$. Therefore we must have that e^π is *not* equal to 23.25.

The careful reader may have noticed that this proof, just as the first, depended on a computation involving the irrational number π. But, of course, if we can explicitly compute the quantity

$$\left|\frac{(4e^z - 93)(16e^z - 93^2)}{(z - \pi)(z - 2\pi)}\right|$$

at $z = \pi + 0.05$, then we might as well simply compute e^π as in our first, quick and slick proof. Thus a proof devoid of numerical computations involving the quantity π still elludes us.

Is there a noncomputational proof ending where we live happily ever after?   If we wish to extend our methods to a large class of exponents, then we must remove all computations involving π. That is, we need to produce an upper bound for

$$\pi \max_{|z|=R}\left\{\left|\frac{(4e^z - 93)(16e^z - 93^2)}{(z - \pi)(z - 2\pi)}\right|\right\},$$

without any subtle computation, that is relatively small. For ease of notation, we write $|G(z)|_R$ for the quantity $\max_{|z|=R}\{|G(z)|\}$. The only upper bound we can produce in general comes from minimizing the denominator and using the triangle inequality on the numerator. Carrying out these steps produces

$$\pi\left|\frac{(4e^z - 93)(16e^z - 93^2)}{(z - \pi)(z - 2\pi)}\right|_R \le \pi \frac{|(4e^z - 93)(16e^z - 93^2)|_R}{\min_{|z|=R}\{|z - \pi||z - 2\pi|\}}$$

$$\le \frac{93^3 e^{2R}\pi}{|R - \pi||R - 2\pi|} = \frac{804357\pi e^{2R}}{|R - \pi||R - 2\pi|},$$

which clearly, for any choice of $R > \pi$, is never less than 769761. Thus this general bound is far too crude to conclude our proof with a happy ending. What we can conclude for certain is that the upper bound depends on R, the degree of the polynomial $P(x)$, and the size of $P(x)$'s maximum coefficient.

The previous proof may frustrate some readers, since we worked so hard to try to prove $e^\pi \neq 23.25$ by some general method only to "cheat" at the end. In reality, through our previous journey we have seen *all* of the major themes to come. In particular, we have foreshadowed the following strategy for a proof of the irrationality of e^π:

- Assume that e^π is rational.
- Find a polynomial with integer coefficients that vanishes at $e^{k\pi}$, for integers $k = 1, 2, \ldots, K$ each with high multiplicity, say, T.
- Use this polynomial to construct an auxilary function $f(z)$ that is entire and satisfies $f^{(t)}(k\pi) = 0$ for $k = 1, 2, \ldots, K$, and $t = 0, 1, \ldots, T - 1$.
- Show there exists a smallest integer $M \geq T$ such that for some k_0, $1 \leq k_0 \leq K$, $f^{(M)}(k_0\pi) \neq 0$. Thus the function $G(z)$ given by

$$G(z) = \frac{f(z)}{(z - \pi)^M (z - 2\pi)^M \cdots (z - K\pi)^M}$$

is entire.
- Verify that for some $n \in \mathbb{Z}$, $\pi^n |G(k_0\pi)|$ is a nonzero rational number, hence for some integer d, $d\pi^n |G(k_0\pi)|$ is a nonzero integer.
- For $R > K\pi$, apply the Maximum Modulus Principle:

$$0 < d\pi^n |G(k_0\pi)| \leq d\pi^n \max_{|z|\leq R}\{|G(z)|\} = d\pi^n \max_{|z|=R}\{|G(z)|\}$$

$$= d\pi^n \max_{|z|=R}\left\{\frac{|f(z)|}{|z - \pi|^M |z - 2\pi|^M \cdots |z - K\pi|^M}\right\}.$$

- Deduce a general upper bound for the above quantity in terms of R, the degree of the polynomial, and the size of the polynomial's coefficients that leads to a contradiction to the Fundamental Principle of Number Theory.

Thus our main mission is clear: We must construct a polynomial with integer coefficients that vanishes with some high multiplicity at $e^{k\pi}$ for $k = 1, 2, \ldots, K$ yet has relatively small coefficients so that we can produce an upper bound that will lead to a contradiction. We are now ready to apply this strategy to prove the following

THEOREM 4.2   *The number e^π is irrational.*

4.3 Our first ill-fated attempt at a proof—A star-crossed relationship between the order of vanishing and the degree

As always, we begin by assuming that e^π is rational, say $e^\pi = \frac{r}{s}$. Given our previous discussion, we first consider the polynomial $P(x) = (sx - r)^T$ and let $f(z) = P(e^z)$.

It is easy to verify that $f(z)$ has a high order of vanishing at $z = \pi$. In particular, we have that $f^{(t)}(\pi) = 0$ for $0 \leq t < T$. Moreover, if we expand $f(z)$ using the Binomial Theorem, then we see that

$$f(z) = \sum_{\ell=0}^{T} \binom{T}{\ell} (se^z)^{\ell} (-r)^{T-\ell}, \qquad (4.5)$$

and thus it follows from repeated differentiation that

$$f^{(t)}(z) = \sum_{\ell=0}^{T} \ell^t \binom{T}{\ell} (se^z)^{\ell} (-r)^{T-\ell}. \qquad (4.6)$$

We notice that given our assumption that $e^{\pi} = \frac{r}{s}$, identity (4.6) implies that $f^{(t)}(\pi)$ is an integer for all t.

We now turn our attention to the value $f^{(T)}(\pi)$ and, in particular, wonder whether $f^{(T)}(\pi)$ is a *nonzero* integer. This issue leads us to a fundamental question: Given that a polynomial $P(x)$ vanishes at $x = x_0$ with multiplicity T, what can we conclude about the order of vanishing of the function $f(z) = P(e^z)$ at $z = z_0$, where $e^{z_0} = x_0$? The answer, as we will discover in the next challenge, is that the two orders of vanishing are equal. Perhaps it is not too surprising that the degree of $P(x)$ is T while the order of vanishing of $f(z)$ at $z = \pi$ is also T. Unfortunately, this seemingly happy coincidence will be the cause of considerable unhappiness ahead.

Challenge 4.2 *Let $P(x) \in \mathbb{C}[x]$ be a nonzero polynomial. If $f(z) = P(e^z)$, then show that $f(z)$ cannot be identically zero. Suppose now that $P(x)$ vanishes at $x = x_0$ with multiplicity T and that $g(z)$ is differentiable to order T at $z = z_0$, where $g(z_0) = x_0$. Furthermore, let $f(z) = P(g(z))$ and assume that $f(z)$ is not identically zero. Show that $f(z)$ vanishes at z_0 to an order greater than T if and only if $g'(z_0) = 0$.*

Challenge 4.2 leads to two important insights. First, since the coefficients of the polynomial $P(x) = (sx - r)^T$ are not all zero, the function $f(z) = P(e^z)$ is not identically zero. This remark implies that the power series expansion for $f(z)$ centered at any point must have some nonzero coefficients. In particular, the power series for $f(z)$ at $z = \pi$,

$$f(z) = \sum_{n=0}^{\infty} \frac{f^{(n)}(\pi)}{n!} (z - \pi)^n,$$

is not identically zero. Since $f^{(t)}(\pi) = 0$ for $0 \leq t < T$, we know that the first T coefficients of this power series are all equal to 0. Thus we have

$$f(z) = \sum_{n=T}^{\infty} \frac{f^{(n)}(\pi)}{n!} (z - \pi)^n = (z - \pi)^T \sum_{n=T}^{\infty} \frac{f^{(n)}(\pi)}{n!} (z - \pi)^{n-T}.$$

As foreshadowed earlier, we now define

$$G(z) = \frac{f(z)}{(z-\pi)^T} = \sum_{n=T}^{\infty} \frac{f^{(n)}(\pi)}{n!} (z-\pi)^{n-T}.$$

In view of its power series above, it follows that $G(z)$ is an entire function, and its value at π is given by

$$G(\pi) = \frac{f^{(T)}(\pi)}{T!}.$$

The second important insight that the previous challenge brings to light is that because $\frac{d}{dz}(e^z)\big|_{z=\pi} = e^\pi \neq 0$, it follows that indeed $f^{(T)}(\pi)$ is a *nonzero* integer. Now using the function $G(z)$, we can provide an upper bound for the integer $|f^{(T)}(\pi)|$ in the hopes of violating the Fundamental Principle of Number Theory.

For a real number $R > \pi$, we apply the Maximum Modulus Principle to deduce that

$$0 < |f^{(T)}(\pi)| = T!|G(\pi)| \leq T! \max_{|z|\leq R}\{|G(z)|\} = T!|G(z)|_R = T! \left| \frac{f(z)}{(z-\pi)^T} \right|_R$$

$$\leq \frac{T!|f(z)|_R}{(\min_{|z|=R}\{|z-\pi|\})^T} = \frac{T!|f(z)|_R}{|R-\pi|^T},$$

where the last equality follows from the fact that π is a positive real number.

We can bound the quantity $|f(z)|_R$ by applying the triangle inequality to (4.5) and recalling that $s < r$, since $1 < e^\pi = \frac{r}{s}$. These two observations yield

$$|f(z)| = \left| \sum_{\ell=0}^{T} \binom{T}{\ell} (se^z)^\ell (-r)^{T-\ell} \right| \leq \sum_{\ell=0}^{T} \binom{T}{\ell} s^\ell |e^z|^\ell r^{T-\ell} < \sum_{\ell=0}^{T} \binom{T}{\ell} r^T |e^z|^\ell.$$

In view of the fact that $\sum_{\ell=0}^{T} \binom{T}{\ell} = 2^T$ and $|e^z|_R = e^R$, the previous inequality implies that

$$|f(z)|_R < \sum_{\ell=0}^{T} \binom{T}{\ell} r^T e^{RT} = (2re^R)^T.$$

Thus we have that

$$0 < |f^{(T)}(\pi)| < \frac{T!(2re^R)^T}{|R-\pi|^T}. \tag{4.7}$$

We remark that we are free to select the parameters T and R. It is often easier to visualize the relationships between the various parameters by estimating $\log|f^{(T)}(\pi)|$ rather than $|f^{(T)}(\pi)|$. Recalling the crude, but difficult to improve upon, estimate $T! < T^T$, inequality (4.7) can be recast to yield

$$\log|f^{(T)}(\pi)| < T\log T + T\log(2r) + TR - T\log|R-\pi|. \tag{4.8}$$

Of course, our desired inequality $|f^{(T)}(\pi)| < 1$ is equivalent to $\log|f^{(T)}(\pi)| < 0$. However, the upper bound of (4.8) is the sum of four terms, only one of which is negative.

Challenge 4.3   *For a positive integer T and a real number $R > \pi$, show that*

$$TR > T\log(R - \pi).$$

Using this inequality, conclude that there are no allowable choices of T and R such that (4.8) leads to a contradiction to the Fundamental Principle of Number Theory.

Looking Back: A major problem inspires a major breakthrough.   There are two important lessons to be taken away from our attempt. The first is that, sadly, the argument does not appear to lead to a complete proof. The second, and perhaps more important, lesson is that we require a different choice of polynomial $P(x)$—but one that satisfies some of the same properties as $(sx - r)^T$. Specifically, imagine that we have a polynomial $P(x) \in \mathbb{Z}[x]$ of degree T such that the function $f(z) = P(e^z)$ vanishes to an *unspecified* order M at $z = \pi$ with $f^{(M)}(\pi) \neq 0$. From Challenge 4.2 we know that $f^{(M)}(\pi)$ is a nonzero rational number and that there is some power δ such that $s^\delta f^{(M)}(\pi)$ is a nonzero integer. Our hope now is that our new choice of $P(x)$ will allow us to conclude that the absolute value of our integer violates the Fundamental Principle of Number Theory.

For a polynomial $P(x) = \sum_{t=0}^T a_t x^t \in \mathbb{C}[x]$, we define the *height of P*, $\mathcal{H}(P)$, by

$$\mathcal{H}(P) = \max\{|a_0|, |a_1|, \ldots, |a_{T-1}|, |a_T|\}.$$

Using this notation, we could apply the Maximum Modulus Principle to $f(z) = P(e^z)$ and take logarithms to deduce an upper bound on the positive integer $|s^\delta f^{(M)}(\pi)|$ of the basic form

$$\log|s^\delta f^{(M)}(\pi)| < \delta\log s + M\log M + \log\mathcal{H}(P) + TR - M\log(R - \pi).$$

Therefore in order to utilize this inequality to deduce $\log|s^\delta f^{(M)}(\pi)| < 0$, we must select the parameters M, T, and R such that the term $M\log(R - \pi)$ dominates the left-hand side; that is, the radius R and the order of vanishing of $f(z)$ at $z = \pi$ must exceed the degree and the height of the polynomial P. In particular, we must have, at least,

$$TR < M\log(R - \pi),$$

which implies that M must be significantly larger than T. In other words, $f(z) = P(e^z)$ must have a higher order of vanishing at $z = \pi$ than $P(x)$ has at $x = e^\pi$. Unfortunately, we saw in Challenge 4.2 that this scenario is impossible.

This problematic situation requires us to modify our approach. That modification, which we explore in the next section, entails replacing the polynomial $P(x)$ with a polynomial of *two* variables. This new important feature will enable us to free the order of vanishing of the function f from the degree of the polynomial P. We will see that this crucial shift in the construction of our polynomial leads not only to a proof of the irrationality and then the transcendence of e^π, but to much much more.

4.4 Polynomials in two variables and the power of i

We come away from our previous attempt to establish the irrationality of e^{π} with an appreciation for the need of a polynomial function with integer coefficients having an enormous number of zeros relative to both its degree and the size of its coefficients. From Challenge 4.2 we see that this desire cannot be attained with a polynomial in one variable. As we will discover, we can realize our goal if we use a polynomial in *two* variables. Thus here we consider polynomials in two variables $P(x, y)$ and construct the auxiliary function $f(z)$.

We let $P(x, y) \in \mathbb{Z}[x, y]$ be a polynomial in two variables with integer coefficients defined by

$$P(x, y) = \sum_{m=0}^{D-1} \sum_{n=0}^{D-1} a_{mn} x^m y^n,$$

where we use $D - 1$ for the partial degrees rather than the perhaps more natural D because it leads to cleaner estimates in our analysis later. We now turn to the subtle issue of using $P(x, y)$ to build the function $f(z)$.

In our first attempt to demonstrate the irrationality of e^{π}, we composed the polynomial in one variable with the function e^z, which enabled us both to use our assumption that $e^{\pi} = \frac{r}{s}$ and to exploit the analytic properties of e^z. Here, if we let $f(z) = P(g(z), h(z))$, then our choice of the functions $g(z)$ and $h(z)$ is not immediately clear. A natural first guess might be to let $g(z) = h(z) = e^z$, which, in view of the identity $e^a e^b = e^{a+b}$, yields

$$f(z) = P(e^z, e^z) = \sum_{m=0}^{D-1} \sum_{n=0}^{D-1} a_{mn} e^{mz} e^{nz} = \sum_{m=0}^{D-1} \sum_{n=0}^{D-1} a_{mn} e^{(m+n)z} = \sum_{m=0}^{D-1} \sum_{n=0}^{D-1} a_{mn} (e^z)^{m+n},$$

which is simply a polynomial in one variable composed with e^z. This observation implies that if we let $g(z) = h(z) = e^z$, then we remain unable to disassociate the degree of P from the order of vanishing of f. The moral here is that we need to select two *different* functions for $g(z)$ and $h(z)$. But how "different" is different enough?

In order to appreciate the expression "*vive la différence*," we consider the highly nonrandom polynomial

$$P(x, y) = y^6 - xy^4 + x^2 y^2 - x^3.$$

Suppose that we select $g(z) = e^z$ and $h(z) = e^{\frac{1}{2}z}$; thus we have two *different* functions. If we now let $f(z) = P(g(z), h(z))$, then we see that

$$f(z) = P\left(e^z, e^{\frac{1}{2}z}\right) = e^{3z} - e^{3z} + e^{3z} - e^{3z},$$

that is, we discover that composing the not-identically-zero polynomial $P(x, y)$ with the distinct nonzero functions $g(z)$ and $h(z)$ results in a function $f(z)$ that is *identically zero*. Thus there is no way to use $f(z)$ or any of its derivatives to construct a *nonzero* integer. Hence we realize that the functions $g(z)$ and $h(z)$ must be *really* different.

The unintentional collapse above leading to $f(z) \equiv 0$ resulted from the fact that $g(z)$ and $h(z)$ are algebraically related by a polynomial equation. Specifically, if we let $p(x, y) = x - y^2$, then we have that $p(g(z), h(z)) \equiv 0$. Therefore, as we saw at the end of the previous chapter, here we see the need for the functions $g(z)$ and $h(z)$ to be algebraically independent as functions. Armed with this new insight, we are ready to hunt down a suitable choice of $g(z)$ and $h(z)$.

As we have already discovered, a reasonable choice for one of our functions is e^z itself. The key to finding the second function is the all-important identity

$$e^{i\pi} = -1.$$

Given this identity, an intriguing choice for our second function is e^{iz}. This possibility is promising, since if we assume that e^{π} is rational, then *both* e^z and e^{iz} will yield rational numbers when evaluated at $z = \pi$. The challenge below is further evidence that we appear to be on the right track.

Challenge 4.4 *Let $P(x, y)$ be a polynomial with integer coefficients and let $f(z) = P(e^z, e^{iz})$. Under the assumption that e^{π} is rational, show that $f(\pi)$ is a rational number. Furthermore, show that for all $t \geq 1$, $f^{(t)}(\pi)$ is a complex number of the form $u + iv$ for some rational numbers u and v. Use this observation to conclude that $|f^{(t)}(\pi)|^2$ is a rational number for all $t \geq 0$.*

We are now faced with the issue of whether the function $f(z) = P(e^z, e^{iz})$ is identically zero. Fortunately, as the next challenge implies, e^z and e^{iz} are sufficiently different so that $f(z)$ does not accidentally disappear.

Challenge 4.5 *Let $P(x, y) \in \mathbb{Z}[x, y]$ be a nonzero polynomial. Prove that the function $f(z) = P(e^z, e^{iz})$ is not identically zero. (Hint: First show that there must exist a complex number z_0 such that the polynomial expression $P(e^z, e^{iz_0})$ in z is not identically zero. Then using the periodicity of e^z, namely that for any integer k, $e^{z + 2\pi i k} = e^z$, conclude that if $f(z)$ is identically zero, then the polynomial $P(x, e^{iz_0})$ in x has infinitely many zeros—in particular, show that $P(x, e^{iz_0})$ vanishes at the distinct values $x_k = e^{z_0 + 2\pi k}$, for each integer k.)*

Thus we close this section with our choice of $f(z)$ firmly in hand, namely $f(z) = P(e^z, e^{iz})$. Next we consider the intricate construction of our polynomial $P(x, y) \in \mathbb{Z}[x, y]$ having a high order of vanishing but comparatively small coefficients.

4.5 The ever-popular polynomial construction

Given our choices of $g(z) = e^z$ and $h(z) = e^{iz}$, we can now restate our goal: Using the assumption that $e^{\pi} = \frac{r}{s}$, find a nonzero integral polynomial

$$P(x, y) = \sum_{m=0}^{D-1} \sum_{n=0}^{D-1} a_{mn} x^m y^n$$

such that the auxilary function $f(z) = P(e^z, e^{iz})$ possesses the necessary properties required to show that some value of one of its derivatives leads to an integer in violation of the Fundamental Principle of Number Theory. As we have seen in Section 4.2, we require the function $f(z)$ to vanish, with some high multiplicity, at $z = \pi$. Thus finding the polynomial $P(x, y)$ is equivalent to determining its coefficients a_{mn}. We can express the high degree of vanishing of $f(z)$ at $z = \pi$ as a system of linear equations having the coefficients a_{mn} as its "unknowns." Specifically, if we write

$$f(z) = \sum_{m=0}^{D-1} \sum_{n=0}^{D-1} a_{mn} e^{mz} e^{niz} \, ,$$

then we wish to determine a solution to the linear system (linear with respect to the "variables" a_{mn})

$$f(\pi) = 0$$
$$f^{(1)}(\pi) = 0$$
$$f^{(2)}(\pi) = 0$$
$$\vdots$$
$$f^{(T-1)}(\pi) = 0,$$

for some integer T yet to be chosen. As was briefly foreshadowed in Section 4.2, if we were to solve this system for the coefficients a_{mn}, then we would find that the function $f(z)$ does not have enough zeros—even when counted with multiplicity—to complete the proof. As we saw in the "proof" of Theorem 4.1, we must construct a polynomial that vanishes at other points. Specifically, we now seek a polynomial function that vanishes to order T not only at $z = \pi$ but at the points $z = 2\pi, z = 3\pi, \ldots, z = K\pi$, for some integer K also to be specified later.

Since we want $f(z)$ to vanish at those K points, and each with multiplicity T, we seek integral coefficients (now viewed as variables) a_{mn} such that the associated function $f(z)$ satisfies the KT linear equations

$$f(\pi) = 0$$
$$f^{(1)}(\pi) = 0$$
$$f^{(2)}(\pi) = 0$$
$$\vdots$$
$$f^{(T-1)}(\pi) = 0$$
$$f(2\pi) = 0$$
$$f^{(1)}(2\pi) = 0$$

$$f^{(2)}(2\pi) = 0$$

$$\vdots$$

$$f^{(T-1)}(2\pi) = 0$$

$$\vdots$$

$$f(K\pi) = 0$$

$$f^{(1)}(K\pi) = 0$$

$$f^{(2)}(K\pi) = 0$$

$$\vdots$$

$$f^{(T-1)}(K\pi) = 0.$$

The first step toward solving this large system of linear equations is to translate the system into one in which the matrix of coefficients is given explicitly. We accomplish this task by exploiting the important properties of e^z. Again using the identity $e^a e^b = e^{a+b}$, we can rewrite $f(z)$ as

$$f(z) = \sum_{m=0}^{D-1}\sum_{n=0}^{D-1} a_{mn} e^{(m+ni)z}.$$

Next, since $\frac{d}{dz}e^{(m+ni)z} = (m+ni)e^{(m+ni)z}$, we can express each derivative of $f(z)$ in a particularly simple form. For example,

$$f^{(1)}(z) = \sum_{m=0}^{D-1}\sum_{n=0}^{D-1} a_{mn}(m+ni)e^{(m+ni)z},$$

and the subsequent derivatives are similarly simple:

$$f^{(t)}(z) = \sum_{m=0}^{D-1}\sum_{n=0}^{D-1} a_{mn}(m+ni)^t e^{(m+ni)z}.$$

Our linear system is formed by evaluating derivatives of $f(z)$ at $z = k\pi$ for $k = 1, 2, \ldots, K$. Given the two key identities $e^{k\pi} = \left(\frac{r}{s}\right)^k$ and $e^{ik\pi} = (-1)^k$, we obtain the following system of KT linear equations in D^2 variables: For each k, $k = 1, 2, \ldots, K$, and for each t, $t = 0, 1, \ldots, T-1$,

$$f^{(t)}(k\pi) = \sum_{m=0}^{D-1}\sum_{n=0}^{D-1}(m+ni)^t \left(\frac{r}{s}\right)^{km}(-1)^{kn} a_{mn} = 0.$$

If we multiply the above identity by $s^{k(D-1)}$, then we clear all denominators, and our linear system becomes

$$\sum_{m=0}^{D-1}\sum_{n=0}^{D-1}(m+ni)^t r^{km} s^{k(D-1-m)}(-1)^{kn} a_{mn} = 0, \quad 1 \le k \le K, \; 0 \le t \le T-1.$$

Thus we wish to find an *integer* solution to a large system of linear equations having unknowns a_{mn} and coefficients of the form

$$(m+ni)^t r^{km} s^{k(D-1-m)}(-1)^{kn}.$$

This issue is made more complex with the appearance of the imaginary number i. We can remove this complexity by simply noting that a complex number $z = x + iy$ equals 0 if and only if *both* its real and imaginary parts, $\mathrm{Re}(z) = x$ and $\mathrm{Im}(z) = y$, respectively, equal 0. Thus the previous linear system of KT equations can be replaced by a linear system of $KT + KT$ equations—the first KT equations arising from the real parts and the second KT equations from the imaginary parts. Specifically, we now have a system of $2KT$ linear equations with *integer* coefficients in D^2 unknowns whose first KT equations are

$$\mathrm{Re}\left(s^{k(D-1)}f^{(t)}(k\pi)\right) = \sum_{m=0}^{D-1}\sum_{n=0}^{D-1}\mathrm{Re}((m+ni)^t)r^{km} s^{k(D-1-m)}(-1)^{kn} a_{mn} = 0,$$

$$(4.9)$$

for $1 \le k \le K$, $0 \le t \le T-1$; and whose last KT equations are

$$\mathrm{Im}\left(s^{k(D-1)}f^{(t)}(k\pi)\right) = \sum_{m=0}^{D-1}\sum_{n=0}^{D-1}\mathrm{Im}((m+ni)^t)r^{km} s^{k(D-1-m)}(-1)^{kn} a_{mn} = 0,$$

$$(4.10)$$

for $1 \le k \le K$, $0 \le t \le T-1$.

We recall that the D^2 unknown "variables" in this impressive linear system, a_{mn}, will be the coefficients for our nonzero polynomial $P(x,y) \in \mathbb{Z}[x,y]$. Thus we need not only a nonzero integer solution to the linear system, but one that is, relatively speaking, "not too big," since the height of that polynomial—the maximum of $|a_{mn}|$— will appear in our upper bound, which we hope will be less than 1. Proving that a "small" nonzero integer solution to our large linear system always exists involves an interesting application of the Pigeonhole Principle. As we will discover later, the usefulness of this result far transcends this one application. Thus we discuss the general principles here before applying the technique to our present situation.

Algebraic Excursion: Solving a system of linear equations in integers

Suppose we wish to find a nonzero integral solution to the homogeneous system of M linear equations in N unknowns given by

$$c_{11}X_1 + c_{12}X_2 + \cdots + c_{1N}X_N = 0$$
$$c_{21}X_1 + c_{22}X_2 + \cdots + c_{2N}X_N = 0$$
$$\vdots$$
$$c_{M1}X_1 + c_{M2}X_2 + \cdots + c_{MN}X_N = 0, \qquad (4.11)$$

where the coefficients c_{mn} are *integers*, not all equal to 0.

We recall from basic linear algebra that if $N > M$, then there is a nonzero *real* solution to the system of equations—in fact, there is an entire subspace of solutions. For our purposes, however, we desire an *integer* solution, and we wish to have an estimate of its size. It is intuitively clear that the size of the coefficients c_{mn} will affect the size of any solution. Thus in order to make this dependency explicit, we let $C = \max\{|c_{mn}| : 1 \le m \le M, \ 1 \le n \le N\}$. If we write $\mathcal{C}$ for the $M \times N$ matrix of coefficients

$$\mathcal{C} = \begin{pmatrix} c_{11} & c_{12} & \cdots & c_{1N} \\ c_{21} & c_{22} & \cdots & c_{2N} \\ \vdots & \vdots & & \vdots \\ c_{M1} & c_{M2} & \cdots & c_{MN} \end{pmatrix},$$

then we are searching for a nonzero vector

$$\vec{X} = \begin{pmatrix} X_1 \\ \vdots \\ X_N \end{pmatrix} \in \mathbb{Z}^N$$

satisfying $\mathcal{C}\vec{X} = \vec{0}$, or equivalently, the system in (4.11).

In fact, we will not directly find a solution to the system of equations (4.11), but instead will view the matrix $\mathcal{C}$ as a map from $\mathbb{Z}^N$ into $\mathbb{Z}^M$. In particular, we will establish the existence of two *distinct* vectors $\vec{x}_1$ and $\vec{x}_2$ in $\mathbb{Z}^N$, each of modest size, satisfying

$$\mathcal{C}\vec{x}_1 = \mathcal{C}\vec{x}_2.$$

Since the mapping defined by $\mathcal{C}$ is linear, the difference $\vec{x}_1 - \vec{x}_2$ will then be a nonzero solution to our original system of equations; that is, $\mathcal{C}(\vec{x}_1 - \vec{x}_2) = \vec{0}$.

Continued

The intuitive idea behind finding integer vectors via the
Pigeonhole Principle

Suppose we take a cube of vectors $\mathcal{D}$ in $\mathbb{Z}^N$ and using the matrix $\mathcal{C}$, map them all into a rectangular box of vectors $\mathcal{R}$ in $\mathbb{Z}^M$. If there are fewer vectors in the range set $\mathcal{R}$ than in the domain set $\mathcal{D}$, then there must exist two *distinct* integer vectors $\vec{x}_1$ and $\vec{x}_2$ in $\mathcal{D}$ that get mapped to the *same* vector in $\mathcal{R}$. That is, $\mathcal{C}\vec{x}_1 = \mathcal{C}\vec{x}_2$. Thus we see that $\vec{X} = \vec{x}_1 - \vec{x}_2$ is a *nonzero* integer solution to $\mathcal{C}\vec{X} = \vec{0}$. Moreover, since the vectors $\vec{x}_1$ and $\vec{x}_2$ are both from the domain cube $\mathcal{D}$, we can bound the size of the largest component of the solution vector $\vec{x}_1 - \vec{x}_2$.

Let $A \geq 1$ be an integer. We define the N-dimensional domain cube $\mathcal{D}(A)$ by

$$\mathcal{D}(A) = \left\{ \begin{pmatrix} x_1 \\ \vdots \\ x_N \end{pmatrix} \in \mathbb{Z}^N : 0 \leq x_n \leq A, \text{ for all } n = 1, 2, \ldots, N \right\}.$$

By simply counting we find that $\text{card}(\mathcal{D}(A)) = (1 + A)^N$, where $\text{card}(S)$ denotes the cardinality of the set S. Describing the M-dimensional range box $\mathcal{R}(A)$ is only slightly more involved.

Given an integer k, we define the integers k^+ and k^- by $k^+ = \max\{0, k\}$ and $k^- = \max\{0, -k\}$. We now define $\mathcal{R}(A) \subseteq \mathbb{Z}^M$ by

$\mathcal{R}(A)$

$$= \left\{ \begin{pmatrix} y_1 \\ \vdots \\ y_M \end{pmatrix} \in \mathbb{Z}^M : -A \sum_{n=1}^{N} c_{mn}^- \leq y_m \leq A \sum_{n=1}^{N} c_{mn}^+, \text{ for all } m = 1, 2, \ldots, M \right\}.$$

Challenge 4.6 *Show that our matrix $\mathcal{C} = (c_{mn})$ maps the finite collection $\mathcal{D}(A)$ into the finite collection $\mathcal{R}(A)$. In addition, for an arbitrary integer k, verify the identity $k^+ + k^- = |k|$. Then apply the identity to prove that*

$$\text{card}(\mathcal{R}(A)) \leq (1 + ACN)^M.$$

(Hint: Recall that $C = \max\{|c_{mn}|\}$.)

Using the previous challenge, we conclude from the Pigeonhole Principle that if

$$(1 + ACN)^M < (1 + A)^N, \tag{4.12}$$

Continued

then $\mathcal{C}$ will map two distinct vectors $\vec{x}_1, \vec{x}_2 \in \mathcal{D}(A)$ to the same vector in $\mathcal{R}(A)$. Thus we have that $\mathcal{C}(\vec{x}_1 - \vec{x}_2) = \vec{0}$, where $\vec{x}_1 - \vec{x}_2$ is a *nonzero* integer vector. Moreover, since each coordinate of both $\vec{x}_1$ and $\vec{x}_2$ is an element of the set $\{0, 1, \ldots, A\}$, we see that the maximum absolute value of the difference of any two of their coordinates must be less than or equal to A. Thus if we define the *height of a vector* $\vec{X} = (X_1 X_2 \cdots X_N)^T \in \mathbb{Z}^N$ by

$$\mathcal{H}(\vec{X}) = \max\{|X_1|, |X_2|, \ldots, |X_N|\},$$

then we conclude that $\vec{X} = \vec{x}_1 - \vec{x}_2$ is a nonzero vector in $\mathbb{Z}^N$ satisfying

$$\mathcal{C}\vec{X} = \vec{0} \text{ and } \mathcal{H}(\vec{X}) \leq A.$$

We are naturally led to the following question: Given that (4.12) must hold, just how small we can take A to be; that is, how small is our bound on the height of our integer solution? The answer is offered as the following challenge.

Challenge 4.7 *Given positive integers C, M, and N, with $N > M$, let*

$$A = \left[(CN)^{\frac{M}{N-M}} \right],$$

where $[x]$ denotes the integer part of x. Verify that $(CN)^M < (1+A)^{N-M}$, and then use the inequality to establish that

$$(1 + ACN)^M < (1 + A)^N.$$

(Hint: First show that $(CN + ACN)^M < (1+A)^N$.)

Challenge 4.7, together with our previous observations, yields the following result, whose usefulness within transcendental number theory cannot be overstated.

THEOREM 4.3 (SIEGEL'S LEMMA) *Let $\mathcal{C} = (c_{mn})$ be a nonzero $M \times N$ matrix having integer entries and let $C = \max\{|c_{mn}| : 1 \leq m \leq M, 1 \leq n \leq N\}$. If $M < N$, then there exists a nonzero vector $\vec{X} \in \mathbb{Z}^N$ satisfying*

$$\mathcal{C}\vec{X} = \vec{0} \text{ and } \mathcal{H}(\vec{X}) \leq (CN)^{\frac{M}{N-M}}.$$

Armed with the power of Siegel's Lemma, we can confidently conquer the complex polynomial construction ahead. With this realization, we now close this important excursion and return to our regularly scheduled polynomial programming already in progress.

Recall that we seek a polynomial $P(x, y)$ with integer coefficients such that the associated function $f(z) = P(e^z, e^{iz})$ vanishes to order T at $z = k\pi$ for $1 \leq k \leq K$. We find $P(x, y)$ by applying Siegel's Lemma to our system of $2KT$ equations in D^2 unknowns appearing in (4.9) and (4.10). Siegel's Lemma requires that $2KT < D^2$, and if this inequality holds, then the lemma gives an upper bound for our integer solution vector in terms of coefficients, which in our case are numbers of the form $\text{Re}((m + ni)^t)r^{km}s^{k(D-1-m)}(-1)^{kn}$ and $\text{Im}((m + ni)^t)r^{km}s^{k(D-1-m)}(-1)^{kn}$. But it is fairly easy to show that the following two estimates hold for any m, n, and k:

$$\left| \text{Re}((m + ni)^t)r^{km}s^{k(D-1-m)}(-1)^{kn} \right| < \frac{t+1}{2}(2mn)^t r^{km}s^{k(D-1-m)}$$

and

$$\left| \text{Im}((m + ni)^t)r^{km}s^{k(D-1-m)}(-1)^{kn} \right| < \frac{t+1}{2}(2mn)^t r^{km}s^{k(D-1-m)} .$$

Challenge 4.8 *Verify the two previous estimates and show that, assuming $2KT < D^2$, they imply the existence of a nonzero integer solution to our linear system having height less than or equal to*

$$\left(D^2(T + 1)(2D)^{2T}r^{KD} \right)^{\frac{2KT}{D^2-2KT}}. \tag{4.13}$$

(Hint: Use the Binomial Theorem to expand $(m + ni)^t$ and recall that $s < r$.)

Certainly, the bound on our integer solution to that linear system—or equivalently, the bound on the size of the coefficients of our polynomial $P(x, y)$—is unwieldy. We close this section by examining the parameters D and T in the hope of replacing the unwieldy bound in (4.13) by one that is more, well, "wieldy."

We have already uncovered a necessary connection between D and T in order to apply Siegel's Lemma: We must have that $2KT < D^2$. In considering the quantity from (4.13), we see that one of the more significant contributions made by D and T is in the awkward exponent $2KT/(D^2 - 2KT)$. Since we will eventually select the parameters D and T to be quite large, a reasonable guess is to connect D and T in sympathy with each other so that the exponent in (4.13) does not dominate our bound. One way to avoid this potential exponent domination is simply to choose the parameters D and T so that they satisfy

$$\frac{2KT}{D^2 - 2KT} = 1. \tag{4.14}$$

That is, we declare that $D = 2\sqrt{KT}$, with the necessary condition that K and T are selected so that KT is a perfect square. We remark that with this choice, $2KT < 4KT = D^2$, and thus we have fulfilled our obligation to satisfy the hypothesis of Siegel's Lemma.

Given that $D = 2\sqrt{KT}$, our unwieldly estimate for the coefficients of $P(x, y)$ in (4.13) becomes

$$4KT(T + 1)(16KT)^T r^{2K^{3/2}T^{1/2}},$$

which implies the only slightly less unwieldy upper bound for (4.13)

$$(4KT)^2 \left(16Kr^{2K^{3/2}} \right)^T T^T .$$

We now can deduce a "wieldy" upper bound for the size of the coefficients of our polynomial $P(x, y)$.

Challenge 4.9 *Verify that regardless of our choice of K, there exists an integer $T_0 = T_0(K)$ such that for all $T > T_0$, we have*

$$(4KT)^2 \left(16Kr^{2K^{3/2}}\right)^T < T^{T/2}. \tag{4.15}$$

Using this upper bound, conclude that by selecting T to be a sufficiently large integer, the coefficients of the polynomial $P(x, y)$ we obtained by applying Siegel's Lemma will have absolute value at most $T^{1.5T}$. (Hint: Show that for all sufficiently large T, the following four inequalities all hold:

$$(4KT)^2 < T^{T/8}, \quad 16^T < T^{T/8}, \quad K^T < T^{T/8}, \quad r^{2K^{3/2}T} < T^{T/8}.$$

Remark: In fact, all these inequalities hold, and hence (4.15) holds, if we take $T > r^{(16K)^8}$.)

4.6 At long last, a proof that e^π is irrational

Suppose that $e^\pi = \frac{r}{s}$. If we let D be the integer defined by $D = 2\sqrt{KT}$, and ensure that the integer T is eventually selected so that $T > r^{(16K)^8}$ and $D \in \mathbb{Z}$, then our previous elaborate polynomial construction gives rise to the nonzero polynomial

$$P(x, y) = \sum_{m=0}^{D-1}\sum_{n=0}^{D-1} a_{mn}x^m y^n \in \mathbb{Z}[x, y],$$

with $\max\{|a_{mn}|\} < T^{1.5T}$ and such that the associated entire function $f(z) = P(e^z, e^{iz})$ satisfies

$$f^{(t)}(k\pi) = 0 \text{ for } k = 1, 2, \ldots, K \quad \text{and} \quad t = 0, 1, \ldots, T - 1.$$

We are now ready to find an integer whose existence contradicts the Fundamental Principle of Number Theory. That integer, as foreshadowed in our overview from Section 4.2, will arise from a derivative of $f(z)$ evaluated at a multiple of π. In this direction we begin with the following challenge.

Challenge 4.10 *Given $f(z)$ as defined above, show that for any positive integers k and M, the quantity $\left|f^{(M)}(k\pi)\right|^2$ is a rational number. (Hint: First show that $f^{(M)}(k\pi)$ can be written as $u + vi$, where u and v are rational numbers. Then verify that $\left|f^{(M)}(k\pi)\right|^2 = u^2 + v^2$.)*

In view of our first attempt at a proof, a natural value to consider in the hope of constructing an integer is $f^{(T)}(k\pi)$ for some choice of k. Unfortunately, our construction does not guarantee that $f^{(T)}(k\pi) \neq 0$ for some k. Fortunately, since the polynomial $P(x,y)$ is not the zero polynomial, the function $f(z)$ is also nonzero. In particular, the power series for $f(z)$ centered at $z = k\pi$ has at least one nonzero coefficient; we let M be the smallest integer such that $f^{(M)}(k_0\pi) \neq 0$ for some k_0 satisfying $1 \leq k_0 \leq K$; so M depends on our as-yet unspecified parameter K. We remark that given the definition of M, we have that

$$T \leq M. \tag{4.16}$$

We observe that as foreshadowed in Challenge 4.4, the nonzero quantity $f^{(M)}(k_0\pi)$ is not a rational number, since it contains an imaginary term. However, in view of Challenge 4.10, we do see that $\left| f^{(M)}(k_0\pi) \right|^2$ is a nonzero rational number. As we will soon discover, it is this quantity that, after clearing denominators, will eventually lead to a nonzero integer failing to satisfy the Fundamental Principle of Number Theory.

From the expression

$$f^{(M)}(z) = \sum_{m=0}^{D-1}\sum_{n=0}^{D-1} a_{mn}(m+ni)^M e^{(m+ni)z}$$

and our assumption about the rationality of e^π, we have that

$$f^{(M)}(k_0\pi) = \sum_{m=0}^{D-1}\sum_{n=0}^{D-1} a_{mn}(m+ni)^M \left(\frac{r}{s}\right)^{k_0 m} (-1)^{k_0 n}.$$

Multiplying the previous identity by $s^{k_0(D-1)}$ reveals that

$$s^{k_0(D-1)}f^{(M)}(k_0\pi) = \sum_{m=0}^{D-1}\sum_{n=0}^{D-1} a_{mn}(m+ni)^M r^{k_0 m} s^{k_0(D-1-m)} (-1)^{k_0 n}.$$

It is easy to see that $s^{k_0(D-1)}f^{(M)}(k_0\pi)$ is a nonzero complex number of the form $A + Bi$, where A and B are *integers*. Unfortunately, $A + Bi$ itself is not an integer. Fortunately, in Chapter 3 we found a method for constructing integers from algebraic numbers. In particular, we introduce symmetry by multiplying by conjugates. In this case, we consider the product $(A + Bi)(A - Bi) = A^2 + B^2$, which not only is an integer but also equals $\left| s^{k_0(D-1)}f^{(M)}(k_0\pi) \right|^2$. Thus we conclude that $\left| s^{k_0(D-1)}f^{(M)}(k_0\pi) \right|^2 = A^2 + B^2$ is a positive integer—the one that will lead us down an impossible road.

In order to show that the positive integer $\left| s^{k_0(D-1)}f^{(M)}(k_0\pi) \right|^2$ is less than 1, we view the quantity $s^{k_0(D-1)}f^{(M)}(k_0\pi)$ as an integer multiple of the function $f^{(M)}(z)$ evaluated at $z = k_0\pi$. Thus we can apply the Maximum Modulus Principle in order to show that $\left| s^{k_0(D-1)}f^{(M)}(k_0\pi) \right|$ is surprisingly small. Hopefully we can select the parameters K, R, and T so that the upper bound is less than 1.

In order to find a small upper bound for the positive integer $\left| s^{k_0(D-1)} f^{(M)}(k_0\pi) \right|^2$, we return to our outline from Section 4.2 and define

$$G(z) = \frac{f(z)}{(z-\pi)^M (z-2\pi)^M \cdots (z-K\pi)^M}.$$

Challenge 4.11 *Given our choice of M, show that $G(z)$ is an entire function and then verify the formula*

$$f^{(M)}(k_0\pi) = M! G(k_0\pi) \prod_{\substack{1 \le k \le K \\ k \ne k_0}} ((k_0 - k)\pi)^M. \tag{4.17}$$

By the previous challenge we conclude that

$$s^{k_0(D-1)} M! |G(k_0\pi)| \prod_{\substack{1 \le k \le K \\ k \ne k_0}} \left| (k - k_0)\pi \right|^M$$

is a positive integer.

We now estimate $|G(k_0\pi)|$ by an application of the Maximum Modulus Principle. If we select R so large that $R > 10K\pi$, then we remark that $0.9R < |R - K\pi|$. Thus by the Maximum Modulus Principle and the previous inequality we conclude that

$$|G(k_0\pi)| \le \max_{|z| \le R}\{|G(z)|\} = |G(z)|_R \le \frac{|f(z)|_R}{\min_{1 \le k \le K}\{|R - k\pi|\}^{KM}}$$

$$= \frac{|f(z)|_R}{|R - K\pi|^{KM}} < \frac{|f(z)|_R}{(0.9R)^{KM}}, \tag{4.18}$$

where we recall that $|G(z)|_R = \max_{|z|=R}\{|G(z)|\}$.

We now require a reasonable estimate for the numerator $|f(z)|_R$, and it is here that the size of the coefficients of the polynomial we constructed through an application of Siegel's Lemma, the degree of that polynomial, and the radius R all play a critical role. We estimate $|f(z)|_R$ by applying the triangle inequality to the expression

$$f(z) = \sum_{m=0}^{D-1} \sum_{n=0}^{D-1} a_{mn} e^{(m+ni)z}.$$

Specifically, we conclude that

$$|f(z)|_R \le D^2 \max\{|a_{mn}|\} e^{2DR},$$

which, in view of our bound on $\max\{|a_{mn}|\}$ from Challenge 4.9 and our choice of $D = 2\sqrt{KT}$, can be expressed as

$$|f(z)|_R \le (4KT)(T^{1.5T}) \left(e^{4R\sqrt{KT}} \right). \tag{4.19}$$

We desire an upper bound for $|f(z)|_R$ that is a simpler function of our three, yet to be specified, parameters: K, R, and T. We recall from Challenge 4.9 that we will select T so large that $r^{(16K)^8} < T$. Thus for any such choice of T, it can be shown that inequality (4.19) implies that

$$|f(z)|_R < e^{1.6T \log T + 4R\sqrt{KT}}. \qquad (4.20)$$

We are now ready to put all the pieces of our upper bound puzzle together. If we let $\mathcal{G} = \left(s^{K(D-1)} f^{(M)}(k_0 \pi)\right)^2$, then we have already seen that $|\mathcal{G}|$ is a positive integer. It now follows from (4.17), (4.18), and (4.20), together with the classic bound $M! < M^M$, that

$$
\begin{aligned}
\log |\mathcal{G}| &< 2K(D-1)\log s + 2M \log M + 2 \log |G(k_0\pi)| + 2MK \log(K\pi) \\
&< 2K(D-1)\log s + 2M \log M + 2 \log |f(z)|_R \\
&\quad - 2KM \log(0.9R) + 2KM \log(K\pi) \\
&< 2K(D-1)\log s + 2M \log M + 3.8T \log T + 8R\sqrt{KT} \\
&\quad - 2KM \log(0.9R) + 2KM \log(K\pi).
\end{aligned}
$$

While that bound appears somewhat daunting, it is easy to see that *if*

$$\max \left\{ K(D-1)\log s, M \log M, 1.6T \log T, 4R\sqrt{KT}, KM \log(K\pi) \right\}$$

$$< \frac{1}{6} KM \log(0.9R), \qquad (4.21)$$

then the previous upper bound on $\log |\mathcal{G}|$ implies the much cleaner inequality

$$\log |\mathcal{G}| < -\frac{1}{3} KM \log(0.9R).$$

Challenge 4.12 *Recall that we have already set $D = 2\sqrt{KT}$ and have noted in (4.16) that $T \le M$. Using these observations, verify that for any integer T satisfying*

$$\max \left\{ 24^3 s^2, (24\pi)^{12} \right\} < T,$$

if we take $R = \frac{10}{9}\sqrt{M}$ and $K = 24$, then inequality (4.21) holds. Conclude that in this case,

$$\log |\mathcal{G}| < -4M \log M.$$

The end is now in sight. We adopt the choices for R and K made in the previous challenge and select the integer T satisfying

$$\max \left\{ r^{16^8 24^8}, 24^3 s^2, (24\pi)^{12} \right\} < T,$$

and such that $D = 2\sqrt{24T}$ is an integer. Thus by the previous challenge we see that

$$\log |\mathcal{G}| < -4M \log M < -4T \log T < 0,$$

which, in turn, implies that $|\mathcal{G}| < 1$. Finally, we remind the weary reader that $|\mathcal{G}|$ is known to be a positive integer, and thus, at long last, we have contradicted the Fundamental Principle of Number Theory. Therefore we conclude that e^π is irrational, which sadly ends our incredibly intricate argument.   ∎

A look back at the proof of the irrationality of e^{π}. It is very easy to focus solely on all the technical details of the previous elaborate argument and miss the overall structure. However, the global structure of the proof is an essential feature, since we will see manifestations of it in several transcendence arguments to come. Thus, in order to appreciate a panoramic view of our previous proof, we close this section with an outline of the entire argument in four "simple" steps.

Step 1. Assume e^{π} is rational, say $e^{\pi} = \frac{r}{s}$.

Step 2. Choose an integer T satisfying $T > \max\left\{r^{16^8 24^8}, 24^3 s^2, (24\pi)^{12}\right\}$ such that $D = 2\sqrt{24T}$ is an integer. Using the assumption from Step 1 and Siegel's Lemma, establish the existence of a nonzero polynomial

$$P(x,y) = \sum_{m=0}^{D-1}\sum_{n=0}^{D-1} a_{mn}x^m y^n \in \mathbb{Z}[x,y],$$

with $\max\{|a_{mn}|\} < T^{1.5T}$ and such that the associated function $f(z) = P(e^z, e^{iz})$ satisfies

$$f^{(t)}(k\pi) = 0 \quad \text{for } k = 1, 2, \ldots, 24, \quad \text{and} \quad t = 0, 1, \ldots, T-1.$$

Step 3. Since the function $f(z)$ in Step 2 is not identically zero, there exists a smallest integer M such that for some k_0, $1 \leq k_0 \leq 24$,

$$f^{(M)}(k_0\pi) \neq 0, \quad \text{while } f^{(m)}(k\pi) = 0 \quad \text{for all } k = 1, 2, \ldots, 24, \quad \text{and all}$$
$$m = 0, 1, \ldots, M-1.$$

Step 4. Given that $P(x,y) \in \mathbb{Z}[x,y]$, together with our choice of $f(z)$ in Step 2 and our choice of M in Step 3, we conclude that $|f^{(M)}(k_0\pi)|^2$ is a nonzero rational number. Moreover, $|\mathcal{G}| = \left|\left(s^{k_0(D-1)}f^{(M)}(k_0\pi)\right)^2\right|$ is a positive integer. A straightforward application of the Maximum Modulus Principle to the function

$$G(z) = \frac{f(z)}{(z-\pi)^M(z-2\pi)^M \cdots (z-K\pi)^M}$$

along a circle of radius $R = \frac{10}{9}\sqrt{M}$ shows us that

$$0 < |\mathcal{G}| < 1.$$

Since no such integer $|\mathcal{G}|$ can exist, our assumption that e^{π} is rational becomes utterly irrational.

4.7 A first glance at the transcendence of e^π—Some algebraic obstacles

All our hard work in establishing the irrationality of e^π will turn out to be extraordinarily valuable as we turn our attention to the transcendence of e^π. In fact, as we will discover in the next section, the reasoning behind the proof of the transcendence of e^π parallels our previous proof. There are only two additional technical complications, each of which was foreshadowed in our proof of Theorem 4.2. In this section we highlight those two points in order to inspire the transcendence proof on the horizon.

In applying Siegel's Lemma in Step 2 of the outline at the close Section 4.6, we were faced with a linear system whose coefficients were of the form $C + iD$, where C and D were integers. Specifically, those complex coefficients were of the form $(m + ni)^t r^{km} s^{k(D-1-m)} (-1)^{kn}$, where the the integers r and s arose from our assumption that $e^\pi = \frac{r}{s}$. In order to utilize Siegel's Lemma, our coefficients are required to be *integers*. We stepped around this complex issue by noting that 1 and i are linearly independent over $\mathbb{R}$—in other words, if a complex number equals zero, then *both* its real and imginary parts must each equal zero. This observation allowed us to create a linear system that had *twice* as many equations as the original one: We separated the real and imaginary parts of the coefficients to produce a larger system having only *integer* coefficients, which enabled us to apply Siegel's Lemma and move forward.

In the proof of the transcendence of e^π, we will assume that e^π is algebraic, say $e^\pi = \alpha$. Given this assumption, it seems reasonable to conjecture that the $r^{km} s^{k(D-1-m)}$ factors from the coefficients of the previous linear system will be replaced by some integer multiple of some power of α in the corresponding linear system in this context. Thus our coefficients will contain not only the algebraic number i, but the assumed-algebraic number α. Hence we will need to expand and develop our idea of linear independence among algebraic numbers in order to break up our linear system into many pieces, each of which contains only *integer* coefficients.

After an elaborate polynomial construction, we were led to an entire function $f(z)$ with the key feature that $f^{(M)}(k_0\pi) \neq 0$, for some integers M and k_0. Unfortunately, the quantity $f^{(M)}(k_0\pi)$ is not an integer, and thus a bound of the form $\left| f^{(M)}(k_0\pi) \right| < 1$ does not immediately lead to a contradiction to the Fundamental Principle of Number Theory. We ironed out this technical wrinkle by observing that the number $s^{k_0(D-1)} f^{(M)}(k_0\pi)$ is a complex number of the form $A + Bi$, for some integers A and B. Thus, introducing some conjugate symmetry, we discovered that

$$(A + Bi)(A - Bi) = A^2 + B^2 = \left| s^{k_0(D-1)} f^{(M)}(k_0\pi) \right|^2$$

is a positive *integer*—in fact, the one that finally led to a contradiction.

Not surprisingly, in proving the transcendence of e^π, the analogous analytic function $f(z)$ satisfying $f^{(M)}(k_0\pi) \neq 0$, for some integers M and k_0, will be such that an appropriate integer multiple of the quantity $f^{(M)}(k_0\pi)$ will involve integer multiples of both i and powers of the assumed-algebraic number α. Therefore in this situation, constructing a nonzero integer from the quantity $f^{(M)}(k_0\pi)$ will require us to consider *all* the conjugates of α. This slightly more complicated integer construction is the only other departure from our previous argument.

We close this section by developing the key ingredients required to overcome our first algebraic obstacle. The fundamental idea revolves around the ability to express higher powers of an algebraic number in terms of its lower powers.

Algebraic Excursion: Re-expressing the powers of an algebraic number

We first inspire the result we require with an illustration. If we let

$$p(x) = 2x^3 - 12x^2 + 24x - 13,$$

then it is a straightforward calculation to verify that the cubic irrational $\alpha = 2 + \left(-\frac{3}{2}\right)^{1/3}$ is a zero of $p(x)$. In fact, $p(x)$ is the minimal polynomial of α; hence we recall that any polynomial $f(x) \in \mathbb{Z}[x]$ for which $f(\alpha) = 0$, must have $p(x)$ as a factor. Implicit in the fact that $p(x)$ is the minimal polynomial for α is the $\mathbb{Q}$-linear independence of the numbers 1, α, and α^2. For if there existed integers c_0, c_1, and c_2, not all equal to zero, satisfying

$$c_2\alpha^2 + c_1\alpha + c_0 = 0,$$

then clearly the second-degree polynomial $c_2x^2 + c_1x + c_0$ would have a cubic polynomial as a factor, which is impossible.

We now consider the number α^3. Given that $p(\alpha) = 0$, we could solve for α^3 and discover that

$$\alpha^3 = 6\alpha^2 - 12\alpha + \frac{13}{2},$$

and thus conclude that α^3 is, in fact, linearly *dependent* on 1, α, and α^2 over $\mathbb{Q}$. Similarly, any higher power of α will also be linearly dependent on 1, α, and α^2. For example,

$$\alpha^4 = \alpha(\alpha^3) = \alpha\left(6\alpha^2 - 12\alpha + \frac{13}{2}\right) = 6\alpha^3 - 12\alpha^2 + \frac{13}{2}\alpha$$

$$= 6\left(6\alpha^2 - 12\alpha + \frac{13}{2}\right) - 12\alpha^2 + \frac{13}{2}\alpha = 24\alpha^2 - \frac{131}{2}\alpha + 39.$$

Continuing in this manner for any $n \geq 3$, it is possible to find rational numbers $c_{0,n}, c_{1,n}$, and $c_{2,n}$ satisfying

$$\alpha^n = c_{2,n}\alpha^2 + c_{1,n}\alpha + c_{0,n}.$$

Our example can be easily generalized. For an arbitrary algebraic number α of degree $\deg(\alpha) = d$, let

$$p(x) = c_dx^d + c_{d-1}x^{d-1} + \cdots + c_0 \in \mathbb{Z}[x]$$

Continued

denote the minimal polynomial for α; thus $\gcd(c_0, c_1, \ldots, c_d) = 1$. The coefficients of $p(x)$ play an important role in representing powers of α, so we define the *height of* α, denoted by $\mathcal{H}(\alpha)$, to be the height of its minimal polynomial. That is,

$$\mathcal{H}(\alpha) = \mathcal{H}(p) = \max\{|c_0|, |c_1|, \ldots, |c_d|\}.$$

We are now ready to express α^n as a linear combination of $1, \alpha, \alpha^2, \ldots, \alpha^{d-1}$ over $\mathbb{Q}$ in the following lemma, which also includes a bound on the size of the coefficients in terms of α and n.

LEMMA 4.4   *Let α be an algebraic number of degree d with minimal polynomial $p(x) = \sum_{j=0}^{d} c_j x^j \in \mathbb{Z}[x]$. Then the numbers $1, \alpha, \alpha^2, \ldots, \alpha^{d-1}$ are linearly independent over $\mathbb{Q}$. Furthermore, for each integer n, $n \geq d$, the quantity α^n can be expressed as a rational linear combination of $1, \alpha, \alpha^2, \ldots, \alpha^{d-1}$. That is, there exist rational numbers $c_{0,n}, c_{1,n}, \ldots, c_{d-1,n}$ such that*

$$\alpha^n = c_{0,n} + c_{1,n}\alpha + c_{2,n}\alpha^2 + \cdots + c_{d-1,n}\alpha^{d-1}. \tag{4.22}$$

Moreover,

$$\max_{0 \leq j \leq d-1}\{|c_{j,n}|\} \leq (1 + \mathcal{H}(\alpha))^{n+1-d},$$

and each rational number $c_{j,n}$ can be expressed as a fraction having a denominator equal to c_d^{n+1-d}.

Proof.   The fact that $1, \alpha, \alpha^2, \ldots, \alpha^{d-1}$ are linearly independent over $\mathbb{Q}$ follows from the observation that the polynomial of minimal degree for which α is a zero has degree d. Next, since $p(\alpha) = 0$, we can solve this identity for α^d to obtain

$$\alpha^d = -\frac{c_0}{c_d} - \frac{c_1}{c_d}\alpha - \cdots - \frac{c_{d-1}}{c_d}\alpha^{d-1},$$

and thus it follows that the lemma holds for $n = d$.

We now proceed by induction and assume that we have established the lemma for some $k \geq d$; that is,

$$\alpha^k = c_{0,k} + c_{1,k}\alpha + c_{2,k}\alpha^2 + \cdots + c_{d-1,k}\alpha^{d-1}, \tag{4.23}$$

where the coefficients have absolute values at most $(1 + \mathcal{H}(\alpha))^{k+1-d}$ and written with a denominator of c_d^{k+1-d}. If we multiply the expression in (4.23) by α, and replace the α^d term by the representation we found in the $k = d$ case, then we find that

$$\alpha^{k+1} = \alpha(\alpha^k) = c_{0,k}\alpha + c_{1,k}\alpha^2 + \cdots + c_{d-2,k}\alpha^{d-1} + c_{d-1,k}\alpha^d$$

$$= c_{0,k}\alpha + c_{1,k}\alpha^2 + \cdots + c_{d-2,k}\alpha^{d-1}$$

$$+ c_{d-1,k}\left(-\frac{c_0}{c_d} - \frac{c_1}{c_d}\alpha - \cdots - \frac{c_{d-1}}{c_d}\alpha^{d-1}\right).$$

Continued

Thus α^{k+1} can be expressed as a linear combination as in (4.22) with $c_{0,k+1} = -\frac{c_{d-1,k}c_0}{c_d}$, and for $j = 1, 2, \ldots, d-1$,

$$c_{j,k+1} = c_{j-1,k} - \frac{c_{d-1,k}c_j}{c_d}.$$

Clearly, each coefficient can be expressed as a rational number having a denominator equal to $c_d^{k+2-d} = c_d^{(k+1)+1-d}$. Finally, the triangle inequality together with our inductive hypothesis implies that

$$|c_{j,k+1}| = \left| c_{j-1,k} - \frac{c_{d-1,k}c_j}{c_d} \right| \leq |c_{j-1,k}| + \left| \frac{c_{d-1,k}c_j}{c_d} \right|$$

$$\leq (1 + \mathcal{H}(\alpha))^{k+1-d} + (1 + \mathcal{H}(\alpha))^{k+1-d}\mathcal{H}(\alpha)$$

$$= (1 + \mathcal{H}(\alpha))^{(k+1)+1-d},$$

which completes the proof and our excursion. ∎

4.8 The transcendence of e^π

In this section we prove the following beautiful theorem which is the main result of this chapter.

THEOREM 4.5 *The number e^π is transcendental.*

Our proof perfectly parallels that of Theorem 4.2 except for the two points outlined in the previous section.

Proof of Theorem 4.5.   We assume that e^π is an algebraic number, say α, of degree d with minimal polynomial $p(x) = \sum_{j=0}^{d} c_j x^j \in \mathbb{Z}[x]$. Since e^π is irrational by Theorem 4.2, we have that $d > 1$. We now follow the same line of reasoning as in the proof of Theorem 4.2.

Building an auxiliary polynomial via a system of linear equations. We wish to construct a nonzero polynomial

$$P(x,y) = \sum_{m=0}^{D-1}\sum_{n=0}^{D-1} a_{mn}x^m y^n \in \mathbb{Z}[x,y],$$

with reasonably small coefficients, and with the additional property that the function $f(z) = P(e^z, e^{iz})$ satisfies

$$f^{(t)}(k\pi) = 0 \quad \text{for } k = 1, 2, \ldots, K \text{ and } t = 0, 1, \ldots, T-1.$$

We could proceed as before and determine an appropriate value for K at the end of this proof, but that approach would only obscure the new features of this argument. Instead, we reveal the punch line up front: We will eventually take $K = 24d$. The parameters D and T, however, will remain a mystery until the dramatic conclusion.

Clearly, we have that

$$f(z) = \sum_{m=0}^{D-1} \sum_{n=0}^{D-1} a_{mn} e^{(m+ni)z},$$

and thus the derivatives of $f(z)$ are simple to express; in particular,

$$f^{(t)}(z) = \sum_{m=0}^{D-1} \sum_{n=0}^{D-1} a_{mn} (m + ni)^t e^{(m+ni)z}.$$

Thus, since $e^{\pi} = \alpha$ and $e^{i\pi} = -1$, we wish to find integers a_{mn}, not all zero, satisfying the following linear system of KT equations:

$$f^{(t)}(k\pi) = \sum_{m=0}^{D-1} \sum_{n=0}^{D-1} (m + ni)^t \alpha^{km} (-1)^{kn} a_{mn} = 0,$$

for $k = 1, 2, \ldots, K$ and $t = 0, 1, \ldots, T - 1$. Unfortunately, the coefficients in the above system, $(m + ni)^t \alpha^{km} (-1)^{kn}$, are algebraic numbers and not integers. Therefore we are unable to apply Siegel's Lemma at this point.

We replace the previous linear system having algebraic coefficients by another system of linear equations having *integral* coefficients in three steps. We first apply our observations from the algebraic excursion in Section 4.7 to replace this system of KT equations by a system of dKT linear equations whose coefficients are free of any α. To accomplish this α-removal we simply replace higher powers of α by rational linear combinations of $1, \alpha, \alpha^2, \ldots, \alpha^{d-1}$, and then group together like-powers of α. We obtain the dKT equations by setting the coefficient of each power of α equal to 0. We next multiply each equation in this system by an appropriate integer to obtain a system having coefficients of the form $C + iD$ for integers C and D. Finally, we replace this system of dKT equations by a system of $2dKT$ equations with integral coefficients, by separating each equation into its real and imaginary parts and setting them equal to zero.

In order to carry out the above outline, we now introduce some notation for the expressions we found in our algebra excursion. Specifically, we write $\alpha^n = p_n(\alpha)$, where $p_n(x) = c_{0,n} + c_{1,n}x + \cdots + c_{d-1,n}x^{d-1} \in \mathbb{Q}[x]$ for $n \geq d$, while for $n = 0, 1, 2, \ldots, d - 1$, $p_n(x) = x^n$. Thus we can rewrite each equation

$$f^{(t)}(k\pi) = \sum_{m=0}^{D-1} \sum_{n=0}^{D-1} (m + ni)^t \alpha^{km} (-1)^{kn} a_{mn} = 0$$

as

$$f^{(t)}(k\pi) = \sum_{m=0}^{D-1}\sum_{n=0}^{D-1}(m+ni)^t p_{km}(\alpha)(-1)^{kn}a_{mn}$$

$$= \sum_{j=0}^{d-1}\left(\sum_{m=0}^{D-1}\sum_{n=0}^{D-1}(m+ni)^t b_{jkm}(-1)^{kn}a_{mn}\right)\alpha^j = 0, \qquad (4.24)$$

for some rational numbers b_{jkm}.

Challenge 4.13 *Verify that each rational number b_{jkm} is the sum of at most D^2 rational numbers, each of whose absolute value is at most $(1+\mathcal{H}(\alpha))^{(D-1)K+1-d}$. Using this observation, establish the following upper bound:*

$$\max\{|b_{jkm}|\} < D^2(2\mathcal{H}(\alpha))^{DK}. \qquad (4.25)$$

Finally, show that the denominators of the numbers b_{jkm} can be taken to be $c_d^{(D-1)K}$.

We now replace our system of KT equations with one consisting of dKT linear equations by setting the coefficients of the α^j terms equal to 0, thereby obtaining the system

$$\sum_{m=0}^{D-1}\sum_{n=0}^{D-1}(m+ni)^t b_{tkm}(-1)^{kn}a_{mn} = 0,$$

for $j = 0, 1, \ldots, d-1$, $k = 1, 2, \ldots, K$, and $t = 0, 1, \ldots, T-1$. To obtain a system of equations with *integral* coefficients, we first clear all denominators of the b_{jkm} terms by multiplying each of the equations above by $c_d^{K(D-1)}$ to produce

$$\sum_{m=0}^{D-1}\sum_{n=0}^{D-1}(m+ni)^t c_d^{K(D-1)} b_{jkm}(-1)^{kn}a_{mn} = 0,$$

for $j = 0, 1, \ldots, d-1$, $k = 1, 2, \ldots, K$, and $t = 0, 1, \ldots, T-1$. We remark that inequality (4.25) implies the following upper bound on the integers $\left|c_d^{K(D-1)}b_{jkm}\right|$:

$$\max\left\{\left|c_d^{K(D-1)}b_{jkm}\right|\right\} \le |c_d|^{K(D-1)}D^2(2\mathcal{H}(\alpha))^{DK} < D^2(2\mathcal{H}(\alpha))^{2KD}.$$

Finally we set the real and imaginary parts of these dKT equations equal to 0, thus obtaining the system of $2dKT$ linear equations

$$\sum_{m=0}^{D-1}\sum_{n=0}^{D-1}\mathrm{Re}((m+ni)^t)c_d^{K(D-1)}b_{jkm}(-1)^{kn}a_{mn} = 0$$

and

$$\sum_{m=0}^{D-1}\sum_{n=0}^{D-1} \mathrm{Im}((m+ni)^t)c_d^{K(D-1)}b_{jkm}(-1)^{kn}a_{mn} = 0,$$

for $j = 0, 1, \ldots, d-1$, $k = 1, 2, \ldots, K$, and $t = 0, 1, \ldots, T-1$. If we recall our estimates for $\mathrm{Re}((m+ni)^t)$ and $\mathrm{Im}((m+ni)^t)$ implicit in Challenge 4.8, then we can bound the integer coefficients of our linear system of equations by

$$\left| \mathrm{Re}((m+ni)^t)c_d^{K(D-1)}b_{jkm}(-1)^{kn} \right| \leq \frac{T+1}{2}(2D^2)^T D^2 (2\mathcal{H}(\alpha))^{2DK}$$

and

$$\left| \mathrm{Im}((m+ni)^t)c_d^{K(D-1)}b_{jkm}(-1)^{kn} \right| \leq \frac{T+1}{2}(2D^2)^T D^2 \mathcal{H}(2(\alpha))^{2DK}.$$

In order to apply Siegel's Lemma to establish the existence of a relatively small nonzero integer solution to the above linear system, we require that

$$D^2 > 2dKT.$$

By Siegel's Lemma we then know that there is a nonzero integer solution a_{mn} to our linear system satisfying

$$\max\{|a_{mn}|\} \leq \left(D^2 \frac{T+1}{2}(2D^2)^T D^2 (2\mathcal{H}(\alpha))^{2KD} \right)^{\frac{2dKT}{D^2-2dKT}}.$$

We can simplify this upper bound by selecting D and T such that $\frac{2dKT}{D^2-2dKT} = 1$. These observations inspire us now to declare that

$$D = 2\sqrt{dKT},$$

where T will be selected such that dKT is a perfect square. With this relationship between D and T, we can apply Siegel's Lemma, since we have more variables than linear equations; that is, $D^2 > 2dKT$. Therefore we conclude that there exist D^2 integers a_{mn}, not all zero, satisfying the above system of linear equations and with

$$\max\{|a_{mn}|\} \leq D^2 \frac{T+1}{2}(2D^2)^T D^2 (2\mathcal{H}(\alpha))^{2KD}.$$

Challenge 4.14 *Show that if we take T to be sufficiently large, then*

$$\max\{|a_{mn}|\} < T^{1.5T}.$$

(Hint: Adopt the same strategy as suggested in Challenge 4.9.)

Hence we have established the existence of a nonzero polynomial

$$P(x,y) = \sum_{m=0}^{D-1}\sum_{n=0}^{D-1} a_{mn}x^m y^n \in \mathbb{Z}[x,y],$$

whose coefficients have absolute value at most $T^{1.5T}$, and such that the associated function

$$f(z) = P(e^z, e^{iz}) \quad \text{satisfies} \quad f^{(t)}(k\pi) = 0 \quad \text{for} \quad k = 1, 2, \ldots, K \quad \text{and}$$

$$t = 0, 1, \ldots, T - 1.$$

Arguing exactly as in the proof of Theorem 4.2, we can discover the analogue of inequality (4.19) in this context:

$$|f(z)|_R \leq (4dKT) \left(T^{1.5T} \right) \left(e^{4R\sqrt{dKT}} \right),$$

which, as before, implies that for all sufficiently large T,

$$|f(z)|_R \leq e^{1.6T \log T + 4\sqrt{dKT}}.$$

Constructing an incredibly small nonzero integer. Since $P(x, y)$ is not the zero polynomial, and e^z and e^{iz} are algebraically independent functions, we know from Challenge 4.5 that $f(z) = P(e^z, e^{iz})$ is not identically zero. It follows that for each $k, k = 1, 2, \ldots, K$, there exists a derivative of $f(z)$ that does not vanish at $k\pi$. That is, for each k, there exists an integer M_k such that

$$f^{(M_k)}(k\pi) \neq 0, \quad \text{while} f^{(m)}(k\pi) = 0 \quad \text{for all} \quad m = 0, 1, \ldots, M_k.$$

If we now let $M = \min\{M_1, M_2, \ldots, M_K\}$, then for any $m < M$, we have that $f^{(m)}(k\pi) = 0$ for $k = 1, 2, \ldots, K$, and there exists some k_0 such that $f^{(M)}(k_0\pi) \neq 0$. We remark that in view of our construction of $P(x, y)$, we have

$$T \leq M. \tag{4.26}$$

The nonzero quantity $f^{(M)}(k_0\pi)$ will eventually lead to an integer violating the Fundamental Principle of Number Theory.

If we let $\mathcal{A} = c_d^{K(D-1)} f^{(M)}(k_0\pi)$, then we note that $\mathcal{A}$ is a nonzero algebraic number. If we now consider a radius $R > 10K\pi$, then we can apply the Maximum Modulus Principle to the auxiliary entire function $G(z)$ exactly as in the proof of Theorem 4.2 and deduce that

$$\log |\mathcal{A}| < K(D - 1) \log c_d + M \log M + 1.6T \log T + 4R\sqrt{dKT}$$

$$- KM \log(0.9R) + 2KM \log(K\pi).$$

Challenge 4.15 *Verify that the previous bound holds and then show that if we set $R = \frac{10}{9}\sqrt{M}$, then for all sufficiently large T, it follows that*

$$\log |\mathcal{A}| < -\frac{1}{6} KM \log M. \tag{4.27}$$

(Hint: Adopt a similar strategy as was suggested in Challenge 4.12.)

Although inequality (4.27) implies that $|\mathcal{A}| < 1$, this observation does not lead us to a contradiction, since $\mathcal{A}$ is merely a nonzero algebraic number and not necessarily a nonzero *integer*. We now adopt the ideas of symmetry and conjugates developed in Chapter 3 to construct an integer using the algebraic number $\mathcal{A}$. In this direction we begin with the important realization that the quantity $c_d^{K(D-1)} f^{(M)}(k_0\pi)$ can be viewed as a polynomial in two variables with integer coefficients evaluated at i and α. That is, there exists a $Q(x, y) \in \mathbb{Z}[x, y]$ such that $c_d^{K(D-1)} f^{(M)}(k_0\pi) = Q(i, \alpha)$.

We now introduce the conjugates of α: $\alpha_1, \alpha_2, \ldots, \alpha_d$, where $\alpha = \alpha_1$. We claim that for any α_l, we have that $Q(i, \alpha_l) \neq 0$. We establish this claim by first observing that $Q(i, \alpha) = \mathcal{A} \neq 0$. If we express the complex number $Q(i, \alpha)$ as $u + iv$ for some real numbers u and v, then since $\alpha = e^\pi$ is a *real* number, it follows that $Q(-i, \alpha) = u - iv$. In particular, we see that $Q(-i, \alpha) \neq 0$, which allows us to conclude that

$$Q(i, \alpha)Q(-i, \alpha) \neq 0. \tag{4.28}$$

We now assume that $Q(i, \alpha_l) = 0$, for some $\alpha_l \in \mathbb{C}$ conjugate to α, which immediately implies that $Q(i, \alpha_l)Q(-i, \alpha_l) = 0$. Since the function $Q(i, y)Q(-i, y)$ is symmetric in the conjugates i and $-i$, it follows from our excursion into symmetric functions in Chapter 3 that $Q(i, y)Q(-i, y)$ is a polynomial with integer coefficients; that is, $Q(i, y)Q(-i, y) = h(y) \in \mathbb{Z}[y]$. Given our assumption, we discover that $h(\alpha_l) = 0$. But since $h(y)$ has integer coefficients, it follows that all the conjugates of α_l are also zeros of h. In particular,

$$h(\alpha) = Q(i, \alpha)Q(-i, \alpha) = 0,$$

which contradicts (4.28). Therefore we must have that $Q(i, \alpha_l) \neq 0$ for all conjugates α_l of α, which establishes our claim.

If we let $\mathcal{Q} = \prod_{l=1}^d Q(i, \alpha_l)$, then from our previous claim we see that

$$\mathcal{Q} = \prod_{l=1}^d Q(i, \alpha_l) \neq 0.$$

We also observe that the previous product is symmetric in the conjugates of α. Therefore by Theorem 3.13, we have that $\mathcal{Q}$ is of the form $U + iV$ for some *rational* numbers U and V. Moreover, in view of (3.15) and Lemma 4.6, it follows that the denominators of U and V can be taken to be c_d^{d-1}. Thus at long last we have that

$$c_d^{2(d-1)}(U + Vi)(U - Vi) = c_d^{2(d-1)}(U^2 + V^2)$$

is a positive integer! We call this much-sought-after integer $\mathcal{N}$. We are now ready to show that $\mathcal{N} < 1$. We begin by observing that

$$\mathcal{N} = c_d^{2(d-1)}(U^2 + V^2) = c_d^{2(d-1)}|U + Vi|^2$$

$$= c_d^{2(d-1)}|\mathcal{Q}|^2 = c_d^{2(d-1)} \prod_{l=1}^d |Q(i, \alpha_l)|^2$$

$$= c_d^{2(d-1)}|Q(i, \alpha)|^2 \prod_{l=2}^d |Q(i, \alpha_l)|^2 = c_d^{2(d-1)}|\mathcal{A}|^2 \prod_{l=2}^d |Q(i, \alpha_l)|^2.$$

Thus we have

$$\log \mathcal{N} = 2(d-1)\log|c_d| + 2\log|\mathcal{A}| + 2\sum_{l=2}^{d}\log|Q(i,\alpha_l)|. \qquad (4.29)$$

By (4.24), we see that

$$Q(i,\alpha_l) = c_d^{K(D-1)}\sum_{j=0}^{d-1}\left(\sum_{m=0}^{D-1}\sum_{n=0}^{D-1}(m+ni)^M b_{jk_0m}(-1)^{k_0n}a_{mn}\right)\alpha_l^j.$$

Now arguing as before, it follows that if we ensure that

$$\max\left\{|\alpha_l|^{d-1}: l = 2,3,\ldots,d\right\} < T,$$

then we can conclude that for $l = 2,3,\ldots,d$,

$$|Q(i,\alpha_l)| < T^{2T}.$$

Therefore (4.29) and (4.26) yield

$$\log \mathcal{N} < 2(d-1)\log|c_d| - \frac{1}{3}KM\log M + 4(d-1)T\log T$$

$$\leq 4(d-1)\log|c_d| - \frac{1}{3}KM\log M + 2(d-1)M\log M. \qquad (4.30)$$

If we further insist that $T > 12(d-1)\log|c_d|$, then since $T \leq M$, it follows that

$$2(d-1)\log|c_d| < \frac{1}{6}KM\log M,$$

which, in turn, allows us to replace (4.30) with

$$\log \mathcal{N} < -\frac{1}{6}KM\log M + 4(d-1)M\log M.$$

But since $K = 24d$, the previous inequality becomes

$$\log \mathcal{N} < -4M\log M < 0,$$

which—after much anticipation—implies that the integer $\mathcal{N}$ satisfies

$$0 < \mathcal{N} < 1,$$

which contradicts the Fundamental Principle of Number Theory. At long last we have earned the right to utter the following sentence: *Therefore we must have that e^π is transcendental.*   ∎

Looking ahead: A final look at e^π.   Given that $e^{\pi i/2} = i$, if we take the principal branch of the logarithm, then we see that $\frac{\pi i}{2} = \log i$. Some basic algebra reveals that $\pi = -2i \log i$, which implies that $e^\pi = e^{-2i \log i}$. Hence we are led to the identity

$$e^\pi = i^{-2i},$$

and discover that the number that was center stage for this entire chapter can be expressed as an algebraic number raised to an algebraic exponent. Inspired by this revelation, in the next chapter we consider quantities of the form α^β where α and β are algebraic. As we will see, the methods we developed here are robust enough to deduce far more general transcendence results that hold within them the transcendence of our hero: e^π.

Number 5

2.665144142690225188650297298 ...

Debunking Conspiracy Theories for Independent Functions:
The Gelfond–Schneider Theorem and the transcendence of $2^{\sqrt{2}}$

5.1 Algebraic exponents and bases—Moving beyond e by focusing on e

In his famous address delivered before the 1900 International Congress of Mathematicians in Paris, David Hilbert outlined 23 open questions that he believed to be the most important mathematical problems that remained unsolved at the dawn of the new century. Hilbert described his 7th problem as follows.

> "Hermite's arithmetical theorems on the exponential function and their extension by Lindemann are certain of the admiration of all generations of mathematicians ... I consider [their work] very difficult; as also the [yet-to-be-found] proof that the expression α^{β}, for an algebraic base α and an irrational algebraic exponent β, e.g., the number $2^{\sqrt{2}}$ or $e^{\pi} = i^{-2i}$, always represents a transcendental or at least an irrational number.
>
> It is certain that the solution of these and similar problems must lead us to entirely new methods and to new insights into the nature of special irrational and transcendental numbers."

In this chapter we explore the solution to Hilbert's 7th problem and its many important implications. The pioneering work in this direction was accomplished in 1934 independently by Aleksandro O. Gelfond and Theodor Schneider. As a consequence of their individual efforts we have the following singular result.

THEOREM 5.1 (THE GELFOND–SCHNEIDER THEOREM) *Suppose α and β are algebraic numbers with $\alpha \neq 0, 1$, and β irrational. Then α^{β} is transcendental.*

Thus we instantly see that $2^{\sqrt{2}}$ (the number appearing in this chapter's title), $\sqrt{3}^{\sqrt{2}}$, and i^i are all transcendental. The transcendence of e^{π} also follows immediately if we view $e^{\pi} = i^{-2i}$; or alternatively if $\alpha = e^{\pi}$ were algebraic, then for $\beta = i$, the Gelfond–Schneider Theorem implies the transcendence of

$$\alpha^{\beta} = (e^{\pi})^i = -1,$$

a flagrant contradiction. Hence we have established the transcendence of e^{π} in one short sentence rather than in one entire chapter. Obviously, the Gelfond–Schneider Theorem packs quite a punch.

Here we establish a particularly appealing formulation of the Gelfond–Schneider Theorem due to Michel Waldschmidt.

THEOREM 5.2 *Given two nonzero complex numbers ξ and ζ, with ξ irrational, at least one of the numbers ξ, e^{ζ}, or $e^{\xi\zeta}$ is transcendental.*

We observe that for algebraic numbers α and β, with $\alpha \neq 0, 1$, and β irrational, if we set $\xi = \beta$ and $\zeta = \log \alpha$, then Theorem 5.2 implies that at least one of the following numbers is transcendental: β, $e^{\log \alpha} = \alpha$, or $e^{\beta \log \alpha} = \alpha^{\beta}$. However since the first two numbers are algebraic, we conclude that α^{β} must be transcendental. Thus Theorem 5.2 immediately implies the validity of the Gelfond–Schneider Theorem.

With essentially no additional effort, we can apply Theorem 5.2 to deduce several other interesting corollaries.

COROLLARY 5.3 *Supppose that α and β are algebraic numbers with $\alpha \neq 0, 1$, and $i\beta$ irrational. Then both $\cos(\beta \log \alpha)$ and $\sin(\beta \log \alpha)$ are transcendental numbers.*

Proof. We begin by recalling the basic identity

$$\cos(\beta \log \alpha) = \frac{e^{i\beta \log \alpha} + e^{-i\beta \log \alpha}}{2}.$$

If we assume that $\cos(\beta \log \alpha)$ is algebraic, then the previous identity implies that $e^{i\beta \log \alpha}$, which equals $\alpha^{i\beta}$, is also algebraic. However given our hypotheses, we see that by Theorem 5.2, $\alpha^{i\beta}$ is, in fact, transcendental. This contradiction leads us to conclude that $\cos(\beta \log \alpha)$ is a transcendental number.

Challenge 5.1 *Applying the identity $\sin z = \frac{e^{iz} - e^{-iz}}{2i}$, prove that $\sin(\beta \log \alpha)$ is also transcendental, thus completing the proof of the corollary.* ∎

The following corollary foreshadows an entire area of study in transcendence theory.

COROLLARY 5.4 *Suppose α_1 and α_2 are nonzero algebraic numbers such that $\log \alpha_1$ and $\log \alpha_2$ are $\mathbb{Q}$-linearly independent. Then the number $\log \alpha_1 / \log \alpha_2$ is transcendental.*

Proof. Let $\xi = \frac{\log \alpha_1}{\log \alpha_2}$. Since $\log \alpha_1$ and $\log \alpha_2$ are $\mathbb{Q}$-linearly independent, it follows that ξ is irrational. If we now set $\zeta = \log \alpha_2$, then Theorem 5.2 implies the transcendence of at least one of the following:

$$\frac{\log \alpha_1}{\log \alpha_2}, \qquad e^{\log \alpha_2} = \alpha_2, \qquad e^{\frac{\log \alpha_1}{\log \alpha_2} \log \alpha_2} = \alpha_1.$$

Since α_1 and α_2 are algebraic, we arrive at the transcendence of $\log \alpha_1 / \log \alpha_2$. ∎

As an aside, it is easy to see that Corollary 5.4 is equivalent to the following statement: *Let α_1 and α_2 be nonzero algebraic numbers such that the numbers $\log \alpha_1$ and $\log \alpha_2$ are $\mathbb{Q}$-linearly independent. Then for any nonzero algebraic numbers β_1 and β_2,*

$$\beta_1 \log \alpha_1 + \beta_2 \log \alpha_2 \neq 0.$$

Phrased in this manner, the result naturally leads to the seminal work of Alan Baker, who in 1966 produced the following important generalization involving linear forms in logarithms, whose statement is reminiscent of the Lindemann–Weierstrass Theorem (see Corollary 3.7).

THEOREM 5.5 *Let $\alpha_1, \alpha_2, \ldots, \alpha_M$ be nonzero algebraic numbers such that the numbers $\log \alpha_1, \log \alpha_2, \ldots, \log \alpha_M$ are $\mathbb{Q}$-linearly independent. Then for any algebraic numbers $\beta_0, \beta_1, \ldots, \beta_M$, not all zero, the number*

$$\beta_0 + \sum_{m=1}^{M} \beta_m \log \alpha_m$$

is transcendental.

While we will not prove this result, we do remark that a quantitative version of Theorem 5.5 has had a significant impact on the theory of Diophantine equations. In recognition of his contributions, Baker was awarded a Fields Medal in 1970.

Returning to the Gelfond–Schneider Theorem, we close this opening section with the observation that both hypotheses of Theorem 5.2, $\zeta \neq 0$ and $\xi \notin \mathbb{Q}$, are necessary.

Challenge 5.2 *Show that if ξ is a nonzero algebraic number, then the conclusion of Theorem 5.2 is false for $\zeta = 0$. Similarly, show that if ζ is the logarithm of an algebraic number different from 0 and 1, then the conclusion of Theorem 5.2 does not hold for $\xi \in \mathbb{Q}$.*

5.2 Some sketchy thoughts on the proof of Theorem 5.2

In the previous chapter we established the transcendence of e^π by employing the assumption that e^π is algebraic to construct a polynomial that led to an integer violating the Fundamental Principle of Number Theory. That argument can be extended to produce a proof of Theorem 5.2.

In Gelfond's proof of Theorem 5.1, the auxiliary function $F(z) = P(e^z, e^{\beta z})$ replaces the function $f(z) = P(e^z, e^{iz})$ from the previous chapter. In both arguments the polynomials $P(x, y)$ are constructed so that their associated auxiliary functions and a large number of their derivatives vanish at certain carefully selected values (for the transcendence of e^{π} those values were integral multiples of π; and for the transcendence of α^{β}, Gelfond's values are integral multiples of $\log \alpha$).

A different approach was offered by Schneider, whose proof does not involve derivatives of the auxiliary function $f(z)$. Instead of constructing a function that has several zeros of relatively large multiplicity, Schneider constructed an auxiliary function that has *many* zeros but each of multiplicity one. Not surprisingly, Schneider's method requires a different choice of auxiliary function, in particular a function of the form $F(z) = P(z, e^{(\log \alpha)z})$. It is this plan of attack that we will adopt in order to establish Theorem 5.2.

The intuitive idea for the proof of Theorem 5.2

In this instance, the clearest way to look ahead is first to look back. Using the broadest of brush strokes to paint an overview of our proof of the transcendence of e^{π}, we recall that we first assumed that $\alpha = e^{\pi}$ is algebraic. We then noticed that for any polynomial $P(x, y) \in \mathbb{Z}[x, y]$, the function $f(z) = P(e^z, e^{iz})$ (and for that matter, each of its derivatives) possesses the property that for any positive integer k, the quantity

$$f(k\pi) = P\left(e^{k\pi}, e^{k\pi i}\right) = P\left(\alpha^k, (-1)^k\right)$$

is an algebraic number. Next, using Siegel's Lemma, we found a polynomial $P(x, y) \in \mathbb{Z}[x, y]$ having coefficients of relatively "modest size" such that the function $f(z) = P(e^z, e^{iz})$ exhibits the following two additional properties:

- $f(z)$ is not identically zero.
- For each $k = 1, 2, \ldots, K$, and $t = 0, 1, \ldots, T - 1$, we have that $f^{(t)}(k\pi) = 0$.

We then found integers M and k_0 such that the algebraic number $f^{(M)}(k_0\pi)$ is nonzero. Using the number $f^{(M)}(k_0\pi)$ together with the properties of $f(z)$, we came upon suitable values for the parameters $\deg_x(P(x, y))$, $\deg_y(P(x, y))$, K, and T, which, in turn, led to an integer $\mathcal{N}$ that violated the Fundamental Principle of Number Theory.

As we now look ahead toward our proof of Theorem 5.2, we find a structure that parallels the above overview for the transcendence of e^{π}. Specifically, using the same big brush strokes, we will begin with the assumption that $\alpha_1 = \xi$, $\alpha_2 = e^{\xi}$, and $\alpha_3 = e^{\xi\zeta}$ are all algebraic numbers. Just as in our previous argument, here we see that for any polynomial $P(x, y) \in \mathbb{Z}[x, y]$, the function $F(z) = P(z, e^{\zeta z})$ possesses the property that for each pair of positive integers (k_1, k_2), the quantity

$$F(k_1 + k_2\xi) = P\left(k_1 + k_2\xi, e^{\zeta(k_1 + k_2\xi)}\right) = P\left(k_1 + k_2\alpha_1, \alpha_2^{k_1}\alpha_3^{k_2}\right)$$

is an algebraic number. Next, we will employ Siegel's Lemma to construct a polynomial $P(x, y) \in \mathbb{Z}[x, y]$ with "modestly sized" coefficients such that the

Continued

associated function $F(z) = P(z, e^{\zeta z})$ satisfies the following two properties:

- $F(z)$ is not identically zero.
- For each pair (k_1, k_2), with $0 \leq k_1 < K$ and $0 \leq k_2 < K$, we have $F(k_1 + k_2 \xi) = 0$.

Then using the hypothesis that ξ is irrational, we will find integers k_1^* and k_2^* such that the algebraic number $F(k_1^* + k_2^* \xi)$ is nonzero. The number $F(k_1^* + k_2^* \xi)$, together with the properties of $F(z)$, will guide us to suitable choices for the parameters $\deg_x(P)$, $\deg_y(P)$, and K, which, in turn, will lead to an integer $\mathcal{N}$ that violates the Fundamental Principle of Number Theory.

In view of our intuitive outline, it may not be too surprising to discover the happy fact that much of the theory required to prove Theorem 5.2 was developed in Chapter 4. There are essentially two new complications—one involves an issue we have already faced, and we will quickly resolve it here. The second wrinkle is more substantial and will require an algebraic excursion in the next section to smooth it out.

A critical step in demonstrating the transcendence of e^{π} was establishing that the functions e^z and e^{iz} are so different that for any nonzero polynomial $P(x, y) \in \mathbb{Z}[x, y]$, the function $f(z) = P(e^z, e^{iz})$ is not identically zero. In other words, we showed that e^z and e^{iz} are algebraically independent. In order to prove Theorem 5.2, we must now show that the functions z and $e^{\zeta z}$ are algebraically independent. The following lemma assures us that there are no conspiracies that would cause $F(z) = P(z, e^{\zeta z})$ to accidentally collapse into the identically zero function for some nonzero polynomial $P(x, y) \in \mathbb{Z}[x, y]$. We note that the proof of this lemma proceeds in the same fashion as the argument behind Challenge 4.5.

LEMMA 5.6 *Let ζ be a nonzero complex number. Then the functions z and $e^{\zeta z}$ are algebraically independent. That is, for any nonzero polynomial $P(x, y) \in \mathbb{Z}[x, y]$, the function $F(z) = P(z, e^{\zeta z})$ is not identically zero.*

Proof. Suppose that $P(x, y) = \sum_{m=0}^{D_1-1} \sum_{n=0}^{D_2-1} a_{mn} x^m y^n \in \mathbb{Z}[x, y]$ is a nonzero polynomial with the property that $P(z, e^{\zeta z}) = 0$ for all $z \in \mathbb{C}$. If we write $P(x, y) = \sum_{m=0}^{D_1-1} p_m(y) x^m$, where $p_m(y) = \sum_{n=0}^{D_2-1} a_{mn} y^n \in \mathbb{Z}[y]$, then we see that there are only finitely many values of y_0 for which the function $g(x) = P(x, y_0)$ is identically zero. In particular, if we select any value for y_0 that is *not* a zero of any polynomial $p_m(y)$, for $m = 1, 2, \ldots, D_1 - 1$, then we see that $P(x, y_0)$ is a nonzero polynomial in x. Thus since $\zeta \neq 0$, we can easily find a complex number z_0 such that the function $G(z) = P(z, e^{\zeta z_0})$ is not identically zero.

Clearly, $G(z)$ is a polynomial in $\mathbb{C}[z]$. However, given our assumption that $P(z, e^{\zeta z}) = 0$ for all $z \in \mathbb{C}$, we see that for each integer k,

$$G(z_0 + (2\pi i k)/\zeta) = P\left(z_0 + (2\pi i k)/\zeta, e^{\zeta z_0}\right) = P\left(z_0 + (2\pi i k)/\zeta, e^{\zeta z_0 + 2\pi i k}\right)$$

$$= P\left(z_0 + (2\pi i k)/\zeta, e^{\zeta(z_0 + (2\pi i k)/\zeta)}\right) = 0.$$

Thus the polynomial $G(z)$ has infinitely many zeros and hence must be the identically zero polynomial—a conclusion that contradicts the last sentence of the previous paragraph. Therefore our assumption that $P(z, e^{\zeta z}) = 0$ for all $z \in \mathbb{C}$ is false, and thus we deduce that the functions z and $e^{\zeta z}$ are algebraically independent.   ∎

In view of the lemma, we can safely move forward comforted by the knowledge that for any nonzero polynomial $P(x, y) \in \mathbb{Z}[x, y]$ we may consider, the associated function $F(z) = P(z, e^{\zeta z})$ will *not* be identically zero.

5.3 Distilling three algebraic numbers down to one primitive element

In this section we face the only significant difference between the proof of the transcendence of e^{π} and the Gelfond–Schneider Theorem. Fortunately, we can get around this potential obstacle quite easily via a brief algebraic detour.

In the proof of the transcendence of e^{π}, we began with the assumption that e^{π} was algebraic of degree d. Thus we immediately were armed with an algebraic number $\alpha = e^{\pi}$, which together with an application of Siegel's Lemma (Theorem 4.3) led us to find modestly sized integers a_{mn}, not all zero, satisfying the homogeneous linear system

$$f^{(t)}(k\pi) = \sum_{m=0}^{D-1}\sum_{n=0}^{D-1}(-1)^{kn}(m+ni)^t\alpha^{km}a_{mn} = 0,$$

for $k = 1, 2, \ldots, K$, and $t = 0, 1, \ldots, T - 1$. However, in order to apply Siegel's Lemma, the coefficients of the linear system need to be integers. Thus we were faced with the task of converting a linear system having coefficients of the form $(-1)^{kn}(m + ni)^t\alpha^{km}$ into one with integer coefficients. We first expressed each α^{km} as $p_{km}(\alpha)$, for some polynomial $p_{km}(x) \in \mathbb{Q}[x]$ of degree at most $d - 1$. We then replaced the above system of KT equations with an associated system of dKT equations in which α no longer appeared in the coefficients. We were then left with only the algebraic number i in the new coefficients. We dispensed with the i's in a similar manner by considering real and imaginary parts of the coefficients. Thus, after clearing out any denominators, we were left with $2dKT$ linear equations with *integer* coefficients and could finally apply Siegel's Lemma.

In our present situation, we have *three* different numbers that are assumed to be algebraic: $\alpha_1 = \xi$, $\alpha_2 = e^{\xi}$, and $\alpha_3 = e^{\xi\zeta}$. Thus we will encounter a homogeneous linear system of K^2 equations of the form

$$F(k_1 + k_2\xi) = \sum_{m=0}^{D_1-1}\sum_{n=0}^{D_2-1}(k_1 + k_2\alpha_1)^m(\alpha_2)^{nk_1}(\alpha_3)^{nk_2}a_{mn} = 0,$$

for $0 \le k_1 < K$ and $0 \le k_2 < K$. We again need to convert this system of linear equations with algebraic coefficients into one that has integral coefficients in order to apply Siegel's Lemma. How can we accomplish this coefficient metamorphosis?

One method is to write the algebraic numbers α_1, α_2, and α_3 in terms of just *one* algebraic number, say α. We could then revert back to the process used in Chapter 4 to dispense with the α's and produce a linear system with integer coefficients.

An illustrative example. In an attempt to inspire what is to come, suppose we have two algebraic numbers, $\alpha_1 = \sqrt{2}$ and $\alpha_2 = \sqrt{3}$, and we wish to express them as $\alpha_1 = p_1(\alpha), \alpha_2 = p_2(\alpha)$ for some algebraic number α and polynomials $p_1(x), p_2(x) \in \mathbb{Q}[x]$. How can we find *one* algebraic number α that simultaneously captures the essence of both $\sqrt{2}$ and $\sqrt{3}$? A natural guess is simply to let $\alpha = \sqrt{2} + \sqrt{3}$. Let us see whether this intuitive hunch turns out to be correct.

We observe that $\alpha^2 = (\sqrt{2} + \sqrt{3})^2 = 5 + 2\sqrt{6}$, so $\frac{\alpha^2 - 5}{2} = \sqrt{2}\sqrt{3}$, and hence

$$\left(\frac{\alpha^2 - 5}{2}\right)\alpha = 2\sqrt{3} + 3\sqrt{2}.$$

Therefore we see that

$$\left(\frac{\alpha^2 - 5}{2}\right)\alpha - 2\alpha = \sqrt{2} \quad \text{and} \quad 3\alpha - \left(\frac{\alpha^2 - 5}{2}\right)\alpha = \sqrt{3};$$

thus we conclude that $\sqrt{2} = p_1(\alpha)$ and $\sqrt{3} = p_2(\alpha)$, where $\alpha = \sqrt{2} + \sqrt{3}$ and where

$$p_1(x) = \frac{1}{2}x^3 - \frac{9}{2}x, \quad \text{and} \quad p_2(x) = -\frac{1}{2}x^3 + \frac{11}{2}x.$$

From an algebraic point of view, what we have discovered is that the smallest field containing $\mathbb{Q}$ and the numbers $\sqrt{2}$ and $\sqrt{3}$, denoted by $\mathbb{Q}(\sqrt{2}, \sqrt{3})$, equals the smallest field containing $\mathbb{Q}$ and the lone number $\sqrt{2} + \sqrt{3}$. In other words, we see that $\sqrt{2} + \sqrt{3}$ can be used to generate the entire field $\mathbb{Q}(\sqrt{2}, \sqrt{3})$. Thus we have that $\mathbb{Q}(\sqrt{2}, \sqrt{3}) = \mathbb{Q}(\sqrt{2} + \sqrt{3})$. We call $\sqrt{2} + \sqrt{3}$ a *primitive element* of the field $\mathbb{Q}(\sqrt{2}, \sqrt{3})$. As we will now see in the algebraic excursion ahead, every finite extension field of $\mathbb{Q}$ has such a primitive element.

Algebraic Excursion: Number fields and primitive elements

Given a field E and an algebraic number α, we write $E(\alpha)$ for the smallest field containing all the elements of E together with the number α. By extending the ideas of Lemma 4.4, we have that if α is algebraic over E of degree d, then $E(\alpha)$ is a finite extension of E of degree d. Specifically, the set $\{1, \alpha, \alpha^2, \ldots, \alpha^{d-1}\}$ forms a basis for the field $E(\alpha)$ over E. That is, any $\beta \in E(\alpha)$ can be expressed as

$$\beta = e_1 + e_2\alpha + e_3\alpha^2 + \cdots + e_d\alpha^{d-1},$$

for some elements $e_1, e_2, \ldots, e_d \in E$.

Given a finite list of algebraic numbers $\alpha_1, \alpha_2, \ldots, \alpha_L$, we write $\mathbb{Q}(\alpha_1, \alpha_2, \ldots, \alpha_L)$ for the smallest field containing the rational numbers and the numbers $\alpha_1, \alpha_2, \ldots, \alpha_L$. We note that we can view $\mathbb{Q}(\alpha_1, \alpha_2, \ldots, \alpha_L)$ as $E(\alpha_L)$,

Continued

where $E = \mathbb{Q}(\alpha_1, \alpha_2, \ldots, \alpha_{L-1})$. An element $\theta \in \mathbb{Q}(\alpha_1, \alpha_2, \ldots, \alpha_L)$ is called a *primitive element* if $\mathbb{Q}(\alpha_1, \alpha_2, \ldots, \alpha_L) = \mathbb{Q}(\theta)$. We call $\mathbb{Q}(\alpha_1, \alpha_2, \ldots, \alpha_L)$ a *number field*.

PROPOSITION 5.7 *Let E be a number field and let α and β be two algebraic numbers. Then there exists an element $\theta \in E(\alpha, \beta)$ such that $E(\alpha, \beta) = E(\theta)$.*

Proof. Let $\alpha_1, \alpha_2, \ldots, \alpha_M$ be the conjugates of α, where $\alpha_1 = \alpha$, and let $\beta_1, \beta_2, \ldots, \beta_N$ be the conjugates of β with $\beta_1 = \beta$. We define $\mathcal{L}$ to be the finite set of algebraic numbers given by

$$\mathcal{L} = \left\{ \frac{\alpha_m - \alpha_1}{\beta_1 - \beta_n} : m = 1, 2, \ldots, M \quad \text{and} \quad n = 2, 3, \ldots, N \right\}.$$

We remark that $\mathcal{L}$ need not be contained in the field $E(\alpha, \beta)$, since $E(\alpha, \beta)$ may not contain all the conjugates of α or β. Since the field E is infinite and $\mathcal{L}$ is a finite collection, it is trivial to select an element $\gamma \in E$ for which $\gamma \notin \mathcal{L}$. We now define $\theta = \alpha + \gamma\beta$ and claim that $E(\alpha, \beta) = E(\theta)$. Certainly, we have that $\theta \in E(\alpha, \beta)$, which implies that $E(\theta) \subseteq E(\alpha, \beta)$. Thus in order to establish that $E(\alpha, \beta) \subseteq E(\theta)$, we need only show that $\alpha \in E(\theta)$ and $\beta \in E(\theta)$.

We write $f_\alpha(x) = (x - \alpha_1)(x - \alpha_2) \cdots (x - \alpha_M) \in \mathbb{Q}[x]$ and $f_\beta(x) = (x - \beta_1)(x - \beta_2) \cdots (x - \beta_N) \in \mathbb{Q}[x]$. If we let $g(x) = f_\alpha(\theta - \gamma x)$, then since $\gamma \in E$, we have that $g(x)$ is a polynomial in $E(\theta)[x]$ with

$$g(\beta) = f_\alpha(\theta - \gamma\beta) = f_\alpha(\alpha) = 0.$$

Thus we see that β is algebraic over the field $E(\theta)$ and its minimal polynomial in $E(\theta)[x]$, let us call it $F_\beta(x)$, must be a factor of $f_\alpha(\theta - \gamma x)$. Hence any x satisfying $F_\beta(x) = 0$ must also satisfy $\theta - \gamma x = \alpha_m$, for some m. We also observe that since $f_\beta(x) \in \mathbb{Q}[x] \subseteq E(\theta)[x]$, the minimal polynomial $F_\beta(x)$ must also be a factor of $f_\beta(x)$. Therefore we discover that the zeros of $F_\beta(x)$ form a subset of $\{\beta_1, \beta_2, \ldots, \beta_N\}$. Let $Z \subseteq \{\beta_1, \beta_2, \ldots, \beta_N\}$ denote the set of all zeros of $F_\beta(x)$. Clearly, Z is nonempty, since we have $\beta = \beta_1 \in Z$.

Suppose now that $\beta_n \in Z$ for some n. Then given our previous observations, we deduce that there exists some m such that $\theta - \gamma\beta_n = \alpha_m$, which, in view of the definition of γ, is equivalent to the expression $\alpha_1 + \gamma(\beta_1 - \beta_n) = \alpha_m$. Now if $\beta_n \neq \beta_1$, then we conclude that $\gamma = \frac{\alpha_m - \alpha_1}{\beta_1 - \beta_n} \in \mathcal{L}$, which, given our choice of γ, is impossible. Hence we see that β_n *must* equal β_1; that is, the minimal polynomial $F_\beta(x) \in E(\theta)[x]$ has only *one* zero and therefore is linear. Thus we conclude that $F_\beta(x) = x - \beta \in E(\theta)[x]$ and so $\beta \in E(\theta)$. Since $\alpha = \theta - \gamma\beta$, we immediately conclude that $\alpha \in E(\theta)$. Therefore we have that $E(\alpha, \beta) \subseteq E(\theta)$, which completes our proof. ∎

Continued

Challenge 5.3   *Prove the following result, known as the Theorem of the Primitive Element:*

THEOREM 5.8   *Let E be a number field and let $\alpha_1, \alpha_2, \ldots, \alpha_L$ be algebraic numbers. Then there exists an element $\theta \in E(\alpha_1, \alpha_2, \ldots, \alpha_L)$ such that $E(\alpha_1, \alpha_2, \ldots, \alpha_L) = E(\theta)$.*

To slightly simplify the seemingly endless stream of complicated inequalities up ahead, we extend the notion of an integer to the realm of algebraic numbers. We begin by noting that we can think of an integer as a rational r/s with $s = 1$. Equivalently, we can say that the integers are precisely those rational numbers whose minimial polynomials are monic; that is, $f(x) = sx - r$ with $s = 1$. An algebraic number is an *algebraic integer* if its minimal polynomial is monic. So, for example, $\sqrt{5}$ is an algebraic integer, since its minimal polynomial, $f(x) = x^2 - 5$, is monic. However, the minimal polynomial for $\sqrt{5}/2$ is $f(x) = 4x^2 - 5$, and thus $\sqrt{5}/2$ is not an algebraic integer.

Of course, multiplying a rational number by its denominator transforms it into an integer. As the previous example suggests, this conversion also carries over for algebraic numbers.

Challenge 5.4   *Let α be an algebraic number. Prove that there exists an integer D such that $D\alpha$ is an algebraic integer. (Hint: Suppose that $f(x) = a_d x^d + a_{d-1} x^{d-1} + \cdots + a_0 \in \mathbb{Z}[x]$ is the minimal polynomial for α. Now consider the polynomial $a_d^{d-1} f(x)$.)*

Thus we can conclude the following slightly stronger formulation of the Theorem of the Primitive Element, which will allow us to avoid dealing with denominators arising from θ in our move toward that elusive integer $\mathcal{N}$.

COROLLARY 5.9   *Let E be a number field and let $\alpha_1, \alpha_2, \ldots, \alpha_L$ be algebraic numbers. Then there exists an algebraic integer $\theta \in E(\alpha_1, \alpha_2, \ldots, \alpha_L)$ such that $E(\alpha_1, \alpha_2, \ldots, \alpha_L) = E(\theta)$.*

Given our assumption that $\alpha_1 = \xi$, $\alpha_2 = e^\zeta$, and $\alpha_3 = e^{\xi\zeta}$ are all algebraic, we can now find an algebraic integer θ of degree d such that $\mathbb{Q}(\alpha_1, \alpha_2, \alpha_3) = \mathbb{Q}(\theta)$. This equality implies that there exist rational numbers r_{lj} such that for $l = 1, 2$, and 3 we have

$$\alpha_l = r_{l1} + r_{l2}\theta + r_{l3}\theta^2 + \cdots + r_{ld}\theta^{d-1}.$$

It is this observation that allows us to transform the linear system of K^2 equations

$$\sum_{m=0}^{D_1-1} \sum_{n=0}^{D_2-1} (k_1 + k_2\alpha_1)^m (\alpha_2)^{nk_1} (\alpha_3)^{nk_2} a_{mn} = 0$$

involving three possibly unrelated algebraic numbers into a system of K^2 linear equations that involves only θ—specifically, a linear system of the form

$$F(k_1 + k_2\xi) = \sum_{m=0}^{D_1-1} \sum_{n=0}^{D_2-1} \left(r_1 + r_2\theta + r_3\theta^2 + \cdots + r_d\theta^{d-1} \right) a_{mn} = 0,$$

where the rational numbers r_j actually depend on our choice of θ, and on m, n, k_1, and k_2; that is, $r_j = r_j(\theta; m, n, k_1, k_2) = r_j(m, n, k_1, k_2)$. If we can bound the size of those r_j's, then after clearing any denominators and applying Siegel's Lemma, we will have our polynomial $P(x, y)$ in hand and will be ready to hunt down an integer that dares to defy the Fundamental Principle of Number Theory.

5.4 Beating the (linear) system—Constructing a polynomial via linear equations

Armed with the primitive element θ, we now wish to construct a nonzero auxiliary polynomial $P(x, y) \in \mathbb{Z}[x, y]$ of the form

$$P(x, y) = \sum_{m=0}^{D_1-1} \sum_{n=0}^{D_2-1} a_{mn} x^m y^n,$$

with the size of the coefficients, $|a_{mn}|$, not too large, so that the associated function

$$F(z) = P(z, e^{\zeta z})$$

vanishes for all $z \in \{k_1 + k_2\xi : k_1, k_2 \in \mathbb{Z}$ with $0 \le k_1 < K$ and $0 \le k_2 < K\}$. Our method mirrors the one employed in Chapter 4: We translate the vanishing of $F(z)$ at all of the desired points $k_1 + k_2\xi$ into a rather explicit homogeneous system of linear equations with integer coefficients. Once we have accomplished this task, we can then find the unknown coefficients of $P(x, y)$, a_{mn}, and obtain an estimate for their absolute values through an application of Siegel's Lemma.

Our first daunting challenge is to devise some clear and concise notation for the coefficients of the system of linear equations corresponding to the conditions

$$F(k_1 + k_2\xi) = 0 \quad \text{for} \quad 0 \le k_1 < K, \quad 0 \le k_2 < K.$$

Given that

$$F(k_1 + k_2\xi) = \sum_{m=0}^{D_1-1} \sum_{n=0}^{D_2-1} a_{mn}(k_1 + k_2\xi)^m (e^{\zeta})^{nk_1} (e^{\xi\zeta})^{nk_2}$$

$$= \sum_{m=0}^{D_1-1} \sum_{n=0}^{D_2-1} (k_1 + k_2\alpha_1)^m (\alpha_2)^{nk_1} (\alpha_3)^{nk_2} a_{mn} = 0,$$

we see that the coefficients of this linear system are of the form $(k_1 + k_2\alpha_1)^m (\alpha_2)^{nk_1}$ $(\alpha_3)^{nk_2}$. Applying the lessons from our algebraic excursion, we find that there exist polynomials $p_{\alpha_1}(z), p_{\alpha_2}(z), p_{\alpha_3}(z) \in \mathbb{Q}[z]$, each of degree less than d, such that $\alpha_1 = p_{\alpha_1}(\theta)$, $\alpha_2 = p_{\alpha_2}(\theta)$, and $\alpha_3 = p_{\alpha_3}(\theta)$. We can now rewrite our linear system as

$$F(k_1 + k_2\xi) = \sum_{m=0}^{D_1-1} \sum_{n=0}^{D_2-1} (k_1 + k_2 p_{\alpha_1}(\theta))^m p_{\alpha_2}(\theta)^{nk_1} p_{\alpha_3}(\theta)^{nk_2} a_{mn} = 0, \qquad (5.1)$$

for $0 \leq k_1 < K$ and $0 \leq k_2 < K$. Thus for each such pair (k_1, k_2), we see that the number $F(k_1 + k_2\xi)$ is an element of the field $\mathbb{Q}(\theta)$, and therefore it can be expressed as a $\mathbb{Q}$-linear combination of $1, \theta, \theta^2, \ldots, \theta^{d-1}$:

$$F(k_1 + k_2\xi) = A_1 + A_2\theta + A_3\theta^2 + \cdots + A_d\theta^{d-1} = 0, \qquad (5.2)$$

where the coefficients $A_1, A_2, \ldots, A_d$ depend on the pair (k_1, k_2) and can be expressed as

$$A_j = A_j(k_1, k_2) = \sum_{m=0}^{D_1-1} \sum_{n=0}^{D_2-1} t_j(m, n, k_1, k_2) a_{mn},$$

where each new coefficient $t_j(m, n, k_1, k_2)$ is the sum of $D_1 D_2$ terms of the form $r_j(m, n, k_1, k_2)$. Since the numbers $1, \theta, \theta^2, \ldots, \theta^{d-1}$ are $\mathbb{Q}$-linearly independent, it follows that $F(k_1 + k_2\xi)$ equals 0 if and only if each of the associated quantities $A_1, A_2, \ldots, A_d$ equals 0. Therefore we can replace the single linear equation $F(k_1 + k_2\xi) = 0$ involving algebraic coefficients with d linear equations involving only rational coefficients, namely,

$$A_1(k_1, k_2) = 0, \quad A_2(k_1, k_2) = 0, \ldots, A_d(k_1, k_2) = 0.$$

After multiplying each equation by a suitable integer to clear denominators, we can assume that the system contains only *integer* coefficients. Of course, we perform this equation transformation for each pair (k_1, k_2). If we wish to follow this approach, then we require an upper bound for the rational coefficients $t_j(m, n, k_1, k_2)$ (and their denominators) of the linear forms $A_j = A_j(k_1, k_2)$, for all $j = 1, 2, \ldots, d, 0 \leq k_1 < K$, and $0 \leq k_2 < K$.

In order to find such a bound, we first recall that for an algebraic number θ, we define the height of θ, $\mathcal{H}(\theta)$, to be the height of its minimal polynomial $p(z) \in \mathbb{Z}[z]$; that is, $\mathcal{H}(\theta) = \mathcal{H}(p)$. We will employ the following generalization of Lemma 4.4 to produce our upper bound.

LEMMA 5.10 *Suppose* $\beta_1, \beta_2, \ldots, \beta_L$ *are elements of* $\mathbb{Q}(\theta)$, *where* θ *is an algebraic integer of degree* d *and of height* $\mathcal{H}(\theta)$. *If for each* $l = 1, 2, \ldots, L$,

$$\beta_l = r_{l1} + r_{l2}\theta + \cdots + r_{ld}\theta^{d-1},$$

where each r_{lj} is a rational number satisfying $|r_{lj}| \leq B_l$ for some bound B_l, then

$$\beta_1 \beta_2 \cdots \beta_L = r_1 + r_2 \theta + \cdots + r_d \theta^{d-1},$$

with rational coefficients r_j satisfying

$$\max_{1 \leq j \leq d} \{|r_j|\} \leq d^L B_1 B_2 \cdots B_L (2\mathcal{H}(\theta))^{dL}. \tag{5.3}$$

Moreover, if $\mathrm{den}(\beta_l)$ denotes the least common multiple of the denominators of the rational coefficients $r_{l1}, r_{l2}, \ldots, r_{ld}$, then each rational number r_j can be expressed with a denominator of the form

$$\mathrm{den}(\beta_1)\mathrm{den}(\beta_2) \cdots \mathrm{den}(\beta_L).$$

Challenge 5.5 *The proof of Lemma 5.10 follows from an argument similar to that used in the proof of Lemma 4.6. Employing the proof of Lemma 4.4 as a guide, convince yourself of the validity of Lemma 5.10.*

We now return to the quantities at hand, namely, the coefficients in (5.1)

$$(k_1 + k_2 p_{\alpha_1}(\theta))^m (p_{\alpha_2}(\theta))^{nk_1} (p_{\alpha_3}(\theta))^{nk_2},$$

where $0 \leq m \leq D_1 - 1$, $0 \leq n \leq D_2 - 1$, and $0 \leq k_1 < K$, $0 \leq k_2 < K$. Given m, n, k_1, and k_2 as above, we apply the lemma with

$$\beta_1 = \beta_2 = \cdots = \beta_m = k_1 + k_2 p_{\alpha_1}(\theta),$$

$$\beta_{m+1} = \beta_{m+2} = \cdots = \beta_{m+n} = p_{\alpha_2}(\theta)^{k_1}, \quad \text{and}$$

$$\beta_{m+n+1} = \beta_{m+n+2} = \cdots = \beta_{m+2n} = p_{\alpha_3}(\theta)^{k_2}.$$

We note that for $l = 1, 2, \ldots, m$, we can select $B_l = 2K\mathcal{H}(p_{\alpha_1})$. If we apply Lemma 5.10 with $l = m+1, m+2, \ldots, m+n$, then by (5.3) we conclude that we can take $B_l = d^K \mathcal{H}(p_{\alpha_2})^K (2\mathcal{H}(\theta))^{dK}$, and for $l = m+n+1, m+n+2, \ldots, m+2n$, we can let $B_l = d^K \mathcal{H}(p_{\alpha_3})^K (2\mathcal{H}(\theta))^{dK}$. Therefore applying all these bounds we see by a final application of Lemma 5.10 that

$$(k_1 + k_2 p_{\alpha_1}(\theta))^m p_{\alpha_2}(\theta)^{nk_1} p_{\alpha_3}(\theta)^{nk_2} = r_1 + r_2 \theta + \cdots + r_d \theta^{d-1},$$

with rational coefficients $r_j = r_j(m, n, k_1, k_2)$ satisfying

$$\max_{1 \leq j \leq d} \{|r_j|\} \leq d^{m+2n} \left(2K\mathcal{H}(p_{\alpha_1})\right)^m \left(d^K \mathcal{H}(p_{\alpha_2})^K (2\mathcal{H}(\theta))^{dK}\right)^n$$

$$\times \left(d^K \mathcal{H}(p_{\alpha_3})^K (2\mathcal{H}(\theta))^{dK}\right)^n (2\mathcal{H}(\theta))^{d(m+2n)}$$

$$\leq d^{D_1 + 2D_2 + 2KD_2} (2K)^{D_1} \left(\mathcal{H}(p_{\alpha_1})\right)^{D_1} \left(\mathcal{H}(p_{\alpha_2})\right)^{KD_2}$$

$$\times \left(\mathcal{H}(p_{\alpha_3})\right)^{KD_2} (2\mathcal{H}(\theta))^{d(D_1 + 2D_2 + 2KD_2)}. \tag{5.4}$$

So the linear system of (5.2) can now be expressed as

$$F(k_1 + k_2\xi) = \sum_{m=0}^{D_1-1} \sum_{n=0}^{D_2-1} (r_1 + r_2\theta + r_3\theta^2 + \cdots + r_d\theta^{d-1})a_{mn} = 0.$$

However, given that our upper bound (5.4) on $|r_j(m, n, k_1, k_2)|$ requires two entire lines, it appears that now is the perfect time to take a moment to introduce an idea that will allow us to shrink that unpleasantly long bound down to a more manageable one that more clearly reflects the critical dependencies.

A brief interlude: The ever-evolving constants

As we have seen throughout this book, our unwieldy upper bounds involve explicitly computable, fixed numbers together with the various parameters whose values are at our disposal. The critical component of each estimate is the contribution of the parameters that we ultimately allow to get large—in the case of the present proof, the integers D_1, D_2, and K. The eventual irrelevance of all other quantities, for example, in this proof, α_1, α_2, α_3, d, θ, and even the functions $p_{\alpha_i}(z)$, in obtaining our final contradiction implies that explicitly displaying them only obscures the relevant activity.

Practitioners of transcendence theory have long recognized this maxim and thus developed a theory of *evolving constants*—complicated, or even simple, expressions involving quantities independent of the free parameter are absorbed into ever changing constants. With each new estimate larger and larger constants are employed. Thus in our proof we will now introduce a sequence of unspecified positive constants $c_1, c_2, \ldots$, each of which, if necessary, *could* be explicitly displayed—although it may require several lines of symbols. The moral is that we will not concern ourselves with those constants that do not depend on any of the free parameters. The only relevant features are the parameters themselves, and our upper bounds will now reflect that relevance.

In light of our interlude, we now see that we can express the multi-lined inequality (5.4) as

$$\max_{1\le j\le d} \{|r_j(m, n, k_1, k_2)|\} \le c_1^{D_1} c_2^{D_2} K^{D_1} c_3^{D_2 K},$$

for suitable positive constants c_1, c_2, and c_3. By writing K as $e^{\log K}$, the previous inequality implies the simpler expression

$$\max_{1\le j\le d} \{|r_j(m, n, k_1, k_2)|\} \le c_4^{D_1 \log K} c_5^{D_2 K},$$

where, to illustrate the theme of evolving constants, we explicitly define the constants: $c_4 = \max\{c_1, e\}^2$ and $c_5 = \max\{c_2, c_3\}^2$. If we let $c_6 = \max\{c_4, c_5\}$, then we can consolidate the upper bound yet further to yield

$$\max_{1\leq j\leq d}\{|r_j(m,n,k_1,k_2)|\} \leq c_6^{D_1 \log K + D_2 K}. \qquad (5.5)$$

Certainly, multiplying through by the integer

$$\delta = \mathrm{den}(p_{\alpha_1}(\theta))^{D_1}\mathrm{den}(p_{\alpha_2}(\theta))^{D_2 K}\mathrm{den}(p_{\alpha_3}(\theta))^{D_2 K}$$

will clear any denominators occurring in the rational numbers r_j. We also note that $\delta \leq c_7^{D_1+D_2 K}$. Thus pulling all our observations together, we see that for each pair of integers (k_1, k_2), we have

$$\delta F(k_1 + k_2 \xi) = A_1 + A_2 \theta + \cdots + A_d \theta^{d-1},$$

where each linear form $A_j = A_j(k_1, k_2)$ can be expressed as

$$A_j = \sum_{m=0}^{D_1-1}\sum_{n=0}^{D_2-1} t_j(m,n,k_1,k_2)a_{mn},$$

and where each coefficient $t_j(m, n, k_1, k_2)$ is an *integer* arising from a sum of $D_1 D_2$ terms of the form $\delta r_j(m, n, k_1, k_2)$. Thus given our previous upper bounds, it follows that each coefficient satisfies

$$|t_j(m,n,k_1,k_2)| \leq D_1 D_2 c_8^{D_1 \log K + D_2 K}.$$

For each pair (k_1, k_2), if we set each of the associated linear forms $A_1, A_2, \ldots, A_d$ equal to 0, then we obtain a homogeneous system of dK^2 linear equations in $D_1 D_2$ unknowns. By Siegel's Lemma (Theorem 4.3), if we take

$$D_1 D_2 > dK^2,$$

then there exist integers a_{mn}, not all zero, that form a solution to the linear system

$$A_j(k_1,k_2) = \sum_{m=0}^{D_1-1}\sum_{n=0}^{D_2-1} t_j(m,n,k_1,k_2)a_{mn} = 0,$$

for $j = 1, 2, \ldots, d$, $k_1 = 0, 1, \ldots, K-1$, and $k_2 = 0, 1, \ldots, K-1$, such that for each m and n,

$$|a_{mn}| < \left((D_1 D_2)^2 c_8^{D_1 \log K + D_2 K}\right)^{\frac{dK^2}{D_1 D_2 - dK^2}}.$$

Just as in the previous chapter, in order to simplify this upper bound, we fix a relationship between the parameters D_1, D_2, and K. It may appear natural to balance the two terms $D_1 \log K + D_2 K$ in the exponent and thus to set $D_1 \log K$ equal to $D_2 K$. Unfortunately, the inclusion of a logarithmic term complicates matters slightly.

Although in some proofs the contribution of that $\log K$ factor is critical, in this some-what less delicate proof we can ignore that relatively slow-growing $\log K$ factor. Thus we now choose D_1 and D_2 such that

$$\frac{dK^2}{D_1 D_2 - dK^2} = 1 \quad \text{and} \quad D_1 = D_2 K.$$

With these declarations in hand we see that

$$D_1 = \sqrt{2d}K^{3/2} \quad \text{and} \quad D_2 = \sqrt{2d}K^{1/2},$$

with the additional understanding that we will henceforth take K such that these quantities are integers. We note that indeed $D_1 D_2 = 2dK^2 > dK^2$ as required in Siegel's Lemma.

Challenge 5.6 *Show that for K sufficiently large, given our choice of parameters, there exist integers a_{mn}, not all zero, satisfying*

$$|a_{mn}| < c_9^{K^{3/2} \log K}, \tag{5.6}$$

so that if $P(x, y) = \sum_{m=0}^{D_1-1} \sum_{n=0}^{D_2-1} a_{mn} x^m y^n$, and $F(z) = P(z, e^{\xi z})$, then $F(z)$ is a nonzero function with the property that for each $k_1 = 0, 1, \ldots, K - 1$ and $k_2 = 0, 1, \ldots, K - 1$,

$$F(k_1 + k_2 \xi) = 0.$$

Before closing our construction, we make the important observation that since ξ is irrational, we see that $k_1 + k_2 \xi = k_1' + k_2' \xi$ if and only if $k_1 = k_1'$ and $k_2 = k_2'$. Therefore we note that $F(z)$ has at least K^2 *distinct* zeros, namely, $z = k_1 + k_2 \xi$, for $0 \leq k_1 < K$ and $0 \leq k_2 < K$.

Thus, at long last, our polynomial construction is complete, and we are armed with an auxiliary function $F(z)$ satisfying our two key conditions. We are now ready to hunt down the integer $\mathcal{N}$.

5.5 The proof of a real special case

In this section we establish Theorem 5.2 with the additional assumption that ξ is a *real* number. This special case would certainly imply the transcendence of $2^{\sqrt{2}}$. We also note that our assumption that ξ and $e^{\xi \zeta}$ are algebraic holds if and only if $-\xi$ and $e^{-\xi \zeta}$ are also algebraic. Thus, without loss of generality, we can assume that the real number ξ is *negative*.

We begin with a hunt for a suitable nonzero value of $F(z)$ that will lead to our contradictory integer. We remark that this step is the second and final point in our proof where the irrationality of ξ plays a role. (Recall that we already used the irrationality of ξ to ensure that $F(z)$ has at least K^2 zeros.) Let $\Xi = \mathbb{Z} + \mathbb{Z}\xi =$

$\{m + n\xi : m \in \mathbb{Z}, n \in \mathbb{Z}\}$. For an integer n, we write ξ_n for the fractional part of $n\xi$; that is, $\xi_n = n\xi - [n\xi]$. Given that ξ is irrational we see that $0 < \xi_n < 1$ for all $n \geq 1$. The necessary groundwork to establish our nonzero value is left as the following extended challenge.

Challenge 5.7 *Show that $\xi_m = \xi_n$ if and only if $m = n$ and conclude that Ξ contains infinitely many distinct elements from the interval $[0, 1]$. Next, using the compactness of $[0, 1]$, argue that the set $\Xi \cap [0, 1]$ has an accumulation point. Then show that there exists a real number ξ^* and a sequence of distinct points $x_i \in \Xi$ of the form $x_i = m_i + n_i\xi$, where m_i and n_i are both positive integers, such that $\lim_{i \to \infty} x_i = \xi^*$. (Remark: This is the only assertion that requires our potentially peculiar assumption that $\xi < 0$.) Lastly, verify that if $F(x_i) = 0$ for all i, then $F(\xi^*) = 0$.*

We now apply the previous observations to prove the following result.

PROPOSITION 5.11 *Suppose that for all positive integers m and n, $F(m + n\xi) = 0$. Then $F(z)$ is the identically zero function.*

Proof. Let $\xi^* = \lim_{i \to \infty} x_i$ be as in Challenge 5.7. If we assume that $F(z)$ is not identically zero, then its power series expanded about the point $z_0 = \xi^*$ is not the identically zero function. Given that we have $F(\xi^*) = 0$, we see that there exists an integer $M \geq 1$ such that

$$F(z) = \sum_{m=M}^{\infty} b_m (z - \xi^*)^m,$$

where $b_M \neq 0$. Thus the function

$$G(z) = \frac{F(z)}{(z - \xi^*)^M} = \sum_{m=M}^{\infty} b_m (z - \xi^*)^{m-M}$$

is entire and $G(\xi^*) = b_M \neq 0$. Since $G(z)$ is a continuous function and is nonzero at $z = \xi^*$, there exists an $\varepsilon > 0$ such that for any $z \in \mathbb{C}$ satisfying $0 < |z - \xi^*| < \varepsilon$, we have that

$$G(z) \neq 0. \tag{5.7}$$

However, from Challenge 5.7 we can find a real number x_I satisfying $0 < |x_I - \xi^*| < \varepsilon$ such that $F(x_I) = 0$. Thus on the one hand, by (5.7) it follows that $G(x_I) \neq 0$, while on the other hand, by the fact that $F(x_I) = 0$, we deduce that $G(x_I) = 0$. This contradiction leads us to the conclusion that $F(z)$ must be identically zero. ∎

From Proposition 5.11 we conclude that there exists some index M, $M > K$, such that for all pairs of integers (k_1, k_2) satisfying $0 \leq k_1 < M$, $0 \leq k_2 < M$, we have that $F(k_1 + k_2\xi) = 0$, while there exists an integer pair (k_1^*, k_2^*), satisfying $0 \leq k_1^* \leq M$, $0 \leq k_2^* \leq M$, with either k_1^* or k_2^* equal to M, for which $F(k_1^* + k_2^*\xi) \neq 0$.

Constructing the infamous integer $\mathcal{N}$. We now transform the nonzero algebraic number $F(k_1^* + k_2^*\xi)$ into a nonzero integer. By Lemma 5.10 and its aftermath, we conclude that there exists a positive integer δ^*, and integers $A_1, \ldots, A_d$, such that

$$\delta^* F(k_1^* + k_2^*\xi) = A_1 + A_2\theta + \cdots + A_d\theta^{d-1}.$$

Moreover, exploiting our previous work on finding upper bounds, we note that $\delta^* \leq c_{10}^{D_1+D_2M}$ and thus in view of inequalities (5.6), (5.5), and our choices for D_1 and D_2, it follows that

$$\max_{1\leq j\leq d}\{|A_j|\} \leq D_1 D_2 \delta^* \max\{|a_{mn}|\} \max\{|t_j(m, n, k_1^*, k_2^*)|\}$$

$$\leq D_1^2 D_2^2 c_{10}^{D_1+D_2M} c_9^{K^{3/2}\log K} c_8^{D_1\log M+D_2M} \leq c_{11}^{M^{3/2}\log M}. \tag{5.8}$$

We now write $\theta_1, \theta_2, \ldots, \theta_d$ for the conjugates of θ, where $\theta = \theta_1$, and define the quantity

$$\mathcal{N} = \prod_{i=1}^{d}\left(A_1 + A_2\theta_i + A_3\theta_i^2 + \cdots + A_d\theta_i^{d-1}\right).$$

As an aside we remark that $\mathcal{N}$ is the "algebraic norm" of $\delta^* F(k_1^* + k_2^*\xi)$. The following lemma shows us that we are heading down the right road for a possible contradiction to the Fundamental Principle of Number Theory.

LEMMA 5.12 *The quantity $\mathcal{N}$ defined above is a nonzero integer.*

Proof. We begin by recalling that θ is an algebraic *integer*. That is, its minimal polynomial is monic. It thus follows that all the elementary symmetric functions in the conjugates of θ are also integers. Hence since $\mathcal{N}$ is a symmetric function in the conjugates of θ with integer coefficients, by Theorem 3.12 we conclude that $\mathcal{N}$ is an integer.

If $\mathcal{N}$ were to equal 0, then there would exist an index, say j, such that

$$A_1 + A_2\theta_j + A_3\theta_j^2 + \cdots + A_d\theta_j^{d-1} = 0.$$

Thus we see that the algebraic number θ_j, which is of degree d, is the zero of a polynomial with integer coefficients having degree at most $d - 1$. Therefore that polynomial must be the identically zero polynomial; that is, $A_1 = A_2 = \cdots = A_d = 0$. However, this conclusion implies that

$$\delta^* F(k_1^* + k_2^*\xi) = A_1 + A_2\theta + \cdots + A_d\theta^{d-1} = 0,$$

which is false by our choice of k_1^* and k_2^*. Hence we conclude that $\mathcal{N} \neq 0$, which completes our proof. ∎

We now wish to show that for all sufficiently large choices of K, we have $|\mathcal{N}| < 1$, which contradicts the Fundamental Principle of Number Theory. Toward this end, in order to produce an upper bound for $|\mathcal{N}|$, we bound each of the d factors of $\mathcal{N}$. These factors come in two flavors: *Vanilla*, those factors involving θ_i for $i = 2, 3, \ldots, d$, and *rocky road*, the rich factor containing $\theta_1 = \theta$. The vanilla factors are quickly bounded by an application of the triangle inequality. For the rocky road factor containing θ, we return to the Maximum Modulus Principle and apply all our previous estimates to produce an *incredible* upper bound.

Bounding the $d - 1$ vanilla factors. For a fixed conjugate θ_i of θ, the triangle inequality together with (5.8) reveals that

$$\left| A_1 + A_2\theta_i + \cdots + A_d\theta_i^{d-1} \right| \leq d \max_{1 \leq j \leq d} \{|A_j|\} \max\{1, |\theta_i|\}^{d-1}$$

$$\leq d c_{11}^{M^{3/2} \log M} \max\{1, |\theta_1|, |\theta_2|, \ldots, |\theta_d|\}^{d-1}$$

$$\leq c_{12}^{M^{3/2} \log M}.$$

Therefore we quickly conclude that

$$\prod_{j=2}^{d} \left| A_1 + A_2\theta_i + A_3\theta_i^2 + \cdots + A_d\theta_i^{d-1} \right| \leq \left(c_{12}^{M^{3/2} \log M} \right)^{d-1}$$

$$= \left(c_{12}^{d-1} \right)^{M^{3/2} \log M} = c_{13}^{M^{3/2} \log M}. \quad (5.9)$$

We now turn our attention to the more decadent factor.

Going down the final rocky road. Here we apply the ideas we developed in the previous chapter involving the Maximum Modulus Principle. In particular, inspired by our previous work, we define

$$G(z) = \frac{\delta^* F(z)}{\prod_{k_1=0}^{M-1} \prod_{k_2=0}^{M-1} (z - (k_1 + k_2\xi))}.$$

In view of our earlier observation that $k_1 + k_2\xi = k_1' + k_2'\xi$ if and only if $k_1 = k_1'$ and $k_2 = k_2'$, we see that the factors of the denominator of $G(z)$ are the distinct zeros of $F(z)$. It now follows that every zero of the denominator is a zero of $F(z)$. Thus we conclude that $G(z)$ is, in fact, an entire function. Moreover, we note that

$$|\delta^* F(k_1^* + k_2^*\xi)| = |G(k_1^* + k_2^*\xi)| \prod_{k_1=0}^{M-1} \prod_{k_2=0}^{M-1} |(k_1^* - k_1) + (k_2^* - k_2)\xi|.$$

Given that $k_1 < M$ and $k_2 < M$, we have that for any $R > M(1 + |\xi|)$, $|k_1^* + k_2^*\xi| < R$, and hence the Maximum Modulus Principle implies that

$$|\delta^* F(k_1^* + k_2^*\xi)| \leq |G|_R \prod_{k_1=0}^{M-1} \prod_{k_2=0}^{M-1} |(k_1^* - k_1) + (k_2^* - k_2)\xi|$$

$$\leq \frac{|\delta^* F|_R}{\left| \prod_{k_1=0}^{M-1} \prod_{k_2=0}^{M-1} (z - (k_1 + k_2\xi)) \right|_R}$$

$$\times \prod_{k_1=0}^{M-1} \prod_{k_2=0}^{M-1} |(k_1^* - k_1) + (k_2^* - k_2)\xi|. \qquad (5.10)$$

Thus in order to give an upper bound for $|\delta^* F(k_1^* + k_2^*\xi)|$, we bound each of the three terms from the right-hand side of (5.10). We bound the first factor, $|\delta^* F|_R$, through an application of the triangle inequality and (5.6):

$$|\delta^* F|_R = \left| \delta^* \sum_{m=0}^{D_1-1} \sum_{n=0}^{D_2-1} a_{mn} z^m \left(e^{\zeta z}\right)^n \right|_R \leq \delta^* D_1 D_2 \max\{|a_{mn}|\} |z|_R^{D_1} |e^{\zeta z}|_R^{D_2}$$

$$\leq \delta^* D_1 D_2 c_9^{K^{3/2} \log K} R^{D_1} \left(e^{|\mathrm{Re}(\zeta)|R}\right)^{D_2}.$$

In view of our choices of D_1 and D_2, our bound on the size of δ^*, and the fact that $R > M > K$, we see that the previous inequality implies

$$|\delta^* F|_R < c_{14}^{K^{3/2} \log R + K^{1/2} R}. \qquad (5.11)$$

It is easy to verify that the second factor in (5.10) satisfies

$$\prod_{k_1=0}^{M-1} \prod_{k_2=0}^{M-1} |(k_1^* - k_1) + (k_2^* - k_2)\xi| \leq (M(1 + |\xi|))^{M^2}.$$

Finally, given that $R > M(1 + |\xi|)$ and $|k_1 + k_2\xi| \leq M(1 + |\xi|)$, it is easy to produce a *lower* bound for the denominator from (5.10): If we take $z = R$, then it follows that

$$\left| \prod_{k_1=0}^{M-1} \prod_{k_2=0}^{M-1} (z - (k_1 + k_2\xi)) \right|_R \geq \left| \prod_{k_1=0}^{M-1} \prod_{k_2=0}^{M-1} (R - (k_1 + k_2\xi)) \right| \geq (R - M(1 + |\xi|))^{M^2}.$$

Therefore if we combine the two bounds we just found for the products appearing in the numerator and denominator of (5.10), we conclude that

$$\frac{\prod_{k_1=0}^{M-1} \prod_{k_2=0}^{M-1} |(k_1^* - k_1) + (k_2^* - k_2)\xi|}{\left| \prod_{k_1=0}^{M-1} \prod_{k_2=0}^{M-1} (z - (k_1 + k_2\xi)) \right|_R} \leq \left(\frac{M(1 + |\xi|)}{R - M(1 + |\xi|)} \right)^{M^2}.$$

In order to simplify this upper bound, let us now declare that $R = M^{1+\varepsilon} + M(1 + |\xi|)$ for some fixed positive number ε to be selected shortly. Thus the previous inequality implies that

$$\frac{\prod_{k_1=0}^{M-1} \prod_{k_2=0}^{M-1} |(k_1^* - k_1) + (k_2^* - k_2)\xi|}{\left| \prod_{k_1=0}^{M-1} \prod_{k_2=0}^{M-1} (z - (k_1 + k_2\xi)) \right|_R} \leq \left((1 + |\xi|)M^{-\varepsilon}\right)^{M^2}$$

$$< \left((1 + |\xi|)e^{-\varepsilon \log M}\right)^{M^2} \leq c_{15}^{-\varepsilon M^2 \log M}. \tag{5.12}$$

Thus putting inequalities (5.10), (5.11), and (5.12) together, we have that

$$|\delta^* F(k_1^* + k_2^* \xi)| < c_{14}^{K^{3/2} \log R + K^{1/2} R} \, c_{15}^{-\varepsilon M^2 \log M}. \tag{5.13}$$

No matter what value we choose for ε, we can select K (and thus indirectly M) large enough so as to ensure that $M^{1+\varepsilon} > M(1 + |\xi|)$, which in turn implies that $R < 2M^{1+\varepsilon}$. From this bound on R and the fact that $K < M$, inequality (5.13) implies that

$$|\delta^* F(k_1^* + k_2^* \xi)| < c_{14}^{K^{3/2} \log(2M^{1+\varepsilon}) + K^{1/2}(2M^{1+\varepsilon})} \, c_{15}^{-\varepsilon M^2 \log M}$$

$$\leq c_{16}^{(1+\varepsilon)M^{3/2} \log M + M^{3/2+\varepsilon} - \varepsilon M^2 \log M}.$$

If we now fix $\varepsilon = \frac{1}{2}$, then combining (5.9) with the previous inequality gives

$$|\mathcal{N}| < c_{13}^{M^{3/2} \log M} \, c_{16}^{\frac{3}{2} M^{3/2} \log M + M^2 - \frac{1}{2} M^2 \log M} \leq c_{17}^{M^{3/2} \log M + M^2 - \frac{1}{2} M^2 \log M}. \tag{5.14}$$

As we let K (and therefore M) become large, we see that the negative term $-\frac{1}{2}M^2 \log M$ dominates the slower-growing positive term $M^{3/2} \log M + M^2$. Thus if we select K sufficiently large, we conclude that the integer $\mathcal{N}$ satisfies

$$0 < |\mathcal{N}| < 1,$$

which, at long last, violates the Fundamental Principle of Number Theory. Therefore our assumption that all three numbers ξ, e^ξ, and $e^{\xi \zeta}$ are algebraic is false, and hence at least one must be transcendental, which completes our proof. ∎

Before rejoicing too much, we recall that we have proved Theorem 5.2 only in the special case of ξ a *real* number. As we will see in the next section, the general case in which ξ is a *complex* irrational number involves only a minor alteration to show the existence of positive integers k_1^* and k_2^* satisfying $F(k_1^* + k_2^* \xi) \neq 0$. Given the fact that the lion's share of the proof is indeed behind us, feel free to roar a bit before facing the slightly complex complication.

5.6 Moving out to the vast complex plane

In our previous argument we made the additional assumption that ξ is a *real* number in order to conclude that there exist positive integers k_1^* and k_2^* such that $F(k_1^* + k_2^*\xi) \neq 0$. But as we will discover in this short section, Theorem 5.2 holds for all *complex* irrational numbers ξ as originally asserted.

For ξ a real number, we used the Diophantine approximation result that the set $\mathbb{Z} + \mathbb{Z}\xi$ contains a convergent sequence of points in order to show the existence of the desired point $k_1^* + k_2^*\xi$. However, for imaginary ξ, for example, $\xi = i$, the set $\mathbb{Z} + \mathbb{Z}\xi$ is a discrete lattice in the complex plane, and thus searching for limit points is futile. In order to find positive integers k_1^* and k_2^* satisfying $F(k_1^* + k_2^*\xi) \neq 0$, for an arbitrary complex irrational number ξ, we must find an alternative proof of Proposition 5.11 that does not require the set $\mathbb{Z} + \mathbb{Z}\xi$ to have an accumulation point.

Our argument in the proof of Proposition 5.11 relied on the analytic fact that the zeros of an entire function are isolated. Our proof in this more general situation also relies on an analytic result, a more powerful one, which we have already put to good use: The Maximum Modulus Principle.

Rounding up the zeros of an entire function.   Proposition 5.11, in essence, asserts that if an entire function has "too many" zeros, then the function must be identically zero. We now produce a quantitative version of this qualitative observation. In order to inspire our result, we begin with an elementary proposition.

PROPOSITION 5.13   *Let $f(z)$ be an entire function. If there exists an unbounded increasing sequence of real numbers $R_1, R_2, \ldots$ such that*

$$\lim_{n \to \infty} |f|_{R_n} = 0, \tag{5.15}$$

then $f(z)$ is the identically zero function.

Proof.   For any $z_0 \in \mathbb{C}$, there exists an integer n_0 such that for all $n \geq n_0$, we have that $|z_0| < R_n$. Therefore by the Maximum Modulus Principle, we conclude that

$$|f(z_0)| \leq |f|_{R_n}.$$

If we now let $n \to \infty$, then we see that $f(z_0) = 0$ for all $z_0 \in \mathbb{C}$, which establishes our modest proposition.   ∎

We are now ready to face the critical question: How many zeros must an entire function have in a sequence of increasingly large disks in order to satisfy the conclusion of Proposition 5.13? The answer, for our purpose here, is captured in the theorem below. Given a function $F(z)$ and a positive real number R, we define the set

$$\mathcal{Z}(F, R) = \{z \in \mathbb{C} : F(z) = 0, \text{ with } |z| \leq R\},$$

with the understanding that if $z_0 \in \mathcal{Z}(F, R)$ is a zero of $F(z)$ with multiplicity m, then z_0 appears m times in the set $\mathcal{Z}(F, R)$. Finally, we denote the cardinality of the set $\mathcal{Z}(F, R)$ by $\text{card}(\mathcal{Z}(F, R))$.

THEOREM 5.14 *Let $F(z)$ be an entire function and suppose that there exists a real number κ such that for all sufficiently large R,*

$$|F|_R \leq e^{R^\kappa}.$$

If there exists an $\varepsilon > 0$ and an unbounded increasing sequence of real numbers $R_1, R_2, \ldots$ such that

$$R_n^{\kappa+\varepsilon} < \text{card}(\mathcal{Z}(F, R_n)),$$

for all $n = 1, 2, \ldots$, then $f(z)$ is the identically zero function.

In other words, Theorem 5.14 asserts that if an entire function has "sufficiently many" zeros in a sequence of ever-increasing disks, then the function must be the identically zero function.

Proof of Theorem 5.14. Given any $z_0 \in \mathbb{C}$, there exists an integer n_0 such that for all $n \geq n_0$, we have $|z_0| < R_n$. For each $n \geq n_0$, we define

$$G_n(z) = \frac{F(z)}{\prod_{\omega \in \mathcal{Z}(F, R_n)}(z - \omega)}.$$

Thus, given that the zeros in $\mathcal{Z}(F, R_n)$ appear with the appropriate multiplicity, we see that $G_n(z)$ is an entire function. Hence by the Maximum Modulus Principle, our hypotheses, and the following figure,

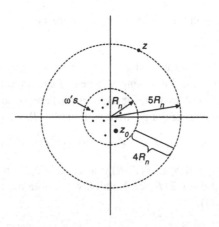

we conclude that for all sufficiently large R_n,

$$|F(z_0)| \leq |G_n|_{5R_n} \prod_{\omega \in \mathcal{Z}(F,R_n)} |z_0 - \omega| = \frac{|F|_{5R_n}}{\left|\prod_{\omega \in \mathcal{Z}(F,R_n)} (z - \omega)\right|_{5R_n}} \prod_{\omega \in \mathcal{Z}(F,R_n)} |z_0 - \omega|$$

$$\leq \frac{e^{(5R_n)^\kappa}}{(4R_n)^{\text{card}(\mathcal{Z}(F,R_n))}} (2R_n)^{\text{card}(\mathcal{Z}(F,R_n))}$$

$$\leq e^{(5R_n)^\kappa} \left(\frac{1}{2}\right)^{\text{card}(\mathcal{Z}(F,R_n))}$$

$$< e^{(5R_n)^\kappa} \left(\frac{1}{2}\right)^{R_n^{\kappa+\varepsilon}}$$

$$= e^{(5R_n)^\kappa} e^{-(\log 2)R_n^{\kappa+\varepsilon}} = e^{(5^\kappa - (\log 2)R_n^\varepsilon)R_n^\kappa}.$$

Since for all sufficiently large n, we have $5^\kappa < (\log 2)R_n^\varepsilon$, we conclude that if we let $n \to \infty$, then $|F(z_0)| = 0$, and hence $F(z_0) = 0$ for all $z_0 \in \mathbb{C}$, which establishes our result. ∎

We now return to the proof of the general case of Theorem 5.2. In particular, we require an estimate for the number of zeros of modulus at most R that the function

$$F(z) = P(z, e^{\zeta z}) = \sum_{m=0}^{D_1-1} \sum_{n=0}^{D_2-1} a_{mn} z^m e^{n\zeta z}$$

can have; that is, we require a bound on $\mathcal{Z}(F,R)$. The first step is to find a value for the exponent κ that fulfills the hypothesis of Theorem 5.14, which is the goal of the following simple challenge.

Challenge 5.8 *Show that for any choice of $\kappa > 1$, we have that for all sufficiently large R,*

$$|F|_R \leq e^{R^\kappa}.$$

Thus, since we have already established that $F(z)$ is not identically zero (since the functions z and $e^{\zeta z}$ are algebraically independent and the coefficients a_{mn} are not all equal to 0), Theorem 5.14 implies that for any choice of $\varepsilon > 0$, for all sufficiently large R, $F(z)$ cannot have more than $R^{\kappa+\varepsilon}$ zeros in $\mathcal{Z}(F,R)$.

More concretely, taking $\kappa = \frac{5}{4}$ and $\varepsilon = \frac{1}{4}$, we see that for all sufficiently large R,

$$\text{card}(\mathcal{Z}(F,R)) \leq R^{3/2}.$$

Challenge 5.9 *Using the above estimate for $\text{card}(\mathcal{Z}(F,R))$ for sufficiently large R, deduce that there exists an integer M, $M \geq K$, such that*

$$F(k_1 + k_2\xi) = 0 \quad \text{for all } 0 \leq k_1 < M \quad \text{and} \quad 0 \leq k_2 < M,$$

while there exists a pair (k_1^*, k_2^*) *with* $0 \le k_1^* \le M$, $0 \le k_2^* \le M$, *such that* $F(k_1^* + k_2^* \xi) \ne 0$. *(Hint: Assume that* $F(k_1 + k_2\xi) = 0$ *for all positive integers* k_1 *and* k_2. *Now show that for all sufficiently large R, a lower bound for the number of complex numbers of the form* $z = k_1 + k_2\xi$, *where* k_1 *and* k_2 *are positive integers, that satisfy* $|z| \le R$ *is proportional to the area of the disk of radius R. Conclude that this assumption implies that* $F(z)$ *is identically zero, which is a contradiction.)*

The remainder of the proof of the general case of Theorem 5.2 is now identical to the one in Section 5.5, where we assumed ξ to be real. Thus our proof of Theorem 5.2, at long *long* last, is complete—Congratulations.   ■

As an entertaining intermission, we suggest the following:

Challenge 5.10   *Use a calculator to confirm that the number* $e^{\pi\sqrt{163}}$, *when rounded to eight digits after the decimal point equals the* integer 262537412640768744. *Despite this calculation, apply the Gelfond–Schneider Theorem to deduce the transcendence of* $e^{\pi\sqrt{163}}$. *Use this transcendental near-integer to impress family and friends.*

5.7 Widening our perspective and extending our results

In all of our efforts thus far we have applied important features of special functions to demonstrate the transcendence of a wide variety of numbers such as e^{π}, $2^{\sqrt{2}}$, and $\log \sqrt{i}$. But Theorem 5.2 offers an alternative point of view for the role of those special functions. That perspective is to view the functions not as necessary tools to study the transcendence of particular numbers, but rather as generators of families of transcendental numbers. Of course, we have already considered this point of view in Chapter 3 when we established the transcendence of e^{α} for any nonzero algebraic number α. In this section we expand this perspective and hence extend our results.

Focusing on this alternative point of view.   We can restate Hermite's result (Theorem 3.1) more in sympathy with Theorem 5.2 as follows: *Given any nonzero complex number* ξ, *at least one of the numbers* ξ *or* e^{ξ} *is transcendental.* This point of view inspires the following theorem, which foreshadows the direction in which we wish to head.

THEOREM 5.15   *Given an irrational* $\xi \in \mathbb{C}$, *for any two* $\mathbb{Q}$-*linearly independent complex numbers* x_1 *and* x_2, *at least one of the four numbers*

$$x_1, \ x_2, \ e^{\xi x_1}, \ e^{\xi x_2}$$

is transcendental.

Proof.   Suppose that the four numbers $x_1, x_2, e^{\xi x_1}$, and $e^{\xi x_2}$ are all algebraic. Thus we easily see that the number x_1/x_2 is algebraic, and that the numbers ξx_1 and ξx_2 are logarithms of algebraic numbers, namely $e^{\xi x_1}$ and $e^{\xi x_2}$, respectively. We also observe that by our hypothesis, ξx_1 and ξx_2 are $\mathbb{Q}$-linearly independent. Thus by Corollary 5.4, we conclude that the ratio $\frac{\xi x_1}{\xi x_2} = \frac{x_1}{x_2}$ is transcendental, which contradicts the fact that x_1/x_2 is algebraic. Therefore at least one of the four numbers in the theorem must be transcendental. ∎

As an illustration, we note that if we let $\xi = e$, $x_1 = \sqrt{2}$, and $x_2 = \sqrt{3}$, then Theorem 5.15 implies the fun fact that at least one of the two numbers $e^{e\sqrt{2}}$ or $e^{e\sqrt{3}}$ is transcendental. How much would you be willing to wager that *either* of these numbers is algebraic?

With these results as inspiration, in this section we deduce a theorem that straddles the line between what we are referring to as *classical* and *modern* transcendental number theory—the so-called *Six Exponentials Theorem*. This result is classical in that its proof is simply an elaboration of our proof of the Gelfond–Schneider Theorem and yet is modern in that it examines the transcendence of values of special functions rather than of particular numbers.

THEOREM 5.16   *Let $\{x_1, x_2\}$ and $\{y_1, y_2, y_3\}$ be two $\mathbb{Q}$-linearly independent sets of complex numbers. Then at least one of the six numbers*

$$e^{x_1 y_1},\ e^{x_1 y_2},\ e^{x_1 y_3},\ e^{x_2 y_1},\ e^{x_2 y_2},\ e^{x_2 y_3}$$

is transcendental.

If we consider the sets $\{1, e\}$ and $\{e, e^2, e^3\}$, then the Six Exponentials Theorem immediately implies the entertaining insight that at least one of the following numbers is transcendental:

$$e^e,\ e^{e^2},\ e^{e^3},\ e^{e^4}.$$

An intuitive 5-step program to the proof of Theorem 5.16

There are two ways to view the conclusion of Theorem 5.16: The two algebraically independent exponential functions $e^{x_1 z}$ and $e^{x_2 z}$ cannot be simultaneously algebraic at three $\mathbb{Q}$-linearly independent values; or the three exponential functions $e^{y_1 z}$, $e^{y_2 z}$, and $e^{y_3 z}$, which can be shown to be algebraically independent, cannot be simultaneously algebraic at two $\mathbb{Q}$-linearly independent values. The theorem can be established using either point of view, and so we adopt the first, since it is more in line with our earlier arguments.

Our proof of Theorem 5.16 mirrors the five key steps involved in proving the Gelfond–Schneider Theorem. Those steps, in this context, are outlined below.

Step 1.   We assume that all of the values $e^{x_i y_j}$ are algebraic. Thus for any $P(x, y) \in \mathbb{Z}[x, y]$, we notice that if $F(z) = P(e^{x_1 z}, e^{x_2 z})$, then for any integers k_1, k_2, and k_3, the quantity $F(k_1 y_1 + k_2 y_2 + k_3 y_3)$ is an algebraic number.

Continued

Step 2. We find a nonzero integral polynomial

$$P(x,y) = \sum_{m=0}^{D_1-1} \sum_{n=0}^{D_2-1} a_{mn} x^m y^n,$$

having "modestly sized" integral coefficients, such that if we let

$$F(z) = P(e^{x_1 z}, e^{x_2 z}),$$

then $F(z) = 0$ for all $z \in \{k_1 y_1 + k_2 y_2 + k_3 y_3 : 0 \le k_j < K\}$. Before proceeding to the next step, we note that since the two functions $e^{x_1 z}$ and $e^{x_2 z}$, which we compose with $P(x,y)$ in order to produce $F(z)$, are so similiar, it is not natural to take $D_1 > D_2$ or $D_2 > D_1$. Thus we now declare that $D_1 = D_2$ and denote this common value by D.

Step 3. We apply Theorem 5.14 to show that the $\mathbb{Q}$-linear independence of y_1, y_2, y_3 implies that $F(z)$ cannot vanish at all of the numbers of the form

$$k_1 y_1 + k_2 y_2 + k_3 y_3, \quad \text{where } k_j \in \mathbb{Z}^+.$$

It then follows that there exists a positive integer M such that

$$F(k_1 y_1 + k_2 y_2 + k_3 y_3) = 0,$$

for all $0 \le k_j < M$, while there exists some triple k_1^*, k_2^*, k_3^*, satisfying $0 \le k_j^* \le M$, where

$$F(k_1^* y_1 + k_2^* y_2 + k_3^* y_3) \ne 0.$$

Steps 4 and 5. We now use the nonzero algebraic number $F(k_1^* y_1 + k_2^* y_2 + k_3^* y_3)$ to obtain a nonzero integer $\mathcal{N}$. We then show that by our construction of the polynomial $P(x,y)$, the integer $\mathcal{N}$ violates the Fundamental Principle of Number Theory.

A slightly more detailed proof of Theorem 5.16.   Since this proof so closely parallels the proof of the Gelfond–Schneider Theorem, we adopt a briefer and slightly more informal style here as we move through the five steps outlined in our overview.

Step 1. We assume that all six numbers $e^{x_1 y_1}, e^{x_1 y_2}, \dots, e^{x_2 y_3}$ are algebraic. Next we let $E = \mathbb{Q}(\theta)$ be the number field generated by $e^{x_1 y_1}, e^{x_1 y_2}, \dots, e^{x_2 y_3}$, and assume that θ is an algebraic integer of degree d. As before, there exists a rational linear combination of $1, \theta, \dots, \theta^{d-1}$ representing each of the numbers $e^{x_i y_j}$. For each i and j, we let $p_{ij}(z) \in \mathbb{Q}[z]$ be the polynomial satisfying $e^{x_i y_j} = p_{ij}(\theta)$. We saw in the proof of the Gelfond–Schneider Theorem that if we let δ denote a common denominator for all of the coefficients of the six polynomials $p_{ij}(z)$, then we could later multiply through by a power of δ to clear all denominators. In order to avoid this distracting complication, we assume that the coefficients of each $p_{ij}(z)$ are all integers.

Step 2. We now wish to construct a nonzero polynomial

$$P(x,y) = \sum_{m=0}^{D-1}\sum_{n=0}^{D-1} a_{mn}x^m y^n \in \mathbb{Z}[x,y]$$

satisfying

$$F(k_1y_1 + k_2y_2 + k_3y_3) = 0, \quad \text{for}\quad 0 \le k_j < K, \qquad (5.16)$$

where the size of the coefficients, $|a_{mn}|$, is under control.

We again view the coefficients a_{mn} of our desired polynomial $P(x,y)$ as the unknowns in a system of linear equations given by (5.16). Explicitly, we consider the linear system

$$\sum_{m=0}^{D-1}\sum_{n=0}^{D-1} e^{mx_1(k_1y_1+k_2y_2+k_3y_3)} e^{nx_2(k_1y_1+k_2y_3+k_3y_3)} a_{mn} = 0,$$

which in view of the ever-important relationship $e^{a+b} = e^a e^b$, is equivalent to

$$\sum_{m=0}^{D-1}\sum_{n=0}^{D-1} \left(e^{k_1x_1y_1} e^{k_2x_1y_2} e^{k_3x_1y_3} \right)^m \left(e^{k_1x_2y_1} e^{k_2x_2y_2} e^{k_3x_2y_3} \right)^n a_{mn}$$

$$= \sum_{m=0}^{D-1}\sum_{n=0}^{D-1} \left(p_{11}(\theta)^{k_1} p_{12}(\theta)^{k_2} p_{13}(\theta)^{k_3} \right)^m \left(p_{21}(\theta)^{k_1} p_{22}(\theta)^{k_2} p_{23}(\theta)^{k_3} \right)^n a_{mn} = 0.$$

We write each of these K^3 linear equations in the form

$$\sum_{m=0}^{D-1}\sum_{n=0}^{D-1} \left(r_1 + r_2\theta + r_3\theta^2 + \cdots + r_d\theta^{d-1} \right) a_{mn} = 0,$$

where the coefficients r_j are integers depending on k_1, k_2, k_3, m, and n. Proceeding as we did in the proof of the Gelfond–Schneider Theorem to deduce inequality (5.5), and recalling that the coefficients of p_{ij} are integral and that θ is an algebraic integer, we can again apply Lemma 5.10 to conclude that those integer coefficients all satisfy

$$|r_j| < c_1^{DK}.$$

If we now combine equal powers of θ, we discover that each linear equation can be expressed as

$$A_1 + A_2\theta + \cdots + A_d\theta^{d-1} = 0,$$

where each A_j is of the form

$$\sum_{m=0}^{D-1}\sum_{n=0}^{D-1} t_j(m,n,k_1,k_2,k_3)a_{mn} = 0,$$

where the integers $t_j(m,n,k_1,k_2,k_3)$ satisfy

$$0 < \max\{|t_j(m,n,k_1,k_2,k_3)|\} < c_2^{DK}.$$

Setting each of the coefficients $A_1, A_2, \ldots, A_d$ equal to zero, for all allowable choices of k_1, k_2, k_3, yields a linear system of dK^3 equations in D^2 unknowns. This system has a nontrivial integral solution provided that $D^2 > dK^3$. If we now fix $D = \sqrt{2dK^3}$, ensuring that K is selected such that D is an integer, then we can apply Siegel's Lemma and find a nonzero integer solution a_{mn} to our linear system satisfying

$$0 < \max\{|a_{mn}|\} < c_3^{K^{5/2}}.$$

Step 3. Applying Theorem 5.14 we deduce that there exists an integer M, $M > K$, such that

$$F(k_1 y_1 + k_2 y_2 + k_3 y_3) = 0$$

for all $0 \le k_j < M$, while

$$F(k_1^* y_1 + k_2^* y_2 + k_3^* y_3) \ne 0,$$

for some triple k_1^*, k_2^*, k_3^* satisfying $0 \le k_j^* \le M$.

Steps 4 and 5. We express the nonzero algebraic number $F(k_1^* y_1 + k_2^* y_2 + k_3^* y_3) \in \mathbb{Z}(\theta)$ as

$$F(k_1^* y_1 + k_2^* y_2 + k_3^* y_3) = A_1 + A_2\theta + \cdots + A_d\theta^{d-1},$$

where the A_j's are *integers*. Arguing as in our previous proof, we find that

$$0 < \max_{1 \le j \le d}\{|A_j|\} < c_4^{M^{5/2}}.$$

We now define the nonzero integer $\mathcal{N}$ by

$$\mathcal{N} = \prod_{i=1}^{d}\left(A_1 + A_2\theta_i + A_3\theta_i^2 + \cdots + A_d\theta_i^{d-1}\right),$$

where $\theta_1, \theta_2, \ldots, \theta_d$ are the conjugates of θ, with $\theta_1 = \theta$.

Applying the Maximum Modulus Principle, we see that if we take $R = M \log M$, then

$$0 < \left|A_1 + A_2\theta + \cdots + A_d\theta^{d-1}\right| < c_5^{-c_6 M^3 \log\log M}.$$

Using the triangle inequality to bound the other factors of $|\mathcal{N}|$, we conclude that

$$0 < |\mathcal{N}| < c_7^{-c_8 M^3 \log\log M},$$

which if we select K (and therefore M) sufficiently large leads us to contradict the Fundamental Principle of Number Theory and brings us to the end of our panoramic proof of the Six Exponentials Theorem.   ∎

5.8 Algebraic values of algebraically independent functions

In this closing section we look beyond the exponential function $f(z) = e^z$ and consider the natural, albeit too-good-to-be-true, conjecture that captures the essence of Hermite's result, the Gelfond–Schneider Theorem, and the Six Exponentials Theorem.

META-CONJECTURE 5.17 *Two algebraically independent functions cannot simultaneously be algebraic at a point. That is, if $f(z)$ and $g(z)$ are distinct functions, then for any $z_0 \in \mathbb{C}$, at least one of the values $f(z_0)$ or $g(z_0)$ must be a transcendental number.*

Although this meta-conjecture might initially sound reasonable, a moment's thought reveals that it is not. In fact, the very results that inspired the ill-fated conjecture actually generate counterexamples. For example, Hermite's Theorem implies that z and e^z are not simultaneously algebraic, *except* at $z = 0$. So the conjecture is false in that two algebraically independent functions can indeed be simultaneously algebraic (although in this particular case the functions are simultaneously algebraic only at $z = 0$). Moreover, it is possible to exhibit algebraically independent functions that are simultaneously algebraic at any prescribed finite number of points. For example, if $\alpha_1, \alpha_2, \ldots, \alpha_L$ are algebraic numbers, then

$$f(z) = z \quad \text{and} \quad g(z) = e^{(z-\alpha_1)(z-\alpha_2)\cdots(z-\alpha_L)} \tag{5.17}$$

both produce algebraic values for $z = \alpha_1, \alpha_2, \ldots, \alpha_L$. These observations lead us to the following refined meta-conjecture.

REFINED META-CONJECTURE 5.18 *Two algebraically independent functions cannot simultaneously be algebraic at very many different complex numbers.*

The Six Exponentials Theorem can be viewed as a special case of this refined meta-conjecture, since it asserts that two algebraically independent exponential functions cannot be simultaneously algebraic at three $\mathbb{Q}$-linearly independent complex numbers. However, such a meta-conjecture is still too vague to be true—we need to specify what kind of functions (continuous? entire? something else?); how many points (a couple? finitely many? infinitely many?); and finally, how "different" must those points be (unequal? $\mathbb{Q}$-linear independent?).

The counterexample in (5.17) leads us to conclude that if such a conjecture is to hold, then the number of points at which the two functions are simultaneously algebraic must depend on some specific properties of the functions themselves. Moreover, if $P(z)$ is any nonzero polynomial with integral coefficients, say of degree $d \geq 1$, then the algebraically independent functions

$$f(z) = z \quad \text{and} \quad g(z) = e^{P(z)}$$

are simultaneously algebraic at the d zeros of $P(z)$. So in this example "too many" must be connected with the degree of $P(x)$.

It is possible to see how the degree of $P(z)$ plays a role in determining an upper bound on the number of simultaneous algebraic points for z and $e^{P(z)}$ by returning to Theorem 5.14 which asserts that if a nonzero entire function $F(z)$ satisfies $|F|_R \leq e^{R^\kappa}$ for all sufficiently large R, then F cannot have more than $R^{\kappa + \varepsilon}$ zeros in the disk of all complex z satisfying $|z| \leq R$. This result implies that for any particular complex number β, such a nonconstant entire function $F(z)$ satisfies

$$\text{card} \{z \in \mathbb{C} : F(z) = \beta \text{ with } |z| \leq R\} < R^{\kappa + \varepsilon}.$$

Challenge 5.11 *Verify the previous inequality by considering the function $G(z) = F(z) - \beta$.*

Hence we discover that the number of times such an entire function can attain any particular algebraic value is bounded by a function of R and κ. Extending this observation, it is not too difficult to imagine that κ might influence how many times a particular entire function takes values in any fixed *set* of algebraic numbers. For example, an immediate consequence of the previous challenge is that if we let $\mathcal{A}$ denote a set of n algebraic numbers, then

$$\text{card} \left\{ z \in \mathbb{C} : F(z) \in \mathcal{A} \text{ with } |z| \leq R \right\} < nR^{\kappa + \varepsilon}. \tag{5.18}$$

Seeing the importance of the exponent κ, we are led to the following notion. Given an entire function $F(z)$, we say that $F(z)$ has *finite order of growth* if there exists a positive constant κ such that for all sufficiently large $|z|$,

$$|f(z)| < e^{|z|^\kappa}.$$

If $f(z)$ has finite order of growth, then we define the *order of $f(z)$* by

$$\rho = \inf \left\{ \kappa > 0 : |f(z)| < e^{|z|^\kappa} \text{ for all sufficiently large } |z| \right\}.$$

When we move beyond the algebraic values of a single function to the simultaneous algebraic values of two algebraically independent functions, the order of growth of each function must play a role. Surprisingly, however, we will see that if we wish to give an upper bound for the number of points in a disk at which two algebraically independent functions simultaneously take values from a prescribed collection of algebraic numbers, neither the radius of the disk nor the cardinality of the set of algebraic values appears in our bound. Instead, the upper bound depends only upon the degree of the field extension of $\mathbb{Q}$ containing the given set of algebraic numbers and the orders of growth of the functions. Stated only slightly more precisely, if $f_1(z)$ and $f_2(z)$ are *sufficiently nice* functions with finite orders of growth ρ_1 and ρ_2, respectively, and E is an algebraic extension of $\mathbb{Q}$ of degree d, then

$$\text{card}\left\{ z \in \mathbb{C} : f_1(z) \in E \text{ and } f_2(z) \in E \right\} \leq (\rho_1 + \rho_2)d.$$

To codify these informal remarks into an actual result, we must determine what types of functions are "*sufficiently nice.*" In order develop an intuitive sense of what that vague phrase should mean, we first reflect back on our previous observations and insights.

A look back. If we revisit the proof of the transcendence of e^π from Chapter 4, we see that we constructed an auxiliary function of the form $f(z) = P(e^z, e^{iz})$ satisfying $f^{(t)}(k\pi) = 0$ for $0 \leq t < T$ and $k = 1, 2, \ldots, K$. In the body of the proof we used the fact that if we assume that e^π is algebraic, then the derivatives of both e^z and e^{iz} evaluated at $z = k\pi$ yield algebraic numbers. Stated more formally, we have that e^z and e^{iz} are each a solution to a polynomial differential equation with algebraic coefficients; specifically, $y = e^z$ is a solution to $\frac{dy}{dz} = y$, while $y = e^{iz}$ is a solution to $\frac{dy}{dz} = iy$.

Our proof of the transcendence of α^β, for algebraic numbers α, β with $\alpha \neq 0, 1$, and β irrational, based on Schneider's argument, proceeded differently than our proof of the transcendence of e^π. Writing $\zeta = \log \alpha$, we see that our proof of the transcendence of α^β relied on the auxiliary function of the form $F(z) = P(z, e^{(\log \alpha)z})$ satisfying $F(k_1 + k_2\beta) = 0$ for a large collection of positive integers k_1 and k_2. Indeed, we could not possibly have used an auxiliary function $F(z) = P(z, e^{(\log \alpha)z})$ together with the differentiation idea we applied in the demonstration of the transcendence of e^π, because even with the assumption that α^β is algebraic,

$$F'(\beta) = \frac{d}{dz} P\left(z, e^{(\log \alpha)z} \right) \bigg|_{z=\beta}$$

is a polynomial expression involving the algebraic number β, the assumed-algebraic number α^β, and the highly transcendental number $\log \alpha$.

However, it *is* possible to establish the transcendence of α^β using the derivatives of an auxiliary polynomial by replacing the functions z and $e^{(\log \alpha)z}$ with suitably defined functions. Indeed, this was precisely Gelfond's approach to proving this result. Specifically, we could construct a polynomial $P(x, y)$ such that the function $f(z) = P(e^z, e^{\beta z})$ satisfies $f^{(t)}(k \log \alpha) = 0$ for an appropriate list of integers t and k. In this case we would see that

$$f^{(t)}(z) = \sum_{m=0}^{D_1-1} \sum_{n=0}^{D_2-1} b_{mn}(\beta)(e^z)^m (e^{\beta z})^n,$$

where each $b_{mn}(\beta)$ is an integral polynomial expression in β. Hence if we assume that α^{β} is algebraic, then so is the quantity

$$f^{(t)}(k \log \alpha) = \sum_{m=0}^{D_1-1} \sum_{n=0}^{D_2-1} b_{mn}(\beta)\alpha^{km}(\alpha^{\beta})^{kn},$$

for any nonnegative integers k and t.

All of these remarks point to a re-refinement of our meta-conjecture that might be within our grasp. In fact we are now able to produce a special case of an important result due to Serge Lang known as the Schneider–Lang Theorem. We restrict our attention to entire functions with finite orders of growth that satisfy polynomial differential equations with algebraic coefficients. We will return to this issue and consider the more general formulation of the Schneider–Lang Theorem in Chapter 7.

THEOREM 5.19 (A WEAK VERSION OF THE SCHNEIDER-LANG THEOREM) *Suppose that $f_1(z)$ and $f_2(z)$ are two algebraically independent entire functions with finite orders of growth, each of which satisfies an algebraic polynomial differential equation. That is, there exists a number field F and a finite collection of functions $f_3(z), f_4(z), \ldots, f_J(z)$ such that the differential operator $\frac{d}{dz}$ maps the ring $F[f_1(z), f_2(z), \ldots, f_J(z)]$ into itself. Then for any number field E containing F,*

$$\mathrm{card}\left\{z \in \mathbb{C} : f_1(z) \in E, f_2(z) \in E, \ldots, f_J(z) \in E\right\} \tag{5.19}$$

is finite.

Moreover, if ρ_1 denotes the order of growth of $f_1(z)$ and ρ_2 denotes the order of growth of $f_2(z)$, then one can give a quantitive version of this result, namely,

$$\mathrm{card}\left\{z \in \mathbb{C} : f_1(z) \in E, f_2(z) \in E, \ldots, f_J(z) \in E\right\} \le (\rho_1 + \rho_2)[E : \mathbb{Q}].$$

Before offering an intuitive sketch of the proof of this result, we remark that the Schneider–Lang Theorem can be applied to produce many of the results we have seen thus far. For example, as we demonstrate below, we can use the Schneider–Lang Theorem to deduce the transcendence of α^{β} for algebraic numbers α, β, with $\alpha \neq 0, 1$, and β irrational.

The deduction of the Gelfond–Schneider Theorem from the Schneider–Lang Theorem. As always, we begin by assuming that α^{β} is algebraic and let E be the number field given by $E = \mathbb{Q}(\alpha, \beta, \alpha^{\beta})$. If we now define $f_1(z) = e^z$ and $f_2(z) = e^{\beta z}$, then in view of the fact that β is irrational, it follows that $f_1(z)$ and $f_2(z)$ are algebraically independent functions. We now observe that each of these functions satisfies an algebraic differential equation,

$$\frac{dy}{dz} = y \quad \text{and} \quad \frac{dy}{dz} = \beta y,$$

respectively. So if we take $F = \mathbb{Q}(\beta) \subseteq E$, then we see that $F[f_1(z), f_2(z)]$ is closed under differentiation. Thus we may apply the Schneider–Lang Theorem and deduce that there are only *finitely* many points $z \in \mathbb{C}$ such that $f_1(z) \in E$ and $f_2(z) \in E$. However, we note that for all integers k,

$$f_1(k \log \alpha) = \alpha^k \in E \quad \text{and} \quad f_2(k \log \alpha) = \left(\alpha^\beta\right)^k \in E,$$

which contradicts the previous sentence. Therefore we conclude that α^β is transcendental.

An overview of the proof of Theorem 5.19.   We fix a number field E and let

$$\Omega = \left\{ z \in \mathbb{C} : f_1(z) \in E \text{ and } f_2(z) \in E \right\}.$$

We wish to show that Ω is a finite set, and we establish this assertion by assuming that $\{w_1, w_2, \ldots, w_L\}$ is an arbitrary set of distinct elements from Ω and showing that L cannot be too large.

We first solve a system of linear equations to find a nonzero polynomial $P(x, y)$ such that the function $f(z) = P(f_1(z), f_2(z))$ vanishes at each of the points $w_1, w_2, \ldots, w_L$, each with multiplicity T. Since $f_1(z)$ and $f_2(z)$ are algebraically independent functions, $f(z)$ is not identically zero. Thus we let t_0 be the smallest positive integer such that there exists an index l_0, $1 \leq l_0 \leq L$, satisfying

$$\frac{d^{t_0}}{dz^{t_0}} f(w_{l_0}) \neq 0.$$

If we let $\gamma = \frac{d^{t_0}}{dz^{t_0}} f(w_{l_0})$, then by the hypothesis it follows that there exists a polynomial $D(x_1, x_2, \ldots, x_J)$ with coefficients in E such that

$$\gamma = D(f_1(w_{l_0}), f_2(w_{l_0}), \ldots, f_J(w_{l_0})). \tag{5.20}$$

Thus we conclude that γ is an algebraic number in E.

If we now consider the nonzero rational number formed by taking the product of all the conjugates of the nonzero algebraic number γ, then after clearing its denominator we are led to a nonzero integer $\mathcal{N}$.

Using (5.20) we can determine upper bounds for the absolute values of the conjugates of γ. Next we again employ the Maximum Modulus Principle to estimate $|\gamma|$. Specifically, we consider the function

$$G(z) = \frac{F(z)}{\prod_{l=1}^{L} (z - w_l)^{t_0 - 1}}$$

on a disk of radius $t_0^{1/(\rho_1 + \rho_2)}$ to get an upper bound for $|\gamma|$ involving ρ_1, ρ_2, and $\deg(\gamma)$. Applying our bounds, we discover that if $L > (\rho_1 + \rho_2)[E : \mathbb{Q}]$, the integer $\mathcal{N}$ satisfies $0 < |\mathcal{N}| < 1$, which contradicts the Fundamental Principle of Number Theory and establishes our result.   ∎

Extending our reach.   If we wish to further extend the Schneider–Lang Theorem, then we must isolate those key features a function should possess in order to generalize the previous argument. In an attempt to inspire the next major advance in transcendental number theory, we return to our short argument that allowed us to deduce the Gelfond–Schneider Theorem from Theorem 5.19. At the heart of that little proof is the following hidden but amazingly powerful observation: $f(z) = e^z$ is a group homomorphism from the additive group of complex numbers into the multiplicative group of nonzero complex numbers. Thus we become conscious of the perhaps-invisible fact that our arguments thus far have exploited the following three fundamental but critical properties of e^z:

- The function e^z is an entire function.
- The function e^z is the solution to an algebraic polynomial differential equation; specifically, $y' = y$.
- The function e^z is a group homomorphism from $\langle \mathbb{C}, + \rangle$ to $\langle \mathbb{C}^*, \cdot \rangle$; in particular, $e^{z_1 + z_2} = e^{z_1} e^{z_2}$.

Thus in order to obtain further insights into the theory of transcendence from such a meta-result as Theorem 5.19, we must search for other functions that satisfy the previous three conditions. As we will discover in Chapter 7, that search begins not within the domain of analysis, but rather in the realm of algebra.

Number 6

2.718281828459 ... + 0.11000100000 ...

Class Distinctions Among Complex Numbers:
Mahler's classification and the transcendence of $e + \sum_{n=1}^{\infty} 10^{-n!}$

6.1 The power of making polynomials nearly vanish

Polynomials with integer coefficients play a central role in the theory of transcendence. In fact, they made their first appearance at the very opening of our story—A number is *transcendental* precisely when it is not a zero of any nonzero polynomial in $\mathbb{Z}[z]$. In this chapter, given an arbitrary complex number ξ, we forgo the fascination of determining whether there exists a nonzero polynomial that vanishes at ξ. Instead, here we will consider those polynomials $P(z) \in \mathbb{Z}[z]$ for which $|P(\xi)|$ is *nonzero* but as small as possible relative to the degree of $P(z)$ and the height of $P(z)$. In some sense, we will play "polynomial limbo" with ξ: How low can $|P(\xi)|$ go with $P(z) \in \mathbb{Z}[z]$ but *without* falling down to zero?

A natural question to ask is: How does a hunt for polynomials that *nearly* vanish at ξ connect to issues of transcendence? Here we will see that an understanding of how small the absolute values of a sequence of polynomials can be made at ξ, in terms of their degrees and heights, not only leads to an alternative definition of a transcendental number, but in fact *refines* the definition, since it partitions transcendental numbers into distinct classes. This classification, discovered by Kurt Mahler in 1932, leads to further insights into transcendental numbers and allows us to produce new transcendence results.

> *An intuitive idea for the connection between near-vanishing and transcendence*
>
> We begin by first considering a fixed rational number r/s. If we try to approximate r/s by rational numbers p/q that *differ* from r/s, then by the Fundamental Principle of Number Theory we discover that the approximation cannot be too close—we
>
> *Continued*

are always bounded away from r/s by a constant multiple of $1/q$. In particular,

$$\frac{1}{sq} \le \frac{|rq - ps|}{sq} = \left| \frac{r}{s} - \frac{p}{q} \right|.$$

Suppose now we are given a real number τ with the property that we can find two positive constants c and ω such that there exists an infinite sequence of rationals p_n/q_n satisfying

$$0 < \left| \tau - \frac{p_n}{q_n} \right| \le \frac{c}{q_n^{1+\omega}}. \tag{6.1}$$

What can we deduce about τ? We claim that τ must be irrational. For if we assume that τ is rational, say $\tau = \frac{r}{s}$, then putting the two previous inequalities together, we see that for all n,

$$\frac{1}{sq_n} \le \frac{c}{q_n^{1+\omega}}.$$

Thus we conclude that for all n,

$$0 < \frac{1}{cs} \le \frac{1}{q_n^{\omega}},$$

which is a contradiction, since $\frac{1}{cs}$ is a positive constant, while the right-hand side approaches 0 as $n \to \infty$. Thus τ is irrational.

The important moral of the previous remark is that rational numbers *cannot* be well-approximated by other rational numbers. In fact, even more is true—rational numbers are precisely those numbers that cannot be well-approximated by other rationals. This assertion is a consequence of Theorem 6.3 from the next section.

If we have a real number that has infinitely many very good approximations by rational numbers as in (6.1), then we see it must be irrational. Those readers who are amused by amazing rational approximations may notice that these ideas parallel those surrounding Liouville's result (Theorem 1.1) from Chapter 1 and our proof of the irrationality of $\sqrt{2}$ long ago in Chapter 0 (Theorem 0.1).

In this chapter we revisit Liouville's work, produce an important generalization, and discover that the rational approximation ideas that lead to irrationality extend to transcendence. Roughly speaking, a complex number ξ is transcendental if it can be approximated "very well" by infinitely many algebraic numbers. Determining exactly how "well" the approximations are required to be, in terms of degrees and heights of the approximates, leads to Mahler's classification of transcendental numbers.

Instead of approximating ξ by algebraic numbers, we will focus on making $|P(\xi)|$ small for infinitely many polynomials $P(z) \in \mathbb{Z}[z]$ satisfying $P(\xi) \ne 0$. These two points of view are connected in a natural way: If $|P(\xi)|$ is "small," then ξ must be "close" to a zero of $P(z)$, and if ξ is "close" to a zero of $P(z)$, then $|P(\xi)|$ must be "small." We will heavily exploit this duality between approximating by algebraic numbers and finding small polynomial values. As a naïve illustration of one direction of this connection, consider the polynomial $P(z) \in \mathbb{Z}[z]$, which we factor as $P(z) = a_N \prod_{n=1}^{N}(z - \alpha_n)$, where $a_N \in \mathbb{Z}$ is the leading coefficient of $P(z)$. If $P(\xi) \neq 0$, then

$$0 < |a_N| \prod_{n=1}^{N} |\xi - \alpha_n| = |P(\xi)|.$$

If we let α^* denote the zero of $P(z)$ that is closest to ξ, that is, $|\xi - \alpha^*| = \min\{|\xi - \alpha_n| : n = 1, 2, \ldots, N\}$, then we immediately observe that

$$0 < |a_N||\xi - \alpha^*|^N \leq |a_N| \prod_{n=1}^{N} |\xi - \alpha_n| = |P(\xi)|,$$

and hence

$$0 < |\xi - \alpha^*| \leq \left(\frac{|P(\xi)|}{|a_N|}\right)^{1/N} \leq |P(\xi)|^{1/N}.$$

If we assume now that $|P(\xi)| < \varepsilon$, then we conclude that ξ is "close"—within $\varepsilon^{1/N}$— to an algebraic number α^* of degree at most N. We will further explore and greatly refine this important connection in Section 6.6.

This broad circle of inquiry is known as *Diophantine approximation*—an area of number theory in which one seeks *integer* solutions to certain inequalities. In our case here, the integers we seek are the coefficients of the polynomials that do not vanish but have very small absolute values at ξ in terms of their degrees and heights.

Before searching for infinitely many polynomials that nearly vanish at the complex number ξ in the hopes of attaining a heightened state of transcendence, we first foreshadow these themes by discovering how nearly-vanishing *linear* polynomials lead to irrationality.

6.2 A rational approach to irrationality

We begin by explicitly stating one of the observations made in our intuitive overview. In fact, it actually is a simple case of Liouville's Theorem (Theorem 1.1) that we did not even bother to consider in Chapter 1.

THEOREM 6.1 (A WATERED-DOWN VERSION OF LIOUVILLE'S THEOREM)   *Given a rational number α, there exists a positive constant $c = c(\alpha)$ such that for all integers p and q satisfying $\alpha \neq \frac{p}{q}$, it follows that*

$$c \leq |\alpha q - p|.$$

Proof. If we write r/s for α with $s > 0$, then we note that for all integer pairs (p, q),

$$|\alpha q - p| = \left| \frac{r}{s} q - p \right| = \frac{|rq - ps|}{s}.$$

If we further assume that $\frac{r}{s} \neq \frac{p}{q}$, then we have that $|rq - ps|$ is a *positive* integer. Therefore by the Fundamental Principle of Number Theory we can conclude that $1 \leq |rq - ps|$, which in view of the previous identity, implies that

$$c \leq |\alpha q - p|,$$

where $c = c(\alpha) = \frac{1}{s}$, and completes our short proof of this tiny result.   ∎

Next we consider how well we can approximate a real number τ by *rational numbers*—or the related question: How small can we make $|P(\tau)|$ when $P(z)$ is a *linear* polynomial with integer coefficients?

THEOREM 6.2 (DIRICHLET'S THEOREM) *Given $\tau \in \mathbb{R}$, there exists a constant $C = C(\tau)$ such that for any positive integer H, there exist integers p and q with $0 < \max\{|p|, |q|\} \leq H$ satisfying the inequality*

$$|\tau q - p| < \frac{C}{H}.$$

Moreover, if $H > C$, then $q \neq 0$.

Proof. Our strategy is to apply the Pigeonhole Principle. For a positive integer H, we let $\mathcal{Q}_H$ denote the set of linear polynomials given by

$$\mathcal{Q}_H = \left\{ P(z) \in \mathbb{Z}[z] : P(z) = a_1 z - a_0 , \ 0 \leq a_0 \leq H \ \text{ and } \ 0 \leq a_1 \leq H \right\}.$$

Given the bounds on the integer coefficients for the elements of $\mathcal{Q}_H$, it follows that the number of polynomials in $\mathcal{Q}_H$ equals $(H + 1)^2$. For each $P(z) \in \mathcal{Q}_H$, the triangle inequality implies that

$$|P(\tau)| = |a_1 \tau - a_0| \leq |a_1||\tau| + |a_0| \leq H|\tau| + H = (1 + |\tau|)H,$$

which, letting $B = (1 + |\tau|)H$, can be expressed simply as: For all $P(z) \in \mathcal{Q}_H$,

$$|P(\tau)| \leq B.$$

Next we let S be the closed interval of length $2B$ defined by $S = [-B, B]$. So each of the $(H + 1)^2$ polynomals $P(z) \in \mathcal{Q}_H$ satisfies $P(\tau) \in S$. If we now partition S into, say, s subintervals of equal lengths, then each $P(\tau)$ will reside in one of these subintervals of length $\frac{2B}{s}$. We now select the number of subintervals, s, so that there are *fewer* subintervals than polynomials in $\mathcal{Q}_H$. Given that $\mathcal{Q}_H$ contains $(H + 1)^2$ polynomials, a reasonable choice for s is $s = H^2$. With this choice of s we can invoke the Pigeonhole Principle and conclude that there must exist two *distinct* polynomials $P_1(z)$ and $P_2(z)$ in $\mathcal{Q}_H$ such that the quantities $P_1(\tau)$ and $P_2(\tau)$ are both contained in the same subinterval of length $\frac{2B}{H^2}$. Thus we see that

$$|P_1(\tau) - P_2(\tau)| \leq \frac{2B}{H^2} = \frac{2(1 + |\tau|)}{H}. \tag{6.2}$$

If we write $P_1(z) = a_1'z - a_0'$ and $P_2(z) = a_1''z - a_0''$, and define $p = a_0' - a_0''$ and $q = a_1' - a_1''$, then, given that each coefficient of $P_1(z)$ and $P_2(z)$ is between 0 and H, inclusively, we see that $\max\{|p|, |q|\} \leq H$. Also, if both p and q were equal to 0, then we would have $P_1(z) = P_2(z)$, which is impossible since $P_1(z)$ and $P_2(z)$ are distinct. Thus we have that $0 < \max\{|p|, |q|\} \leq H$ and by (6.2),

$$|\tau q - p| \leq \frac{C}{H}, \tag{6.3}$$

where $C = C(\tau) = 2(1 + |\tau|)$. Finally, we note that if $C < H$ and $q = 0$, then (6.3) would reveal that the integer p satisfies $|p| < 1$. Thus by the Fundamental Principle of Number Theory we must have that $p = 0$ and hence $P_1(z) = P_2(z)$, which again is a contradiction. Therefore if $C < H$, then we see that $q \neq 0$, which completes our proof. ∎

We now search for a Diophantine condition for irrationality. Given a real number τ, we now wish to find the linear polynomial $P(z)$ satisfying $P(\tau) \neq 0$ having its coefficients bounded by some fixed number, and minimizing the *positive* quantity $|P(\tau)|$: The "linear polynomial limbo champion" so to speak. Toward this end, for any positive integer H, we define

$$\Omega(\tau, H) = \min\{|P(\tau)| : P(z) = a_1 z + a_0 \in \mathbb{Z}[z] \text{ with } P(\tau) \neq 0 \text{ and } \mathcal{H}(P) \leq H\},$$

where we recall that the height of P, $\mathcal{H}(P)$, is the maximum absolute value of its coefficients.

If τ is irrational, then for any nonzero $P(z) = a_1 z + a_0 \in \mathbb{Z}[z]$, we are guaranteed that $P(\tau) \neq 0$. Therefore by Theorem 6.2 we see that

$$\Omega(\tau, H) \leq C(\tau)H^{-1}.$$

In order to better compare these values, we define the quantity $\omega(\tau, H)$, which clearly depends on τ and H, to be the exponent satisfying

$$\Omega(\tau, H) = H^{-\omega(\tau, H)}.$$

We recall that our goal is to determine how small $\Omega(\tau, H)$ can be for infinitely many H's. This desire is therefore equivalent to making the exponent $\omega(\tau, H)$ as *large* as possible for infinitely many H's. Unfortunately, we cannot simply compute the limit as H approaches infinity, since $\lim_{H \to \infty} \omega(\tau, H)$ may not exist. Hence we seek the largest accumulation point of $\omega(\tau, H)$ as $H \to \infty$; that is, we wish to find the limit superior, $\lim \sup_{H \to \infty} \omega(\tau, H)$, with the usual understanding that this quantity is declared to equal infinity if there is no largest accumulation point.

Challenge 6.1 *Prove that for all $H > C(\tau)$, we have that $\omega(\tau, H) \geq 0$.*

Thus we see that for all sufficiently large H,

$$H^{-\omega(\tau,H)} \leq CH^{-1},$$

which, after taking logarithms, reveals that

$$1 \leq \frac{\log C}{\log H} + \omega(\tau, H). \tag{6.4}$$

If we let $\omega(\tau) = \lim \sup_{H \to \infty} \omega(\tau, H)$, then, in view of (6.4), we have $1 \leq \omega(\tau)$. Thus we conclude that if τ is irrational, then $\omega(\tau)$ is nonzero.

On the other hand, if τ is rational, then by Theorem 6.1, we have for all H,

$$0 < c(\tau) \leq H^{-\omega(\tau,H)}.$$

Thus, since $\omega(\tau, H) \geq 0$, we discover that $\omega(\tau) = 0$. Hence we have just established the following Diophantine irrationality result:

THEOREM 6.3   *Given a real number τ and positive integer H, let*

$$\Omega(\tau, H) = \min\{|P(\tau)| : P(z) = a_1 z + a_0 \in \mathbb{Z}[z] \quad \text{with } P(\tau) \neq 0 \text{ and } \mathcal{H}(P) \leq H\},$$

and define $\omega(\tau, H)$ by

$$\Omega(\tau, H) = H^{-\omega(\tau,H)}$$

and $\omega(\tau)$ by $\omega(\tau) = \lim \sup_{H \to \infty} \omega(\tau, H)$. Then τ is irrational if and only if $\omega(\tau)$ is nonzero.

With this result as our inspiration, we are ready to extend our analysis to arbitrary polynomials and introduce a dependency on the degrees of the polynomials. Such an extension not only leads to a necessary and sufficient Diophantine condition for transcendence, but, also to a refined measure of transcendence.

6.3 An algebraic approach to transcendence

We begin by extending the first two results from Section 6.2 to polynomials of higher degrees. We note that while the proofs become somewhat more elaborate, the basic ideas and structure are identical to their earlier, simpler counterparts.

THEOREM 6.4 (A GENERALIZED VERSION OF LIOUVILLE'S THEOREM)   *Let α be an algebraic number of degree d, and let N be a positive integer. Then there exists a* *positive constant* $c = c(\alpha, N)$ *such that for all* $P(z) \in \mathbb{Z}[z]$ *satisfying* $\deg(P) \leq N$ *and* $P(\alpha) \neq 0$, *it follows that*

$$\frac{c}{\mathcal{H}(P)^{d-1}} \leq |P(\alpha)|.$$

Proof. Let $f(z) = \sum_{m=0}^{d} a_m z^m \in \mathbb{Z}[z]$ be the minimal polynomial for α, that is, the irreducible polynomial for α with $a_d > 0$, and $\gcd(a_0, a_1, \ldots, a_d) = 1$. If we let $\alpha_1, \alpha_2, \ldots, \alpha_d$ denote the conjugates of α, where $\alpha_1 = \alpha$, then we note that

$$f(z) = a_d \prod_{m=1}^{d} (z - \alpha_m). \tag{6.5}$$

Let $P(z) = \sum_{k=0}^{K} b_k z^k$ be an integral polynomial with $K \leq N$, and let $\beta_1, \beta_2, \ldots, \beta_K$ denote its zeros. Thus we have that

$$P(z) = b_K \prod_{k=1}^{K} (z - \beta_k). \tag{6.6}$$

We now claim that the assumption $P(\alpha) \neq 0$ implies that $\alpha_m \neq \beta_k$ for all indices m and k. To establish this assertion, we assume that for some m and k, $\alpha_m = \beta_k$. Thus we see that $P(\alpha_m) = 0$, which, in turn, reveals that the minimal polynomial for α_m, $f(z)$, must be a factor of the polynomial $P(z)$. Therefore we conclude that $P(\alpha) = P(\alpha_1) = 0$, which is a contradiction and thus confirms our claim that $\alpha_m \neq \beta_k$.

By the previous claim, we see that $\alpha_m - \beta_k \neq 0$ for all m and k. In order to study these differences simultaneously, we examine their product. In particular, we consider the *positive* quantity

$$\prod_{k=1}^{K} \prod_{m=1}^{d} |\alpha_m - \beta_k|.$$

In view of (6.5), we see that

$$\prod_{k=1}^{K} \prod_{m=1}^{d} |\alpha_m - \beta_k| = \prod_{k=1}^{K} \frac{1}{a_d} |f(\beta_k)| = \frac{1}{a_d^K} \prod_{k=1}^{K} |f(\beta_k)|.$$

On the other hand, we can deduce a similar identity by applying (6.6):

$$\prod_{k=1}^{K} \prod_{m=1}^{d} |\alpha_m - \beta_k| = \prod_{m=1}^{d} \prod_{k=1}^{K} |\beta_k - \alpha_m| = \prod_{m=1}^{d} \left| \frac{1}{b_K} P(\alpha_m) \right| = \frac{1}{|b_K|^d} \prod_{m=1}^{d} |P(\alpha_m)|.$$

Putting these two observations together we have that since $\alpha = \alpha_1$,

$$|P(\alpha)| = \frac{|b_K|^d \left| \prod_{k=1}^{K} f(\beta_k) \right|}{a_d^K \prod_{m=2}^{d} |P(\alpha_m)|}. \tag{6.7}$$

Challenge 6.2 *Prove that $\left| b_K^d \prod_{k=1}^{K} f(\beta_k) \right|$ is a positive integer by arguing that the quantity $\prod_{k=1}^{K} f(\beta_k)$ is a symmetric function in $\beta_1, \beta_2, \ldots, \beta_K$ and thus can be expressed as a rational number having a denominator of b_K^d. (Hint: Revisit Lemma 3.14 for inspiration.)*

In view of the previous challenge and the Fundamental Principle of Number Theory we conclude that

$$1 \leq \left| b_K^d \prod_{k=1}^{K} f(\beta_k) \right|,$$

which together with (6.7) and the fact that $K \leq N$ implies

$$\frac{1}{a_d^N \prod_{m=2}^{d} |P(\alpha_m)|} \leq \frac{1}{a_d^K \prod_{m=2}^{d} |P(\alpha_m)|} \leq |P(\alpha)|. \qquad (6.8)$$

Since we desire a lower bound that depends only on the height of the polynomial $P(z)$, we wish to find an *upper* bound for the $|P(\alpha_m)|$ terms appearing in the denominator of the lower bound of (6.8). By the triangle inequality we have

$$|P(\alpha_m)| \leq \sum_{k=0}^{K} |b_k||\alpha_m|^k \leq \mathcal{H}(P) \left(1 + |\alpha_m| + |\alpha_m|^2 + \cdots + |\alpha_m|^K \right),$$

where we recall that the height of P is given by $\mathcal{H}(P) = \max\{|b_0|, |b_1|, \ldots, |b_K|\}$. If we let $\mathcal{A} = \max\{|\alpha_1|, |\alpha_2|, \ldots, |\alpha_d|\}$, then the previous inequality together with the fact that $K \leq N$ implies

$$|P(\alpha_m)| \leq \mathcal{H}(P) \left(1 + \mathcal{A} + \mathcal{A}^2 + \cdots + \mathcal{A}^N \right).$$

Combining this inequality with (6.8), we conclude that

$$\frac{c}{\mathcal{H}(P)^{d-1}} \leq |P(\alpha)|,$$

where

$$c = c(\alpha, N) = a_d^{-N} \left(1 + \mathcal{A} + \mathcal{A}^2 + \cdots + \mathcal{A}^N \right)^{-(d-1)},$$

which completes our proof.   ∎

Now that we have found a lower bound for $|P(\alpha)|$ that generalizes Liouville's result, we turn our attention to an extension of Theorem 6.2 in this polynomial context.

THEOREM 6.5 (AN EXTENSION OF DIRICHLET'S THEOREM) *Given a complex number ξ and a positive integer N, there exists a constant $C = C(\xi, N)$ such that for any* *positive integer H, there exists a nonzero polynomial $P(z) \in \mathbb{Z}[z]$ with $\deg(P) \leq N$ and $\mathcal{H}(P) \leq H$ satisfying the inequality*

$$|P(\xi)| < \frac{C}{H^{\frac{1}{2}(N-1)}}.$$

Proof. We first remark that if $N = 1$, then we can simply select $C = C(\xi, 1) = 2|\xi|$ and take $P(z) = z$. Clearly, with this choice of $P(z)$, $\mathcal{H}(P) = 1 \leq H$ and

$$|P(\xi)| = |\xi| < 2|\xi| = C(\xi, 1) = \frac{C}{H^0}.$$

Thus we see that the theorem holds for $N = 1$.

We now consider the more interesting case of $N \geq 2$. We again apply the Pigeonhole Principle: For positive integers N and H, we let $\mathcal{Q}_{N, H}$ denote the set of polynomials given by

$$\mathcal{Q}_{N, H} = \left\{ P(z) \in \mathbb{Z}[z] : P(z) = \sum_{n=0}^{N} a_n z^n \quad 0 \leq a_n \leq H \quad \text{for all } n = 0, 1, \ldots, N \right\}.$$

Challenge 6.3 *Verify that the set $\mathcal{Q}_{N, H}$ contains $(H + 1)^{N+1}$ polynomials.*

Applying the triangle inequality as in the proof of Theorem 6.2, it follows that for any polynomial $P(z) \in \mathcal{Q}_{H,N}$,

$$|P(\xi)| \leq B,$$

where

$$B = \left(1 + |\xi| + |\xi|^2 + \cdots + |\xi|^N \right) H.$$

Thus we note that since $|P(\xi)| \leq B$, it follows that $|\mathrm{Re}(P(\xi))| \leq B$ and $|\mathrm{Im}(P(\xi))| \leq B$. If we define S to be the square in the complex plane given by $S = \{x + iy : \max\{|x|, |y|\} \leq B\}$, then it follows that for every $P(z) \in \mathcal{Q}_{N, H}$, the quantity $P(\xi)$ resides in the square S. If we now subdivide S into s^2 subsquares, then the length of the sides of each subsquare would equal $2B/s$. In order to apply the Pigeonhole Principle, we wish the number of polynomials in $\mathcal{Q}_{N, H}$ to exceed the number of subsquares in S. That is, we wish to select the positive integer s so that

$$s^2 < (H + 1)^{N+1}.$$

If we let $s = \left[(H + 1)^{\frac{1}{2}(N+1)} \right] - 1$, where $[x]$ denotes the integer part of x, then indeed we have that s^2 is less than the number of polynomials in $\mathcal{Q}_{N, H}$, since

$$s^2 < \left((H + 1)^{\frac{1}{2}(N+1)} \right)^2 = (H + 1)^{N+1}.$$

Moreover, since $N \geq 2$ and $H \geq 1$, we see that

$$s \geq \left[\sqrt{2}^3 \right] - 1 = 3 - 1 = 2,$$

and thus s is a *positive* integer. By the Pigeonhole Principle, there must exist two distinct polynomials $P_1(z)$ and $P_2(z)$ in $\mathcal{Q}_{N,H}$ such that $P_1(\xi)$ and $P_2(\xi)$ reside in the same $\frac{2B}{s} \times \frac{2B}{s}$ subsquare. Measuring the length of the diagonal of this small subsquare, we see that

$$|P_1(\xi) - P_2(\xi)| \leq \sqrt{2}\, \frac{2B}{s}.$$

The following challenge completes our proof.

Challenge 6.4 *Let $P(z) = P_1(z) - P_2(z)$. Prove that $P(z) \in \mathbb{Z}[z]$ is a nonzero polynomial having $\deg(P) \leq N$ and $\mathcal{H}(P) \leq H$ and satisfying the inequality*

$$|P(\xi)| < \frac{C}{H^{\frac{1}{2}(N-1)}},$$

where $C = C(\xi, N) = 6\sqrt{2}\left(1 + |\xi| + |\xi|^2 + \cdots + |\xi|^N\right)$. ∎

We now wish to examine how small the nonzero quantity $|P(\xi)|$ can be among all polynomials $P(z)$ of degrees bounded by N and heights bounded by H. It this direction, we let

$$\mathcal{P}_{N,H} = \{P(z) \in \mathbb{Z}[z] : \deg(P) \leq N \text{ and } \mathcal{H}(P) \leq H\}$$

and define

$$\Omega(\xi, N, H) = \min\{|P(\xi)| : P(z) \in \mathcal{P}_{N,H}, \; P(\xi) \neq 0\}.$$

In order to better compare $\Omega(\xi, N, H)$ to the exponent $-\frac{1}{2}(N-1)$ in the upper bound of Theorem 6.4—essentially a constant multiple of N—we wish to express $\Omega(\xi, N, H)$ as $H^{-\Theta N}$ for a suitable choice of Θ. This remark leads us to define the exponent $\omega(\xi, N, H)$ so that it satisfies

$$\Omega(\xi, N, H) = H^{-\omega(\xi, N, H)N}.$$

Understanding how *small* the quantity $\Omega(\xi, N, H)$ can be for infinitely many H's and N's is equivalent to determining how *large* the exponent $\omega(\xi, N, H)$ can be made for infinitely many H's and N's.

If ξ is transcendental, then $P(\xi) \neq 0$ for all nonzero polynomials $P(z) \in \mathbb{Z}[z]$. So by Theorem 6.5,

$$H^{-\omega(\xi, N, H)N} < C(\xi, N)H^{-\frac{1}{2}(N-1)},$$

which, after taking logarithms, yields

$$\frac{N-1}{2} < \frac{\log C}{\log H} + \omega(\xi, N, H)N. \qquad (6.9)$$

As in our previous investigation of linear polynomials, here we wish to deduce how large $\omega(\xi, N, H)$ can be for infinitely many H's. However, as before, we do not know that $\lim_{H\to\infty}\omega(\xi, N, H)$ even exists. Thus we again look for the largest accumulation point of the values $\omega(\xi, N, H)$ as $H \to \infty$ and N is held fixed; that is, we consider $\limsup_{H\to\infty}\omega(\xi, N, H)$. If we let

$$\omega(\xi, N) = \limsup_{H\to\infty}\omega(\xi, N, H),$$

then it follows from (6.9) that

$$\frac{N-1}{2} \leq \omega(\xi, N)N,$$

or equivalently,

$$\frac{N-1}{2N} \leq \omega(\xi, N).$$

If we now let $\omega(\xi)$ be the largest accumulation point of $\omega(\xi, N)$ as $N \to \infty$; that is, $\omega(\xi) = \limsup_{N\to\infty}\omega(\xi, N)$, then we discover that

$$\frac{1}{2} \leq \omega(\xi).$$

Thus we have just discovered that if ξ is transcendental, then $\omega(\xi)$ is nonzero (in particular, it must be at least $1/2$).

We now investigate the quantity $\omega(\xi)$ in the case that ξ is an algebraic number, say $\xi = \alpha$, where α is of degree d. Let $P(z) \in \mathscr{P}_{N,H}$ be a polynomial satisfying

$$|P(\alpha)| = \Omega(\alpha, N, H) = H^{-\omega(\alpha, N, H)N}.$$

Applying Theorem 6.4, we conclude that

$$\frac{c(\alpha, N)}{H^{d-1}} \leq \frac{c(\alpha, N)}{\mathcal{H}(P)^{d-1}} < |P(\alpha)| = H^{-\omega(\alpha, H, N)N},$$

which, in turn, implies that

$$\omega(\xi, N, H) < \frac{d-1}{N} - \frac{\log c(\alpha, N)}{N\log H}.$$

Hence we have

$$\omega(\xi, N) = \limsup_{H\to\infty}\omega(\xi, N, H) \leq \frac{d-1}{N},$$

which yields

$$\omega(\xi) = \limsup_{N \to \infty} \omega(\xi, N) = 0.$$

The previous bounds for $\omega(\xi)$ imply the following criterion for transcendence, which perfectly parallels the irrationality result of Theorem 6.3.

THEOREM 6.6 *Given a complex number ξ and positive integers H and N, let*

$$\Omega(\xi, N, H) = \min \left\{ |P(\xi)| : P(z) \in \mathcal{P}_{N, H}, P(\xi) \neq 0 \right\},$$

and define $\omega(\xi, N, H)$ to be the exponent satisfying

$$\Omega(\xi, N, H) = H^{-\omega(\xi, N, H)N}.$$

If $\omega(\xi, N) = \limsup_{H \to \infty} \omega(\xi, N, H)$ and $\omega(\xi) = \limsup_{N \to \infty} \omega(\xi, N)$, then the number ξ is transcendental if and only if $\omega(\xi)$ is nonzero.

We are now ready to study the critical quantity $\omega(\xi)$ in greater detail. This delicate Diophantine value not only will allow us to distinguish different classes of transcendental numbers, but also will enable us to discover new transcendence results.

6.4 Detecting subtle distinctions among the transcendent

We open with a fundamental question: What are the possible values for $\omega(\xi)$? We have just discovered that $\omega(\xi) = 0$ is equivalent to the statement that ξ is algebraic. If $\omega(\xi)$ is nonzero, then we conclude that ξ is transcendental, but in this situation we have two natural subcases: Since $\omega(\xi)$ is a limit superior, we could have $0 < \omega(\xi) < \infty$ or $\omega(\xi) = \infty$.

We note that there are two different ways of finding ourselves in the case of $\omega(\xi) = \infty$. We recall that

$$\omega(\xi) = \limsup_{N \to \infty} \omega(\xi, N),$$

where the term $\omega(\xi, N)$ is itself a limit superior, namely

$$\omega(\xi, N) = \limsup_{H \to \infty} \omega(\xi, N, H).$$

The following challenge shows that $\omega(\xi) = \infty$ in one of two ways: Either some term $\omega(\xi, N)$ is infinite or the sequence $\omega(\xi, 1), \omega(\xi, 2), \ldots$ has no accumulation points.

Challenge 6.5 *Show that if for some integer N_0, $\omega(\xi, N_0) = \infty$, then $\omega(\xi, N) = \infty$ for infinitely many N.*

Thus our simple classification of all complex numbers into two classes,

- $\omega(\xi) = 0$ (in which case we know that ξ is algebraic)
- $\omega(\xi) \neq 0$ (in which case we know that ξ is transcendental)

has blossomed into a more robust classification consisting of four classes,

- $\omega(\xi) = 0$
- $0 < \omega(\xi) < \infty$
- $\omega(\xi) = \infty$ and there exists an N_0 for which $\omega(\xi, N_0) = \infty$
- $\omega(\xi) = \infty$ and for all N, $\omega(\xi, N) \neq \infty$

where the last three classes form a partition of the set of transcendental numbers.

In order to analyze the last three cases, we first introduce a useful auxiliary quantity $\nu(\xi)$. We write $\nu(\xi)$ for the least positive integer N for which $\omega(\xi, N) = \infty$, with the understanding that if $\omega(\xi, N)$ is finite for all N, then we declare $\nu(\xi) = \infty$. Thus if $\omega(\xi)$ is finite, then certainly $\nu(\xi) = \infty$. However, if $\omega(\xi) = \infty$, then there are two possibilities: $\nu(\xi)$ is either finite or infinite. Thus we have four possibilities corresponding to the four classes above, and they give rise to the following classification, due to Mahler, for a complex number ξ:

If $\quad \omega(\xi) = 0 \quad$ and (thus) $\quad \nu(\xi) = \infty, \quad$ then ξ is called an *A-number*.

If $\quad 0 < \omega(\xi) < \infty \quad$ and (thus) $\quad \nu(\xi) = \infty, \quad$ then ξ is called an *S-number*.

If $\quad \omega(\xi) = \infty \quad$ and $\quad \nu(\xi) < \infty, \quad$ then ξ is called a *U-number*.

If $\quad \omega(\xi) = \infty \quad$ and $\quad \nu(\xi) = \infty, \quad$ then ξ is called a *T-number*.

We remark that Mahler's classification is a subtle refinement of Theorem 6.6, which states that ξ is transcendental if and only if $\omega(\xi)$ is nonzero. While the name "A-number" seems appropriate given that we have already established that A-numbers are precisely the set of algebraic numbers, the significance of the letters S, T, and U appears less clear. Thus, to offer a taste of some transcendence trivia, we now share the following rumor. Mahler selected the letter S to honor his teacher C.L. Siegel—the same Siegel as in Siegel's Lemma (Theorem 4.3). And what is the meaning of the two remaining letters? Absolutely nothing—Mahler simply chose the next two letters in the alphabet.

An intuitive feel for Mahler's classification

Here we strip away all the "lim sups" and try to make sense out of what it means for a number to be an A-, S-, T-, or U-number. The A-numbers are the most easily understood. By Theorem 6.6, we have that $\omega(\xi) = 0$ if and only if α is algebraic, and thus the A-numbers are precisely the algebraic numbers.

We now consider the case $0 < \omega(\xi) < \infty$ and attempt to develop a naïve sense for what it means to be an S-number. If ξ is an S-number, then we have that $\omega(\xi, N)$ is a bounded function of N. That is, there exists a real number $\rho = \rho(\xi)$ such that $\omega(\xi, N) < \rho$ for all N. Thus for any fixed choice of N, we have that $\omega(\xi, N, H) \leq \rho$ for all but finitely many values of H. Hence for all but finitely

Continued

many H's, we conclude that

$$H^{-\rho N} \leq H^{-\omega(\xi, N, H)N} = \Omega(\xi, N, H) = \min\{|P(\xi)| : P(z) \in \mathcal{P}_{N, H}, P(\xi) \neq 0\}.$$
(6.10)

We formalize this insight in the following result.

THEOREM 6.7 *Suppose that $\xi \in \mathbb{C}$ is a transcendental number. Then $0 < \omega(\xi) < \infty$ if and only if there exists a real number $\rho > 0$ such that for each integer $N \geq 1$, there exists a constant $c' = c'(\xi, N) > 0$ such that for all integers $H \geq 1$ and all polynomials $P(z) \in \mathcal{P}_{H,N}$, the inequality*

$$\frac{c'}{H^{\rho N}} < |P(\xi)|$$
(6.11)

holds.

In other words, this theorem asserts that the growth rate of the exponent in Theorem 6.5 cannot be improved. That is, a constant multiple of N is the best possible exponent on H if and only if we have $0 < \omega(\xi) < \infty$. Therefore S-numbers are precisely those transcendental numbers ξ for which $|P(\xi)|$ cannot be made substantially smaller than the upper bound in Theorem 6.5 found by an application of the Pigeonhole Principle.

Proof of Theorem 6.7.   We begin by assuming that $\omega(\xi)$ is finite. Thus there exists a real number ρ such that $\omega(\xi, N) < \rho$ for all N. From our previous observations together with (6.10), we have that for any fixed N, for all but finitely many H's,

$$\frac{1}{H^{\rho N}} \leq \Omega(\xi, N, H) \leq |P(\xi)|,$$
(6.12)

where $P(z)$ is any polynomial in $\mathcal{P}_{H,N}$. If we write $H_1 < H_2 < \cdots < H_L$ for the *finite* list of exceptional H's for the given N, then we note that the number

$$m = \min\{\Omega(\xi, N, H_l) : l = 1, 2, \ldots, L\}$$

is positive.

We now claim that (6.11) holds with the explicit constant

$$c' = c'(\xi, N) = \min\left\{1, \frac{m}{2}H_L^{\rho N}\right\}.$$

Given the definition of c' and inequality (6.12), we see that for any H different from the finite list of H_l's, (6.11) holds. In view of the definition of m and c', it follows that (6.11) also holds for the remaining H's: $H_1, H_2, \ldots, H_L$.

Next we assume that inequality (6.11) holds for some complex number ξ and positive constants ρ and c'. We wish to show that $\omega(\xi)$ is finite and nonzero and so ξ is an S-number. Since ξ is transcendental, we know that $0 < \omega(\xi)$. On the other hand, by (6.11), we have

$$\frac{c'}{H^{\rho N}} < H^{-\omega(\xi, N, H)N},$$

which implies that $\omega(\xi) \leq \rho < \infty$ and completes our proof. ∎

Continued

Now that we have developed a better sense of the two classes for which $\omega(\xi) \neq \infty$, we turn our attention to the remaining two classes—those for which $\omega(\xi) = \infty$. If $\omega(\xi) = \infty$, then the set $\{\omega(\xi, N) : N = 1, 2, 3, \ldots\}$ contains an unbounded subsequence; that is, given any positive real number ρ, there exists a subsequence of positive integers $N_1, N_2, N_3, \ldots$, depending on ρ, such that

$$\omega(\xi, N_k) > \rho, \tag{6.13}$$

for all $k = 1, 2, 3, \ldots$. We now claim that the previous sentence implies that for each N_k, the inequality

$$\omega(\xi, N_k, H) > \rho$$

is satisfied for infinitely many different values of H. To establish this claim, we suppose the contrary; namely, we assume that there exists an N_K satisfying $\omega(\xi, N_K, H) \leq \rho$ for all but finitely many different values of H. Thus the largest accumulation point as $H \to \infty$ must be bounded above by ρ; that is, $\omega(\xi, N_K) \leq \rho$, which contradicts (6.13) and establishes our claim.

Thus we conclude that for any ρ and any N_k satisfying (6.13), for each of infinitely many choices of H associated with N_K, there exists a $P(z) \in \mathcal{P}_{N_k, H}$ such that

$$|P(\xi)| = \Omega(\xi, N_k, H) = H^{-\omega(\xi, N_k, H)N_k} < H^{-\rho N_k}. \tag{6.14}$$

Challenge 6.6 *Prove that in the case $\omega(\xi) = \infty$, Theorem 6.5 can be improved as follows:*

THEOREM 6.8 *Suppose that $\xi \in \mathbb{C}$. Then $\omega(\xi) = \infty$ if and only if given any positive, unbounded function $f : \mathbb{N} \to \mathbb{R}^{+}$, there exists an infinite sequence of polynomials $P_k(z) \in \mathcal{P}_{N_k, H_k}$ having increasing heights such that for each index k,*

$$|P_k(\xi)| < \frac{1}{H_k^{f(k)N_k}}.$$

Moreover, if $\nu(\xi) = \infty$, then for each k, $P_k(z)$ can be selected such that $N_k < \log \mathcal{H}(P_k)$.

We now consider the two *subcases* in which $\omega(\xi) = \infty$:

$$\nu(\xi) \text{ finite; and } \nu(\xi) \text{ infinite.}$$

If $\nu(\xi) < \infty$, then there exists an $\tilde{N}$ such that for any $\rho > 0$, the inequality $\omega(\xi, \tilde{N}, H) > \rho$ is satisfied for infinitely many different values of H. Thus in this case we can find an infinite sequence of positive integers $\{H_l\}$ such that $\omega(\xi, \tilde{N}, H_l) \to \infty$. That is, the previous theorem can be refined as follows.

Continued

THEOREM 6.9 *Suppose that $\xi \in \mathbb{C}$ with $\omega(\xi) = \infty$. Then $\nu(\xi) < \infty$; that is, ξ is a U-number, if and only if there exists a positive integer $\tilde{N}$ such that given any positive, unbounded function $f : \mathbb{N} \to \mathbb{R}^+$, there exists an infinite sequence of polynomials $P_k(z) \in \mathcal{P}_{\tilde{N}, H_k}$ having increasing heights such that for each index k,*

$$|P_k(\xi)| < \frac{1}{H_k^{f(k)\tilde{N}}}.$$

Therefore the U-numbers are precisely those ξ's for which we can find infinitely many polynomials $P(z) \in \mathbb{Z}[z]$ all of degrees bounded by some universal constant such that $|P(\xi)|$ can be made *substantially* smaller than the upper bound of Theorem 6.5.

Finally, we remark that if $\omega(\xi) = \nu(\xi) = \infty$, then for all N, we have that $\omega(\xi, N, H)$ is a bounded function of H. That is, for each N, there exists a $\rho > 0$ such that for all H, $\omega(\xi, N, H) \leq \rho$. Thus T-numbers are precisely those ξ's for which we have infinitely many polynomials $P(z) \in \mathbb{Z}[z]$ such that $|P(\xi)|$ can be made *substantially* smaller than the upper bound of Theorem 6.5 but the degrees of those amazing polynomials *must* be unbounded.

As foreshadowed in our informal discussion from Section 6.1, we now revisit the deep connection between Mahler's classification and the notion of how well we can approximate a complex number by algebraic numbers. Thus we return to our first example of transcendental numbers—the Liouville numbers from Chapter 1. Since these are numbers that have amazing approximations by "modest" rational numbers, we see that they nearly vanish at linear polynomials at an incredible rate; thus it appears that they should be U-numbers. We now verify this hunch.

THEOREM 6.10 *Liouville numbers are U-numbers.*

Remark.   We note that Liouville numbers are precisely those for which the integer $\tilde{N}$ in Theorem 6.9 equals 1.

Proof of Theorem 6.10.   Let ξ be a Liouville number. Thus by definition there exists an infinite sequence of rational numbers p_n/q_n, with $q_n \geq 1$, satisfying

$$\left| \xi - \frac{p_n}{q_n} \right| < \frac{1}{q_n^n}.$$

If we write $P_n(z) = q_n z - p_n$, then we see that

$$|\xi q_n - p_n| = |P_n(\xi)| < \frac{1}{q_n^{n-1}} \leq 1.$$

In order to introduce the height $\mathcal{H}(P_n)$ into the upper bound, we require a bound on $|p_n|$. However, using the previous inequality together with the triangle inequality, we observe that

$$|p_n| = |-\xi q_n + p_n + \xi q_n| \le |\xi q_n - p_n| + |\xi q_n| \le 1 + |\xi q_n| \le (1 + |\xi|)|q_n|.$$

Therefore, if we write H_n for $\mathcal{H}(P_n)$, then we have that for all n,

$$|P_n(\xi)| < \frac{(1 + |\xi|)^{n-1}}{H_n^{n-1}}.$$

Given that $H^{-\omega(\xi,1,H)} = \min\{|P(\xi)| : P(z) \in \mathcal{P}_{1,H}, P(\xi) \ne 0\}$, we see that

$$H_n^{-\omega(\xi,1,H_n)} < (1 + |\xi|)^{n-1} H_n^{-(n-1)},$$

which, after taking logarithms, implies that

$$n - 1 < \frac{\omega(\xi, 1, H_n) \log H_n}{\log H_n - \log(1 + |\xi|)}.$$

If we now let $n \to \infty$ and look for the largest accumulation point of $\omega(\xi, 1, H_n)$, we discover that $\omega(\xi, 1) = \limsup_{H \to \infty} \omega(\xi, 1, H) = \infty$. Thus clearly $\omega(\xi) = \limsup_{N \to \infty} \omega(\xi, N) = \infty$ and $\nu(\xi) = 1$. Hence ξ is a U-number.   ∎

6.5 A critical consequence of the classification

While Mahler's classification, upon first inspection, does not appear to be particularly useful in practice, here we will discover that it is, in fact, natural and provides the means to establish the transcendence of a wide array of numbers. The importance of Mahler's partition of complex numbers is illustrated in the following important result.

THEOREM 6.11 *If two numbers are algebraically dependent, then they are elements of the same Mahler class.*

We note that the contrapositive of this theorem enables us to produce the following attractive transcendence result.

COROLLARY 6.12 *Suppose that ξ and ζ are two transcendental numbers having different Mahler classifications. If $F(x, y) \in \mathbb{Z}[x, y]$ is a nonzero polynomial, then the number $F(\xi, \zeta)$ is transcendental.*

Proof. Suppose that the quantity $F(\xi, \zeta)$ is an algebraic number. By Theorem 6.11 we know that ξ and ζ are algebraically independent. Thus we know that $F(\xi, \zeta)$ is a *nonzero* algebraic number. Therefore there exists a nonzero irreducible polynomial $p(z) \in \mathbb{Z}[z]$ such that $p(F(\xi, \zeta)) = 0$.

Challenge 6.7 *Show that the polynomial $p(F(x, y)) \in \mathbb{Z}[x, y]$ is nonzero.*

Hence by the previous challenge we conclude that ξ and ζ are algebraically dependent, which contradicts that fact that they are in different Mahler classes. Thus $F(\xi, \zeta)$ is a transcendental number.   ∎

The intuitive idea behind the proof of Theorem 6.11

The basic idea behind the proof of this theorem is fairly simple. We suppose that ξ and ζ are algebraically dependent; so there exists a nonzero polynomial $Q(x, y) \in \mathbb{Z}[x, y]$ satisfying $Q(\xi, \zeta) = 0$. We wish to show that ξ and ζ have the same Diophantine approximation properties.

Suppose that $\alpha_1, \alpha_2, \ldots$ is a sequence of algebraic numbers that approximate ξ. If we take one of these algebraic values, say α_j, then $|Q(\alpha_j, \zeta)|$ is small. Factoring the polynomial $P(y) = Q(\alpha_j, y)$ over $\mathbb{Q}(\alpha_j)$ allows us to conclude that ζ must be close to one of the zeros of $P(y)$, say β_j. We are now faced with the challenging task of showing that the algebraic numbers β_j approximate ζ neither better nor worse than how well the algebraic numbers α_j approximate ξ.

Proof of Theorem 6.11.   Suppose that ξ and ζ are algebraically dependent and let $Q(x, y) = \sum_{l=0}^{L} \sum_{m=0}^{M} c_{lm} x^l y^m \in \mathbb{Z}[x, y]$ be a nonzero irreducible polynomial satisfying $Q(\xi, \zeta) = 0$.

We first dispense with the simple case in which either ξ or ζ is algebraic. If ξ is an algebraic number, then we write $\xi_1, \xi_2, \ldots, \xi_K$ for its conjugates, where $\xi = \xi_1$. Thus by the hypothesis of the theorem we clearly have that

$$\prod_{k=1}^{K} Q(\xi_k, \zeta) = Q(\xi, \zeta) \prod_{k=2}^{K} Q(\xi_k, \zeta) = 0.$$

We also observe that $\prod_{k=1}^{K} Q(\xi_k, \zeta)$ is a symmetric function in the conjugates $\xi_1, \xi_2, \ldots, \xi_K$. Thus by Theorem 3.13, there exists a nonzero polynomial $P(z) \in \mathbb{Z}[z]$ satisfying $P(\zeta) = 0$. Hence ζ is also algebraic, and therefore both ξ and ζ are A-numbers.

We now assume that both ξ and ζ are transcendental numbers. If we define the polynomial $F(x) = Q(x, \zeta) \in (\mathbb{Z}[\zeta])[x]$, then we see that ξ is a zero of F. We caution that F plainly is not in $\mathbb{Z}[x]$. Let $\xi_1, \xi_2, \ldots, \xi_L$ be the zeros of $F(x)$, with $\xi_1 = \xi$. We now claim that ξ_l is a transcendental number for each l. To establish this claim, we assume that for some l, ξ_l is algebraic and let $f(x) \in \mathbb{Z}[x]$ be its minimal polynomial. We will soon show that $f(x)$ is a factor of $Q(x, y)$, which would contradict the fact that $Q(x, y)$ is irreducible.

Let $\xi_{l1}, \xi_{l2}, \ldots, \xi_{lK}$ be the conjugates of the presumed algebraic number ξ_l, where $\xi_l = \xi_{l1}$. If we now write

$$Q(x, y) = \sum_{m=0}^{M} \left(\sum_{l=0}^{L} c_{lm} x^l \right) y^m = \sum_{m=0}^{M} B_m(x) y^m,$$

then we note that $B_m(x) \in \mathbb{Z}[x]$ and

$$Q(\xi_l, \zeta) = Q(\xi_{l1}, \zeta) = \sum_{m=0}^{M} B_m(\xi_{l1})\zeta^m = 0. \tag{6.15}$$

Next we observe that the function

$$\prod_{k=1}^{K} Q(\xi_{lk}, y)$$

is symmetric in the conjugates $\xi_{l1}, \xi_{l2}, \ldots, \xi_{lK}$. Thus again by Theorem 3.13, we conclude that $\prod_{k=1}^{K} Q(\xi_{lk}, y) \in \mathbb{Q}[y]$, and therefore after multiplying by a suitable positive integer to clear any denominators of the coefficients, we have that

$$G(y) = d \prod_{k=1}^{K} Q(\xi_{lk}, y)$$

is a polynomial in $\mathbb{Z}[y]$. In view of the fact that $Q(\xi, \zeta) = 0$, we note that $G(\zeta) = 0$, which given that ζ is transcendental, implies that the polynomial $G(y)$ must be the identically zero polynomial. Thus we have that $Q(\xi_{lk}, y)$ must be identically zero for some k; that is, every coefficient of $Q(\xi_{lk}, y)$ must equal zero. Hence we conclude that $B_m(\xi_{lk}) = 0$ for all $m = 0, 1, \ldots, M$. Since $f(x)$ is the minimal polynomial for ξ_{lk}, we discover that $f(x)$ is a factor of $B_m(x)$ for all m. Thus there exist polyonimals $B_m^*(x)$ in $\mathbb{Z}[x]$ such that $B_m(x) = f(x)B_m^*(x)$. Therefore we discover that

$$Q(x, y) = \sum_{m=0}^{M} B_m(x)y^m = f(x)\left(\sum_{m=0}^{M} B_m^*(x)y^m\right),$$

which contradicts the fact that $Q(x, y)$ is irreducible. Thus we have established our claim that ξ_l is transcendental for each l.

Now given positive integers N and H, let $P(z) = \sum_{n=0}^{N} a_n z^n \in \mathcal{P}_{N,H}$ be a polynomial satisfying

$$\Omega(\xi, N, H) = |P(\xi)| = H^{-\omega(\xi, N, H)N}. \tag{6.16}$$

If we write $\mathcal{A} = \prod_{l=1}^{L} P(\xi_l)$, then in view of (6.16) and the triangle inequality we see that

$$|\mathcal{A}| = \prod_{l=1}^{L} |P(\xi_l)| = |P(\xi_1)| \prod_{l=2}^{L} |P(\xi_l)| = |P(\xi)| \prod_{l=2}^{L} \left| \sum_{n=0}^{N} a_n \xi_l^n \right|$$

$$\leq |P(\xi)| \prod_{l=2}^{L} \left(H \sum_{n=0}^{N} |\xi_l|^n \right) = H^{-\omega(\xi, N, H)N + L - 1} \prod_{l=2}^{L} \left(\sum_{n=0}^{N} |\xi_l|^n \right)$$

$$= cH^{-\omega(\xi, N, H)N + L - 1},$$

where c is defined by $c = c(\xi, \zeta, Q, N) = \prod_{l=2}^{L} \left(\sum_{n=0}^{N} |\xi_l|^n \right)$. On the other hand, $\mathcal{A}$ is a symmetric function of $\xi_1, \xi_2, \ldots, \xi_L$, which we recall are the zeros of $F(x) =$

$Q(x, \zeta)$. The function $F(x) = Q(x, \zeta)$ can be expressed as $F(x) = \sum_{l=0}^{L} D_l(\zeta)x^l$, where $D_l(y) = \sum_{m=0}^{M} c_{lm}y^m$. We note that $D_0(\zeta) \neq 0$, since if $D_0(y)$ vanished at the transcendental number ζ, then $D_0(y)$ would have to be the identically zero polynomial, and thus $Q(x, y)$ could be factored as $Q(x, y) = x \sum_{l=1}^{L} D_l(y)x^{l-1}$, which again contradicts the fact that $Q(x, y)$ is irreducible. Thus we observe that

$$F(x) = D_0(\zeta) \sum_{l=0}^{L} \left(\frac{D_l(\zeta)}{D_0(\zeta)} \right) x^l = D_0(\zeta) \prod_{l=1}^{L} (x - \xi_l).$$

Therefore by Theorems 3.12 and 3.13, we conclude that the quantity $\mathcal{A}$ is a polynomial with integer coefficients of degree at most N evaluated at $D_l(\zeta)/D_0(\zeta)$ for $l = 1, 2, \ldots, L$.

Challenge 6.8 *Show that there exists a polynomial $R(y) \in \mathbb{Z}[y]$ with $\deg(R) \leq MN$ and $\mathcal{H}(R) \leq \kappa_1 H^L$ satisfying $R(\zeta) = D_0(\zeta)^N \mathcal{A}$, where the constant $\kappa_1 = \kappa_1(\xi, \zeta, Q, N)$ does not depend upon H.*

We note that $R(\zeta) \neq 0$, since otherwise, $\mathcal{A}$ would equal zero, which is impossible, since we have established that the numbers ξ_l are all transcendental. Thus we have that

$$\Omega(\zeta, MN, \kappa_1 H^L) \leq |R(\zeta)| = |D_0(\zeta)^N \mathcal{A}| \leq |D_0(\zeta)|^N c H^{-\omega(\xi, N, H)N + L - 1}.$$

By definition we have

$$\Omega(\zeta, MN, \kappa_1 H^L) = (\kappa_1 H^L)^{-\omega(\zeta, MN, \kappa_1 H^L)MN};$$

thus the previous inequality, after taking logarithms and dividing by $\log H$, implies that

$$\omega(\xi, N, H)N - L + 1 \leq \frac{\log \left(|D_0(\zeta)|^N c \right)}{\log H} + \omega(\zeta, MN, \kappa_1 H^L)LMN$$

$$+ \frac{\omega(\zeta, MN, \kappa_1 H^L) \log \kappa_1}{\log H}.$$

Taking the limit superior as $H \to \infty$ reveals that

$$\omega(\xi, N)N - L + 1 \leq \omega(\zeta, MN)LMN,$$

which after dividing by N yields

$$\omega(\xi, N) + \frac{1 - L}{N} \leq \omega(\zeta, MN)LM.$$

Now taking the limit superior as $N \to \infty$ implies that

$$\omega(\xi) \leq \omega(\zeta)LM.$$

Similiarly, reversing the roles of ξ and ζ above leads to

$$\omega(\zeta) \le \omega(\xi)LM.$$

Therefore we see that $\omega(\xi)$ is finite if and only if $\omega(\zeta)$ is finite.

Challenge 6.9   *Show that $v(\xi)$ is finite if and only if $v(\zeta)$ is finite, thus completing the proof of the theorem.*   ∎

6.6 Which is the most popular class? Granting "Most favored number status"

We have already seen that the A-numbers are precisely the algebraic numbers, while Theorem 6.10 established that all Liouville numbers are U-numbers. What about familiar transcendental numbers such as e and π? It turns out that e is an S-number, while the classification of π remains unknown. We will return to these remarks in Section 6.8, but here we discover that we should not be too shocked by the assertion that e is an S-number, since we will now see that "almost all" numbers are S-numbers.

A set $Z \subseteq \mathbb{C}$ is said to have *measure zero* if given any $\varepsilon > 0$, there exists a collection of disks each having a positive radius such that Z is contained in the union of all these disks and the total area of the union of the disks is less than ε. Roughly speaking, the probablity of selecting a random complex number and having it be an element of Z is Zero. Thus "almost all" complex numbers are *not* in Z. This inspires us to say that *almost all* complex numbers are in a particular set X if the the complement $\mathbb{C} \setminus X$ is a set of measure zero.

Intuitively, almost all numbers are in a particular set X precisely when the probability that a randomly selected number is an element of X equals 1. Thus we will discover that if we pick a random number, then it is with probablistic certainty that we have selected an S-number. We warm up to this highly nontrivial observation by first establishing a far more famous but also far weaker result that foreshadows the basic ideas to come.

THEOREM 6.13   *Almost all numbers are transcendental.*

Proof.   The statement of the theorem is equivalent to establishing that the set of all algebraic numbers is a set of measure zero. That is, given $\varepsilon > 0$, we wish to find a collection of disks each having a positive radius such that all algebraic numbers are contained in the union of these disks and the area of this union of the disks is less than ε.

For positive integers N and H, we define the set

$$\mathcal{A}_{N,H} = \{\alpha \in \mathbb{C} : \alpha \text{ is algebraic and } \deg(\alpha) \le N \text{ and } \mathcal{H}(\alpha) \le H\}.$$

Thus given any algebraic number α, we see that for all sufficiently large values of N and H, $\alpha \in \mathcal{A}_{N,H}$. This last remark is crucial in demonstrating that all algebraic numbers can be encircled in disks whose total area is arbitrarily small. Clearly, for fixed integers N and H, any $\alpha \in \mathcal{A}_{N,H}$ is contained in the disk centered at α of radius H^{-3N}. How many such disks are there? Precisely as many as there are algebraic numbers of degree not exceeding N having height not exceeding H. As we have already seen, the number of polynomials in $\mathcal{P}_{N,H}$ is $(2H + 1)^{N+1}$. Each of those polynomials has at most N zeros. Thus the total number of algebraic numbers in $\mathcal{A}_{N,H}$ is bounded from above by

$$N(2H + 1)^{N+1} \leq 3^{N+1} N H^{N+1} \leq 3^{N+1} N H^{2N}. \tag{6.17}$$

Introducing some new notation we can recast our previous observation. We write $\mathcal{D}(\alpha, H^{-3N})$ for the open disk centered at α having radius H^{-3N}; that is,

$$\mathcal{D}\left(\alpha, H^{-3N}\right) = \left\{ z \in \mathbb{C} : |z - \alpha| < H^{-3N} \right\}.$$

For positive integers N and H, we define

$$\mathcal{U}(N, H) = \bigcup_{\alpha \in \mathcal{A}_{N,H}} \mathcal{D}\left(\alpha, H^{-3N}\right).$$

Thus we note that for every $\alpha \in \mathcal{A}_{N,H}$, we have that $\alpha \in \mathcal{D}(\alpha, H^{-3N})$, and our previous observation can be reformulated as: For any $h \geq H$, if $\alpha \in \mathcal{A}_{N,H}$, then $\alpha \in \mathcal{U}(N, h)$. Thus we have that for any positive integer h^*, each algebraic number α satisfies

$$\alpha \in \bigcup_{H=h^*}^{\infty} \bigcup_{N=2}^{\infty} \mathcal{U}(N, H).$$

We now wish to find an upper bound for the area of the (implicit) triple union above and show that for a sufficiently large choice of h^*, that area is less than ε. We begin by providing an upper bound for the area of $\mathcal{U}(N, H)$ by ignoring all intersections and merely summing up the areas of these disks. Thus in view of (6.17) we have

$$\text{Area}(\mathcal{U}(N, H)) \leq \sum_{\alpha \in \mathcal{A}_{N,H}} \text{Area}\left(\mathcal{D}(\alpha, H^{-3N})\right) \leq 3^{N+1} N H^{2N} \text{Area}\left(\mathcal{D}(\alpha, H^{-3N})\right)$$

$$\leq \frac{3^{N+1} N H^{2N} \pi}{H^{6N}} \leq \left(\frac{3^{N+1} N \pi}{H^{2N}}\right) \frac{1}{H^{2N}} \leq \left(\frac{3^{N+1} N \pi}{H^{2N}}\right) \frac{1}{H^2}.$$

Challenge 6.10 *Show that for any $N \geq 2$ and any $H \geq 9$, we have that*

$$\frac{3^{N+1} N \pi}{H^{2N}} \leq \frac{1}{N^2}.$$

Thus for all $N \geq 2$ and $H \geq 9$, we conclude that

$$\text{Area}(\mathcal{U}(N, H)) \leq \frac{1}{N^2 H^2}.$$

In view of the famous identity $\sum_{N=1}^{\infty} \frac{1}{N^2} = \frac{\pi^2}{6}$, we see that

$$\text{Area}\left(\bigcup_{N=2}^{\infty} \mathcal{U}(N,H)\right) \leq \sum_{N=2}^{\infty} \text{Area}(\mathcal{U}(N,H)) \leq \sum_{N=1}^{\infty} \frac{1}{N^2 H^2} = \frac{\pi^2}{6}\frac{1}{H^2}. \qquad (6.18)$$

We now select $h^* > 9$ so large that $\sum_{H=h^*}^{\infty} \frac{1}{H^2} < \frac{6}{\pi^2}\varepsilon$. Thus we see that

$$\text{Area}\left(\bigcup_{H=h^*}^{\infty}\bigcup_{N=2}^{\infty} \mathcal{U}(N,H)\right) \leq \sum_{H=h^*}^{\infty} \text{Area}\left(\bigcup_{N=2}^{\infty} \mathcal{U}(N,H)\right) \leq \sum_{H=h^*}^{\infty} \frac{\pi^2}{6}\frac{1}{H^2} < \varepsilon.$$

Therefore given any $\varepsilon > 0$, we have found that all algebraic numbers are contained within a union of disks having nonzero radii such that the union has area less than ε. Hence we discover that the collection of A-numbers is a set of measure zero, and we conclude that almost all complex numbers are transcendental numbers.   ■

The simplicity of the previous argument is due to the basic fact that the collection of algebraic numbers is countably infinite, and therefore we were able to represent the total area of the disks centered at each algebraic number as an infinite series. The central step in the proof was, given a specified $\varepsilon > 0$, choosing the radii of these disks so that the infinite series converged to a value less than ε.

We now wish to extend the ideas of the previous proof to show that in the world of complex numbers, not only are almost all numbers transcendental, but in fact almost all numbers are S-numbers.

Since the complement of the collection of S-numbers, that is, the union of the collections of all A-, U-, and T-numbers, is an uncountably infinite set, we cannot simply place a small disk around each of these numbers and then show that this encirclement can be acheived in such a manner that the total area contained in all of the disks is smaller than any given $\varepsilon > 0$. Instead, we must demonstrate that since U- and T-numbers are especially well-approximated by algebraic numbers, they each lie in infinitely many of the disks constructed in our proof of Theorem 6.13.

Unfortunately, Mahler's classification scheme is based on the smallness of $|P(\xi)|$ among integral polynomials of a given degree and height, rather than on the quality of approximations of the number ξ by algebraic values. But as we have noted earlier, these two ideas are related. Intuitively it is clear that if $|P(\xi)|$ is small, then ξ should be reasonably close to a zero of $P(z)$. To establish a quantitative version of the previous vague assertion, we require the following two important results, whose elaborate proofs we postpone to the next algebraic excursion.

LEMMA 6.14   *Given a polynomial $P(z) \in \mathbb{Z}[z]$, let $\sigma(P) = \max\{\deg(P), \log \mathcal{H}(P)\}$. Suppose that ξ is a complex number such that there exists a real number $\rho > 0$ satisfying*

$$|P(\xi)| < e^{-\rho \deg(P)\sigma(P)}.$$

Then there is an irreducible factor $Q(z) \in \mathbb{Z}[z]$ of $P(z)$ satisfying

$$|Q(\xi)| \leq e^{-\frac{1}{3}\rho \deg(Q)\sigma(Q)}.$$

LEMMA 6.15 *Suppose that $Q(z)$ is an irreducible polynomial in $\mathbb{Z}[z]$ and $\xi \in \mathbb{C}$. Let α be a zero of $Q(z)$ that is closest to ξ. If $2 \leq \deg(Q) \leq \mathcal{H}(Q)$, then*

$$|\xi - \alpha| \leq \mathcal{H}(Q)^{6 \deg(Q)} |Q(\xi)|.$$

For now, we assume the validity of both lemmas and move closer to our main result. As a first step, we demonstrate that almost all numbers are *not* U-numbers.

THEOREM 6.16 *The set of U-numbers is a set of measure zero.*

Proof. Let $\xi \in \mathbb{C}$ be a U-number and $f(k)$ a positive, increasing, unbounded function. Then by Theorem 6.9, there exist a positive integer $\tilde{N}$ and an infinite sequence of polynomials $P_k(z) \in \mathcal{P}_{N, H_k}$ with increasing heights such that for each index k,

$$|P_k(\xi)| < H_k^{-f(k)\tilde{N}}.$$

By Lemma 6.14 we see that there must exist an irreducible factor $Q_k(z) \in \mathbb{Z}[z]$ of $P_k(z)$ satisfying

$$|Q_k(\xi)| \leq e^{-\frac{1}{3} f(k) \deg(Q_k) \sigma(Q_k)}. \tag{6.19}$$

Since Mahler's classification involves $\mathcal{H}(Q_k)$ rather than $\sigma(Q_k)$, our intermediate goal is to find an infinite sequence of irreducible polynomials $Q_k(z)$ such that (6.19) holds with $\sigma(Q_k)$ replaced by $\log \mathcal{H}(Q_k)$. Toward this end, we now claim that there exists a subsequence $\{k_i\}$ such that

$$\mathcal{H}(Q_{k_1}) < \mathcal{H}(Q_{k_2}) < \mathcal{H}(Q_{k_3}) < \cdots.$$

To verify this claim, we assume that the assertion is false. Thus there exists a constant H such that $\mathcal{H}(Q_k) \leq H$ for all sufficiently large k. Recalling that $\deg(Q_k) \leq \deg(P_k) \leq \tilde{N}$, we can then conclude that for all sufficiently large k, $Q_k(z) \in \mathcal{P}_{N, H}$. Since $\mathcal{P}_{N, H}$ is a finite set of polynomials, by the Pigeonhole Principle we deduce that there must exist a nonzero polynomial $Q(z) \in \mathcal{P}_{N, H}$ such that for an infinite subsequence $\{k_i\}$,

$$|Q(\xi)| \leq e^{-\frac{1}{3} f(k_i) \deg(Q) \sigma(Q)}.$$

Letting $i \to \infty$ reveals that $Q(\xi) = 0$ and thus ξ is algebraic. However, ξ is a U-number and thus is transcendental. This contradiction establishes our claim.

Given the previous claim, we now see that there exists a subsequence $\{k_i\}$ such that for each i,

$$e^{\tilde{N}} < \mathcal{H}(Q_{k_1}) < \mathcal{H}(Q_{k_2}) < \mathcal{H}(Q_{k_3}) < \cdots.$$

Thus since $e^{\deg(Q_{k_i})} \leq e^{\tilde{N}} < \mathcal{H}(Q_{k_i})$, we have that

$$\sigma(Q_{k_i}) = \max\{\deg(Q_{k_i}), \log \mathcal{H}(Q_{k_i})\} = \log \mathcal{H}(Q_{k_i}).$$

In view of (6.19), we have the inequalities that realize our intermediate goal, namely,

$$|Q_{k_i}(\xi)| \leq \mathcal{H}(Q_{k_i})^{-\frac{1}{3}f(k_i)\deg(Q_{k_i})}. \tag{6.20}$$

We now apply Lemma 6.15 to deduce that for each index k_i, there exists a zero of $Q_{k_i}(z)$, say α_{k_i} satisfying $\alpha_{k_i} \in \mathcal{A}_{N,\mathcal{H}(Q_{k_i})}$ and

$$|\xi - \alpha_{k_i}| \leq \mathcal{H}(Q_{k_i})^{6\deg(Q_{k_i})}\,\mathcal{H}(Q_{k_i})^{-\frac{1}{3}f(k_i)\deg(Q_{k_i})}.$$

Given that $f(k)$ is increasing and unbounded, we have that for all sufficiently large i, $f(k_i) > 27$ and hence

$$|\xi - \alpha_{k_i}| < \mathcal{H}(Q_{k_i})^{-3\deg(Q_{k_i})},$$

which in view of the fact that $Q_{k_i}(z)$ is the minimal polynomial for α_{k_i}, can be rewritten as

$$|\xi - \alpha_{k_i}| < \mathcal{H}(\alpha_{k_i})^{-3\deg(\alpha_{k_i})}.$$

As in the proof of Theorem 6.13, for $\zeta \in \mathbb{C}$ and $r > 0$, we again define $\mathcal{D}(\zeta, r)$ to be the open disk in the complex plane centered at ζ and having radius r. Thus our previous observation can be recast as follows: For each U-number ξ, there exist an integer $\tilde{N} = \tilde{N}(\xi)$ and infinitely many distinct algebraic numbers $\alpha_{k_i} \in \mathcal{A}_{N,H_{k_i}}$ such that for all k_i,

$$\xi \in \mathcal{D}\left(\alpha_{k_i}, H_{k_i}^{-3\deg(\alpha_{k_i})}\right).$$

Since for all k_i, $\deg(\alpha_{k_i}) \leq \tilde{N}$, it follows by the Pigeonhole Principle we conclude that there must be an infinite subsequence of $\{\alpha_{k_i}\}$ for which all the elements are of the same algebraic degree. Thus redefining the quantity $\tilde{N}$ and re-indexing the sequence α_{k_i} if necessary, we now assume that for all k_i, $\deg(\alpha_{k_i}) = \tilde{N}$ and

$$\xi \in \mathcal{D}\left(\alpha_{k_i}, H_{k_i}^{-3\tilde{N}}\right).$$

Again, for positive integers N and H, we let

$$\mathcal{U}(N,H) = \bigcup_{\alpha \in \mathcal{A}_{N,H}} \mathcal{D}\left(\alpha, H^{-3N}\right).$$

We remark that given a U-number ξ, for any $N \geq \tilde{N}(\xi)$ and for any sufficiently large H, $\xi \in \mathcal{U}(N,H)$. Thus we have that for any positive integer h^*, each U-number ξ satisfies

$$\xi \in \bigcup_{H=h^*}^{\infty} \bigcup_{N=2}^{\infty} \mathcal{U}(N,H).$$

As we saw in the proof of Theorem 6.13, we can select h^* such that

$$\text{Area}\left(\bigcup_{H=h^*}^{\infty}\bigcup_{N=2}^{\infty}\mathcal{U}(N,H)\right) < \varepsilon.$$

Thus we conclude that the set of U-numbers is a set of measure zero.   ∎

Since we have shown that the set of A-numbers has measure zero and the set of U-numbers has measure zero, then so does their union. Thus almost all numbers are either S- or T-numbers. We now prove more, which establishes our main result of this section.

THEOREM 6.17 *Almost all numbers are S-numbers.*

Proof. Proving this theorem is equivalent to establishing that the union of all A-, T-, and U-numbers is a set of measure zero. Thus we need only show that the T-numbers form a set of measure zero. That is, given $\varepsilon > 0$, we must find a collection of disks each having a positive radius such that all T-numbers are contained in the union of these disks and the area of this union of disks is less than ε. Just as we discovered in the case of U-numbers, here we will see that because T-numbers are sufficiently well-approximated by algebraic numbers, they too are contained in the sets introduced in the proof of Theorem 6.13.

Given a T-number, say $\xi \in \mathbb{C}$, we know by Theorem 6.8 that for any given positive, unbounded function $f(k)$, there exists an infinite sequence of polynomials $P_k(z) \in \mathcal{P}_{N_k, H_k}$ having increasing heights such that for each index k, $N_k < \log \mathcal{H}(P_k)$ and

$$|P_k(\xi)| < \frac{1}{H_k^{f(k)N_k}}. \tag{6.21}$$

Furthermore, we can assume that the sequence of integers H_k and the associated sequence of polynomials $P_k(z) \in \mathcal{P}_{N_k, H_k}$ have been selected so that

$$\mathcal{H}(P_1) < \mathcal{H}(P_2) < \cdots < \mathcal{H}(P_k) < \cdots.$$

Given the previous assumptions, certainly we could weaken inequality (6.21) to deduce that

$$|P_k(\xi)| < \mathcal{H}(P_k)^{-f(k)N_k} \le e^{-f(k)\deg(P_k)\sigma(P_k)}, \tag{6.22}$$

where we recall that $\sigma(P_k) = \max\{\deg(P_k), \log\mathcal{H}(P_k)\}$. Therefore by Lemma 6.14 we conclude that for each k, there exists an irreducible polynomial $Q_k(z) \in \mathcal{P}_{N_k, H_k}$ satisfying the inequality

$$|Q_k(\xi)| \le e^{-\frac{1}{3}f(k)\deg(Q_k)\sigma(Q_k)}. \tag{6.23}$$

We now claim that after finding a suitable subsequence and re-indexing, if necessary, we can assume that

$$\deg(Q_1) < \deg(Q_2) < \deg(Q_3) < \cdots.$$

Here we briefly sketch the idea behind the proof of this claim. If the claim were false, then there must exist a constant M such that $\deg(Q_k) \leq M$ for all k. We first consider the possibility that $\mathcal{H}(Q_k)$ is bounded for all k; that is, for all k, $\mathcal{H}(Q_k) \leq H$ for some H. Then in view of the fact that $f(k) \to \infty$ as $k \to \infty$, we could argue as we did in the proof of Theorem 6.16 to deduce that (6.23) implies that ξ must be algebraic, which again is impossible. Thus we conclude that $\mathcal{H}(Q_k)$ is unbounded, and hence we can find a subsequence $Q_{k_i}(z)$ such that $\mathcal{H}(Q_{k_i}) \to \infty$ as $i \to \infty$. In this case, inequality (6.23) together with Theorem 9.9 can be employed to show that ξ must be a U-number, which is again a contradiction and thus establishes the claim. Moreover, we remark that after finding another suitable subsequence and re-indexing yet again, we can assume that the $Q_k(z)$'s have been selected such that

$$\mathcal{H}(Q_1) < \mathcal{H}(Q_2) < \cdots < \mathcal{H}(Q_k) < \cdots .$$

In view of the previous claims, we can assume that the indices have been selected such that the irreducible polynomials $Q_k(z)$ satisfy

$$|Q_k(\xi)| \leq \mathcal{H}(Q_k)^{-\frac{1}{3}f(k)N_k^*},$$

for some integers $N_1^* < N_2^* < N_3^* < \cdots$, and where $\mathcal{H}(Q_1) < \mathcal{H}(Q_2) < \mathcal{H}(Q_3) < \cdots$. Thus we now conclude that there exist strictly increasing sequences of natural numbers N_k^* and H_k^* such that for each k, there exists an *irreducible* polynomial $Q_k(z) \in \mathscr{P}_{N_k^*, H_k^*}$ satisfying

$$|Q_k(\xi)| \leq H_k^{*-\frac{1}{3}f(k)N_k^*}, \tag{6.24}$$

which is comparable to inequality (6.21) with the additional feature of irreducibility.

We worked so hard to obtain (6.24) for *irreducible* polynomials because we can now apply Lemma 6.15 to deduce that for all sufficiently large k, there exists an algebraic number $\alpha_k \in \mathcal{A}_{N_k^*, H_k^*}$ satisfying

$$|\xi - \alpha_k| \leq H_k^{*6N_k^*}|Q_k(\xi)|,$$

which, in view of inequality (6.24) and the fact that $f(k) \to \infty$ as $k \to \infty$, implies that for all sufficiently large k,

$$|\xi - \alpha_k| \leq H_k^{*-3N_k^*}.$$

Thus we see that for each T-number ξ, there exist infinitely many H's and N's such that for any such pair H and N, there exists an algebraic number $\alpha \in \mathcal{A}_{N, H}$ satisfying

$$\xi \in \mathcal{D}\left(\alpha, H^{-3N}\right).$$

As we have done before, for positive integers N and H, we write

$$\mathcal{U}(N, H) = \bigcup_{\alpha \in \mathcal{A}_{N, H}} \mathcal{D}\left(\alpha, H^{-3N}\right),$$

and remark that our previous observations imply that for each T-number ξ,

$$\xi \in \bigcup_{H=h^*}^{\infty} \bigcup_{N=2}^{\infty} \mathcal{U}(N,H),$$

for all sufficiently large integers h^*. Again, as we established in the proof of Theorem 6.13, we can select h^* such that

$$\text{Area}\left(\bigcup_{H=h^*}^{\infty} \bigcup_{N=2}^{\infty} \mathcal{U}(N,H) \right) < \varepsilon.$$

Therefore given any $\varepsilon > 0$, we have found that all T-numbers are contained within a union of disks having nonzero radii such that the union has area less than ε. Hence we discover that the union of all A-, U-, and T-numbers is a set of measure zero, and thus we conclude that almost all complex numbers are S-numbers. ■

6.7 The proofs of Lemmas 6.14 and 6.15

We now return to the heart of the matter: Proving Lemmas 6.14 and 6.15. Before we turn our attention to the justification of those lemmas, we inspire the development to come through a simple example. If we let

$$P(z) = z^8 - 7z^6 - 4z^5 + 28z^3 + 4z^2 - 28,$$

then we note that

$$\left| P\left(\frac{82}{31}\right) \right| = 0.85070\ldots \quad \text{and} \quad \left| P\left(\frac{4}{3}\right) \right| = 0.71635\ldots.$$

Thus it seems reasonable to assume that 82/31 and 4/3 are near zeros of $P(z)$. This assumption is quickly confirmed once we factor $P(z)$ and observe that, in fact,

$$P(z) = (z^2 - 7)(z^3 - 2)^2,$$

and

$$\left| \frac{82}{31} - \sqrt{7} \right| < 0.00059, \quad \text{while} \quad \left| \frac{4}{3} - \sqrt[3]{2} \right| < 0.07342.$$

A careful examination of the previous estimates offers the following important insight. If ξ is *very* close to a zero of $P(z)$, then $|P(\xi)|$ is small; but if ξ is merely *fairly* close to a zero of $P(z)$ having higher multiplicity, then it still follows that $|P(\xi)|$ is small. In order to quantify this observation, we begin with the following lemma.

LEMMA 6.18 *Let $P(z)$ be a polynomial with integer coefficients and ξ a complex number. Suppose that α is the zero of $P(z)$ closest to ξ. Then*

$$2^{-(\deg(P)-1)}|P'(\alpha)||\xi - \alpha| \le |P(\xi)|.$$

Proof.   If we let N denote the degree of $P(z)$, then factoring $P(z)$ over the complex numbers yields an expression of the form

$$P(z) = a_N \prod_{i=1}^{N} (z - \alpha_i), \qquad (6.25)$$

where, without loss of generality, we assume that α_1 is the zero of $P(z)$ closest to ξ. Thus it follows from the triangle inequality that for each index i, $|\xi - \alpha_i| \geq \frac{1}{2}|\alpha_1 - \alpha_i|$. Therefore we conclude that

$$|P(\xi)| = \left| a_N \prod_{i=1}^{N} (\xi - \alpha_i) \right| \geq |\xi - \alpha_1| |a_N| \prod_{i \neq 1} \frac{|\alpha_1 - \alpha_i|}{2}$$

$$= 2^{-(N-1)} |\xi - \alpha_1| |a_N| \prod_{i \neq 1} |\alpha_1 - \alpha_i|.$$

However, applying the familiar product rule from calculus to (6.25), we find that the product $|a_N \prod_{i \neq 1} (\alpha_1 - \alpha_i)|$ is precisely $|P'(\alpha_1)|$, which completes our proof. ∎

In view of the lemma, if $|P'(\alpha)|$ is nonzero, then we see that

$$|\xi - \alpha| \leq 2^{\deg(P)-1} \frac{|P(\xi)|}{|P'(\alpha)|}.$$

Thus if $|P(\xi)|$ is *very* small as compared to $2^{\deg(P)-1}/|P'(\alpha)|$, then ξ is near a zero α of $P(z)$. We will see below that if $P(z)$ is an *irreducible* polynomial with integer coefficients having α as a zero, then $|P'(\alpha)|$ is neither zero nor particularly small.

Specifically, our program is to first establish Lemma 6.14; that is, to show that if $|P(\xi)|$ is sufficiently small, then $P(z)$ has an irreducible factor $Q(z)$, having a degree and height that we can bound, for which $|Q(\xi)|$ remains relatively small. Since $Q(z)$ is irreducible, we are certain that it has no multiple zeros. Next, we establish a lower bound for the nonzero value $|Q'(\alpha)|$, where α is the zero of $Q(z)$ closest to ξ. Finally, we deduce the inequality of Lemma 6.15 and conclude that ξ is well-approximated by α. We carry out this plan of attack in the following fairly lengthy excursion. Some readers may wish to postone the details contained in the excursion for another day.

Algebraic Excursion: Algebraic approximations and polynomials with small moduli

Our first goal is to apply the fact that the quantity $|P(\xi)|$ is relatively small as compared to the degree and height of $P(z)$ to find an irreducible polynomial $Q(z)$ such that $|Q(\xi)|$ is also small when measured in terms of the degree and height of $Q(z)$. Since the irreducible polynomial $Q(z)$ will be a factorof $P(z)$, we first

Continued

require a relationship between the degree and height of $P(z)$ and those of $Q(z)$. Clearly, $\deg(Q) \le \deg(P)$, but $Q(z)$ might have larger coefficients than $P(z)$. The following result provides an upper bound for the height of a factor $Q(z)$ in terms of the height of $P(z)$.

PROPOSITION 6.19 *Suppose $P(z) \in \mathbb{Z}[z]$ is a polynomial of degree N. If $Q(z) \in \mathbb{Z}[z]$ is an irreducible factor of $P(z)$, then*

$$\deg(Q) \le N \quad \text{and} \quad \mathcal{H}(Q) \le 4^N \mathcal{H}(P).$$

Preparing for the proof.   The proof of this simply stated proposition relies on the observation made in Chapter 3 that the coefficients of a monic polynomial can be expressed in terms of the elementary symmetric functions evaluated at the zeros of the polynomial. Thus, perhaps not surprisingly, we require a deeper understanding of the relationship between the height of a polynomial and the absolute value of the product of any number of its zeros. Toward this end, we have the following strong result that applies not only to the product of all the zeros of $P(z)$ but also to a product of *any* number of zeros of $P(z)$.

LEMMA 6.20 *Let $P(z) \in \mathbb{C}[z]$ be a polynomial written in a factored form as $P(z) = a_N \prod_{i=1}^{N}(z - \alpha_i)$. If $\mathcal{H}(P) \ge 1$, then*

$$|a_N| \prod_{i=1}^{N} \max\{1, |\alpha_i|\} \le 2^N \mathcal{H}(P). \tag{6.26}$$

Proof.   Without loss of generality we may assume that $a_N = 1$. We establish this lemma by induction on the number of zeros of $P(z)$ that have absolute value greater than 2. If $P(z)$ has no such zeros, then for each zero α_i of $P(z)$, $\max\{1, |\alpha_i|\} \le 2$, and so inequality (6.26) holds. Thus we suppose that $P(z)$ has at least one zero α with $|\alpha| > 2$ and factor $P(z)$ as

$$P(z) = R(z)(z - \alpha),$$

where $R(z) = z^{N-1} + r_{N-2}z^{N-2} + \cdots + r_0 \in \mathbb{C}[z]$.

Since we are applying induction on the number of zeros of $P(z)$ having absolute value greater than 2, and $R(z)$ has fewer such zeros than $P(z)$, our first goal is to relate $\mathcal{H}(P)$ with $\mathcal{H}(R)$. This relationship is not too difficult to establish, since every coefficient of $P(z)$ is of the form $r_{j-1} - \alpha r_j$, with the understanding that we declare $r_{-1} = 0$. If we focus on the coefficient of $R(z)$ having the smallest subscript with $\mathcal{H}(R) = |r_j|$, then $|r_j| > |r_{j-1}|$, and from the factorization of $P(z)$ we conclude that

$$\mathcal{H}(P) \ge |r_{j-1} - \alpha r_j| \ge |\alpha r_j| - |r_{j-1}| > |\alpha r_j| - |r_j| > |r_j|(|\alpha| - 1) \ge \frac{1}{2}|\alpha||r_j|.$$

Continued

This sequence of inequalities implies that

$$2\mathcal{H}(P) > |\alpha|\mathcal{H}(R).$$

We now apply our induction hypothesis to $R(z)$ and multiply the inequality

$$\prod_{\alpha_j \text{ a root of } R} \max\{1, |\alpha_j|\} \le 2^{N-1}\mathcal{H}(R)$$

by the quantity $\max\{1, |\alpha|\}$ to obtain the inequality

$$\prod_{j=1}^{N} \max\{1, |\alpha_j|\} \le \max\{1, |\alpha|\}2^{N-1}\mathcal{H}(R) \le 2^N \mathcal{H}(P),$$

which completes the proof.   ∎

Proof of Proposition 6.19.   Let $d_Q = \deg(Q)$ and $H_Q = \mathcal{H}(Q)$, and for notational simplicity let us suppose that we have ordered the zeros of $P(z)$ so that $Q(z) = b_{d_Q} \prod_{j=1}^{d_Q}(z - \alpha_j)$. We seek to estimate the absolute value of an arbitrary coefficient b_j of $Q(z)$, which we express in terms of the zeros of $Q(z)$. Explicitly, if we write $Q(z) = b_{d_Q}z^{d_Q} + \cdots + b_1 z + b_0$, then b_j/b_{d_Q} is ± 1 times the sum of all possible products of the zeros of $Q(z)$ taken $d_Q - j$ at a time. If we let

$$M_j = \max\{|\alpha_{i_1}\alpha_{i_2}\cdots\alpha_{i_{d_Q-j}}| : 1 \le i_1 < i_2 < \cdots < i_{d_Q-j} \le d_Q\},$$

then

$$\left|\frac{b_j}{b_{d_Q}}\right| \le \binom{d_Q}{d_Q - j}M_j \le 2^{d_Q}M_j \le 2^{d_Q} \prod_{\alpha_i \text{ a zero of } Q(z)} \max\{1, |\alpha_i|\}$$

$$\le 2^{d_Q} \prod_{i=1}^{N} \max\{1, |\alpha_i|\}.$$

In view of Lemma 6.20 and the inequality $|b_{d_Q}| \le |a_N|$, we conclude that

$$\mathcal{H}(Q) \le |b_{d_Q}|2^{N+d_Q}\frac{1}{|a_N|}\mathcal{H}(P) \le 4^N \mathcal{H}(P),$$

which completes our proof.   ∎

Thus we have established an upper bound for the height of an irreducible factor of a given polynomial with integer coefficients in terms of the degree and height of the given polynomial. We are now ready to prove Lemma 6.14, which connects the size of $|P(\xi)|$ with the size of $|Q(\xi)|$. For clarity, we restate the result below.

Continued

LEMMA 6.14   *Given a polynomial $P(z)$ in $\mathbb{Z}[z]$, let $\sigma(P) = \max\{\deg(P),$ $\log \mathcal{H}(P)\}$. Suppose that ξ is a complex number such that there exists a real number $\rho > 0$ satisfying*

$$|P(\xi)| < e^{-\rho \deg(P)\sigma(P)}.$$

Then there is an irreducible factor $Q(z) \in \mathbb{Z}[z]$ of $P(z)$ satisfying

$$|Q(\xi)| \le e^{-\frac{1}{3}\rho \deg(Q)\sigma(Q)}.$$

Challenge 6.11   *Prove Theorem 6.14. (Hint: First note that if $P(z)$ is irreducible, then the result is immediate. Thus we may assume that $P(z)$ is not irreducible and factor it into irreducible integral polynomials $P(z) = Q_1(z) \cdots Q_s(z)$. Now apply Proposition 6.19 to deduce the following inequality: $\sigma(Q_i) \le 3\sigma(P)$. Next suppose that for each index i, $|Q_i(\xi)| > e^{-\frac{1}{3}\rho d_{Q_i}\sigma(Q_i)}$. Finally, use these assumed inequalities and the fact that $|P(\xi)| = |Q_1(\xi)| \cdots |Q_s(\xi)|$ to obtain a contradiction to the known upper bound on $|P(\xi)|$.)*

As we turn our attention to Lemma 6.15, we see that by Lemma 6.18, for an irreducible polynomial $Q(z) \in \mathbb{Z}[z]$ and a complex number $\xi \in \mathbb{C}$, if α is the zero of $Q(z)$ that is closest to ξ, then

$$|\xi - \alpha| \le 2^{\deg(Q)-1} \frac{|Q(\xi)|}{|Q'(\alpha)|}.$$

Thus in order to produce our upper bound in Lemma 6.15, we need a lower bound for $|Q'(\alpha)|$.

A lower bound for $|Q'(\alpha)|$—A result of the resultant of two polynomials.   We begin with a useful condition for when two polynomials share a common zero and then discover that this condition can be verified by computing a determinant. We show that when this determinant is nonzero, the Fundamental Principle of Number Theory allows us to produce a lower bound for the quantity $|Q'(\alpha)|$.

LEMMA 6.21   *Suppose that $P(z)$ and $Q(z)$ are polynomials with integer coefficients having degrees N and M, respectively. Then $P(z)$ and $Q(z)$ share a common zero if and only if there exist nonzero polynomials $U(z)$ and $V(z)$, with $\deg(U) \le M - 1$, $\deg(V) \le N - 1$, satisfying*

$$U(z)P(z) = V(z)Q(z).$$

Proof.   We first suppose that $P(z)$ and $Q(z)$ share a common zero, say α, whose minimal polynomial is $G(z)$. Then there exist polynomials $U(z)$ and $V(z)$ in $\mathbb{Z}[z]$ such that

$$G(z)V(z) = P(z) \quad \text{and} \quad G(z)U(z) = Q(z).$$

Thus we have $U(z)P(z) = V(z)Q(z)$, as desired.

Continued

Conversely, we now assume that there exist nonzero polynomials $U(z)$ and $V(z)$, with $\deg(U) \leq M - 1$, $\deg(V) \leq N - 1$, such that $U(z)P(z) = V(z)Q(z)$. Clearly, each of the N zeros of $P(z)$ is a zero of the product $V(z)Q(z)$. However, since $V(z)$ has fewer than N zeros, it follows that one of the zeros of $P(z)$ must be a zero of $Q(z)$, which completes our proof.   ∎

We next define the *resultant* of the polynomials $P(z) = \sum_{i=0}^{N} a_i z^i$ and $Q(z) = \sum_{j=0}^{M} b_j z^j$, denoted by $\mathrm{Res}(P, Q)$, to be the determinant of the $(N + M) \times (N + M)$ matrix whose first M rows are the coefficients of $P(z)$, each subsequent row shifted one unit to the right, and whose last N rows are the coefficients of $Q(z)$, similarly shifted. For example, if $M = N$, then the matrix is:

$$
\mathcal{R} = \begin{pmatrix}
a_N & a_{N-1} & \cdots & a_0 & & & & \\
 & a_N & a_{N-1} & \cdots & a_0 & & & \\
 & & a_N & a_{N-1} & \cdots & a_0 & & \\
 & & & \ddots & & & \ddots & \\
 & & & & a_N & a_{N-1} & \cdots & a_0 \\
b_M & b_{M-1} & \cdots & b_0 & & & & \\
 & b_M & b_{M-1} & \cdots & b_0 & & & \\
 & & & \ddots & & & & \ddots \\
 & & & & b_M & b_{M-1} & \cdots & b_0
\end{pmatrix}.
$$

Challenge 6.12 *By treating the coefficients of the polynomials* $U(z) = \sum_{i=0}^{M-1} u_i z^i$ *and* $V(z) = \sum_{j=0}^{N-1} v_j z^j$ *as unknowns, show that* $U(z)P(z) = V(z)Q(z)$ *if and only if* $\mathrm{Res}(P, Q) = 0$. *(Hint: Show that the existence of the nonzero polynomials* $U(z)$ *and* $V(z)$ *is equivalent to the matrix equation*

$$
(u_{M-1} \; u_{M-2} \; \cdots \; u_0 \; -v_{N-1} \; -v_{N-2} \; \cdots \; -v_0)\mathcal{R} = (0 \; 0 \; \cdots \; 0).)
$$

Thus from the two previous results we conclude the important fact that the polynomials $P(z)$ and $Q(z)$ share a common zero if and only if $\mathrm{Res}(P, Q) = 0$.

Yet another critical application of the Fundamental Principle of Number Theory. We are finally able to establish an arithmetic lower bound for $|Q'(\alpha)|$ and thereby deduce a good algebraic approximation to ξ.

PROPOSITION 6.22 *If* $Q(z) \in \mathbb{Z}[z]$ *is an irreducible polynomial of degree* d_Q *having* α *as a zero, then*

$$
\left(d_Q \mathcal{H}(Q)\right)^{-3d_Q} < |Q'(\alpha)|.
$$

Continued

Proof. We consider two cases based on the size of α: $|\alpha| \leq 1$ and $|\alpha| > 1$. We first consider the case $|\alpha| \leq 1$. Since $Q(z)$ is an irreducible polynomial, $Q(z)$ and $Q'(z)$ do not share a common zero, and thus $\mathrm{Res}(Q, Q') \neq 0$. By the Fundamental Principle of Number Theory, since $Q(z)$ has integer coefficients, we have

$$1 \leq |\mathrm{Res}(Q, Q')|.$$

On the other hand, it is possible to produce an upper bound for $|\mathrm{Res}(Q, Q')|$ that involves $|Q'(\alpha)|$. To realize this bound, we write $Q(z) = \sum_{j=0}^{d_Q} b_j z^j$ and consider the $(2d_Q - 1) \times (2d_Q - 1)$ matrix

$$
\begin{pmatrix}
b_{d_Q} & b_{d_Q-1} & \cdots & b_2 & b_1 & b_0 & & & \\
 & b_{d_Q} & b_{d_Q-1} & \cdots & b_2 & b_1 & b_0 & & \\
 & & b_{d_Q} & \cdots & \cdots & b_2 & b_1 & b_0 & \\
 & & & \ddots & & & & & \ddots \\
 & & & & b_{d_Q} & b_{d_Q-1} & \cdots & \cdots & b_0 \\
d_Q b_{d_Q} & (d_Q-1)b_{d_Q-1} & \cdots & 2b_2 & b_1 & & & & \\
 & d_Q b_{d_Q} & (d_Q-1)b_{d_Q-1} & \cdots & 2b_2 & b_1 & & & \\
 & & & \ddots & & & \ddots & & \\
 & & & & d_Q b_{d_Q} & \cdots & & 2b_2 & b_1
\end{pmatrix}.
$$

By adding z-multiples of each column to the last column of the previous impressive matrix, it is easy to show that the determinant of that large matrix is equal to the determinant of the matrix

$$
\begin{pmatrix}
b_{d_Q} & b_{d_Q-1} & \cdots & b_2 & b_1 & b_0 & & z^{d_Q-2}Q(z) \\
 & b_{d_Q} & b_{d_Q-1} & \cdots & b_2 & b_1 & b_0 & z^{d_Q-3}Q(z) \\
 & & b_{d_Q} & \cdots & \cdots & b_2 & b_1 & b_0 & \\
 & & & \ddots & & & & & \ddots \\
 & & & & b_{d_Q} & b_{d_Q-1} & \cdots & \cdots & Q(z) \\
d_Q b_{d_Q} & (d_Q-1)b_{d_Q-1} & \cdots & 2b_2 & b_1 & & & z^{d_Q-1}Q'(z) \\
 & d_Q b_{d_Q} & (d_Q-1)b_{d_Q-1} & \cdots & 2b_2 & b_1 & & z^{d_Q-2}Q'(z) \\
 & & & \ddots & & & \ddots & & \\
 & & & & d_Q b_{d_Q} & \cdots & & 2b_2 & Q'(z)
\end{pmatrix}.
$$

Evaluating this matrix at $z = \alpha$, using the fact that $Q(\alpha) = 0$, expanding the determinant along its last column, and applying *Hadamard's inequality* for the

Continued

determinant of a matrix, we obtain

$$1 \le |\text{Res}(Q, Q')| \le d_Q \max\{1, |\alpha|\}^{d_Q-1} |Q'(\alpha)| \left((d_Q + 1)\mathcal{H}(Q)^2\right)^{\frac{1}{2}(d_Q-1)}$$

$$\times \left(d_Q(d_Q\mathcal{H}(Q))^2\right)^{\frac{1}{2}d_Q}.$$

Since we are assuming that $|\alpha| \le 1$, the proposition, in this case, follows from the previous inequality.

Challenge 6.13 *Establish the proposition in the case $|\alpha| > 1$ by modifying the previous argument by adding multiples of the columns to the first column—thus completing the proof.* ∎

We are finally in a position to establish the second key lemma required in showing that almost all numbers are S-numbers. We restate this result here.

LEMMA 6.15 *Suppose that $Q(z)$ is an irreducible polynomial in $\mathbb{Z}[z]$ and $\xi \in \mathbb{C}$. Let α be the zero of $Q(z)$ that is closest to ξ. If $2 \le \deg(Q) \le \mathcal{H}(Q)$, then*

$$|\xi - \alpha| \le \mathcal{H}(Q)^{6\deg(Q)}|Q(\xi)|.$$

Challenge 6.14 *You guessed it: Give a proof of this lemma.*

We close this lengthy section with one final remark. Given Lemmas 6.14 and 6.15, it may appear as if we could produce an alternate classification into four classes by replacing $|P(\xi)|$, for $P(z) \in \mathcal{P}_{N, H}$, by $|\xi - \alpha|$, for $\alpha \in \mathcal{A}_{N, H}$. Such a classification was discovered and studied by Jurjen Koksma in 1939 and our analysis indirectly shows that these two classification schemes are equivalent.

6.8 Declassified quantities: e, π, and the elusive T-numbers

We begin with some reflections on e. By Hermite's result (Theorem 3.1), we know that for any nonzero polynomial $P(z) \in \mathbb{Z}[z]$, and any nonzero algebraic number α,

$$0 < \left|P\left(e^\alpha\right)\right|.$$

A refinement to the proof we gave for the Lindemann–Weierstrass Theorem (Theorem 3.4) would allow us to deduce a much stronger result. In particular, we state the following theorem without proof.

THEOREM 6.23 *If α is a nonzero algebraic number, then there exists a positive constant $c = c(\alpha)$ such that for any nonzero polynomial $P(z) \in \mathbb{Z}[z]$ whose height is* *sufficiently large as compared to its degree, the inequality*

$$\mathcal{H}(P)^{-c \deg(P)} \leq \left| P\left(e^{\alpha}\right) \right|$$

holds.

The exponent in the lower bound of an inequality such as the one in Theorem 6.23 is what is known as a *transcendence measure*, in this case, for e^{α}.

Challenge 6.15 *Using Theorem 6.23, prove that for any nonzero algebraic number α, e^{α} is an S-number.*

The previous challenge reveals that e is an S-number. Moreover, as an immediate consequence of this challenge, Theorem 6.10, and Corollary 6.12, we have

COROLLARY 6.24 *Let α be a nonzero algebraic number and $\mathcal{L}$ a Liouville number. Then $e^{\alpha} + \mathcal{L}$ is transcendental.*

In particular, the number in this chapter's title, $e + \sum_{n=1}^{\infty} 10^{-n!}$, is transcendental.

Much less is known about π. Transcendence measures for π are known but they are weaker than the ones known for e and only lead to the conclusion that π is not a U-number. Of course, in light of Theorem 6.17, we might guess that π is an S-number, but the question remains open.

We close this chapter with one final observation. We note that we have given examples of A-, U-, and S-numbers. Yet we have not exhibited a T-number. We do know that T-numbers form a set of measure zero, but a more fundamental question remains: Do T-numbers even exist? This question remained unanswered for over 35 years after Mahler introduced his classification scheme. Finally, in 1968, Wolfgang Schmidt gave a very delicate construction showing that T-numbers do, in fact, exist.

The difficulty in finding T-numbers was foreshadowed in our intuitive look at Theorem 6.8. There we remarked that T-numbers are precisely those ξ's for which we have infinitely many polynomials $P(z) \in \mathbb{Z}[z]$ such that $|P(\xi)|$ can be made *substantially* smaller than the upper bound of Theorem 6.5 *but* the degrees of those amazing polynomials *must* be unbounded. It is ensuring that the degrees of the best polynomial approximations are unbounded that poses the greatest challenge in generating a T-number.

Number 7

Extending Our Reach Through Periodic Functions:
The Weierstrass ℘-function and the transcendence of $\frac{\Gamma(1/4)^2}{\sqrt{\pi}}$

7.1 Transcending our beloved e^z and challenging its centrality

In this penultimate chapter we look beyond transcendental numbers associated with the familiar function $f(z) = e^z$ and examine values associated with a function that has taken center stage in modern number theory—the Weierstrass elliptic function, denoted by $\wp(z)$. Here we will not only introduce the function $\wp(z)$ and apply it to obtain an attractive transcendence result, but through that extended development we endeavor to highlight the function's importance. The scope of this development involves both algebraic and analytic aspects of $\wp(z)$. Toward these ends, our approach will be somewhat more relaxed than in the previous chapters. While some readers may classify our treatment here, at moments, as "sketchy," we would argue that we provide sufficient depth so that the casual reader will be able to appreciate the beauty of the theory, while the motivated reader will be able to fill in any details we may have suppressed.

As foreshadowed at the close of Chapter 5, we inspire the construction of the Weierstrass $\wp$-function through a journey that begins with the theory of homomorphisms of abelian groups. However, in this opening section, we offer a panoramic overview of the analytic vistas to which that algebraic journey will eventually lead. The central analytic feature of $\wp(z)$, which we will discover and then exploit, is its periodicity in the complex plane.

Of course, we have already exploited the fact that e^z is periodic modulo the lattice $2\pi i \mathbb{Z}$. More generally, for any nonzero $w \in \mathbb{C}$, we know that there exists an entire function that is periodic modulo $\mathbb{Z}w$; namely the function $f(z) = e^{\frac{2\pi i}{w}z}$. As we will soon realize, a critical difference between the function $\wp(z)$ and e^z is that $\wp(z)$ has *two* $\mathbb{Q}$-linearly independent periods (such a function is said to be *doubly periodic*). Moreover, just as there is an exponential function periodic modulo $\mathbb{Z}w$ for any nonzero $w \in \mathbb{C}$, here we will show that given any two $\mathbb{Q}$-linearly independent complex numbers w_1 and w_2 satisfying $w_2/w_1 \notin \mathbb{R}$, there exists a Weierstrass $\wp$-function that is periodic modulo the lattice $W = \mathbb{Z}w_1 + \mathbb{Z}w_2 \subseteq \mathbb{C}$.

Such robust periodicity may sound promising and exciting from a transcendence point of view—for with respect to e^z, we found that all of its nonzero periods, $2\pi i n$, for $n \in \mathbb{Z} \setminus \{0\}$, are transcendental. Those early discoveries arose from exploiting the power series associated with e^z. In our present situation, we will come to the unfortunate realization that the only doubly periodic functions that can be represented by an everywhere-convergent power series are the constant functions. This observation will follow from the fact that a non-constant, doubly periodic function cannot be defined at all points in the complex plane. Thus there cannot exist as attractive a power series for $\wp(z)$ as there is for e^z.

However, all is not lost, since we will also discover that the complex numbers for which $\wp(z)$ is not defined form a *discrete* subset in the complex plane. In fact, we will be able to normalize the Weierstrass $\wp$-function so that the points at which it is not defined are precisely its periods. Moreover, the behavior of $\wp(z)$ at the periods $w \in W$ will be well understood—we will see that it has essentially the same behavior as the function $1/(z - w)^2$ for z near w. These observations will allow us to demonstrate the transcendence of the nonzero periods $w \in W$ by extending the arguments used to deduce the transcendence of the nonzero periods of e^z.

7.2 A circle of ideas behind basic elliptic curves

As we noted earlier, our development of the Weierstrass $\wp$-function begins not with analysis but with algebra. In order to inspire the ideas to come, we first introduce a geometrically defined binary operation on the points of a circle and show that this operation transforms those points into an abelian group.

Given a circle $\mathcal{C}$ in the plane, we select and fix a point $\mathcal{O}$ on $\mathcal{C}$. The point $\mathcal{O}$ will be the identity element of an abelian group. We now define the binary operation $\oplus$ on $\mathcal{C}$ as follows: Given two points P and Q on $\mathcal{C}$, let L_{PQ} denote the chord between P and Q and let L denote the line through $\mathcal{O}$ that is parallel to L_{PQ}. Since the circle is a conic section, we know that L will intersect $\mathcal{C}$ at a second point (with the understanding that if L is tangent to $\mathcal{C}$, then the second point of intersection is also $\mathcal{O}$). We define $P \oplus Q$ to be this second point of intersection of L and $\mathcal{C}$.

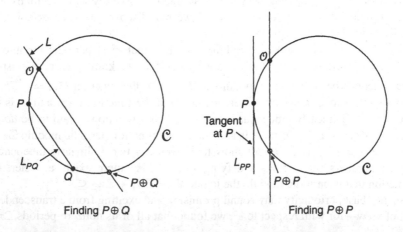

Finding $P \oplus Q$ Finding $P \oplus P$

In the case that P and Q are the same point, the chord L_{PQ} is replaced by the line tangent to the circle at P.

Given a point $P \in \mathcal{C}$, we associate with it the unique point $Q \in \mathcal{C}$ such that the line through $\mathcal{O}$ parallel to the chord L_{PQ} is tangent to $\mathcal{C}$. It is easy to verify that this point Q is the *inverse* of P, denoted by P^{-1}, for the binary operation $\oplus$.

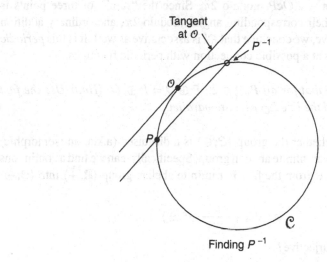

Finding P^{-1}

Summarizing these remarks, the chord $L_{PP^{-1}}$ is parallel to the tangent line at $\mathcal{O}$; or equivalently, the chord $L_{PP^{-1}}$ is perpendicular to the diameter of the circle that passes through $\mathcal{O}$.

It is an interesting exercise in geometry to verify that the points on $\mathcal{C}$ under the binary operation $\oplus$ form an abelian group. The most challenging step of this exercise is the verification that $\oplus$ is associative. Fortunately, the associativity of $\oplus$ will follow from an alternative description of this binary operation that depends on angular measurement.

We let $\mathbf{c}$ denote the center of the circle $\mathcal{C}$. For any point P on the circle, we let $\angle\mathcal{O}\mathbf{c}P$ denote the angle $\theta, 0 \leq \theta < 2\pi$, formed by moving counterclockwise from the radius $\overline{\mathbf{c}\mathcal{O}}$ to the radius $\overline{\mathbf{c}P}$.

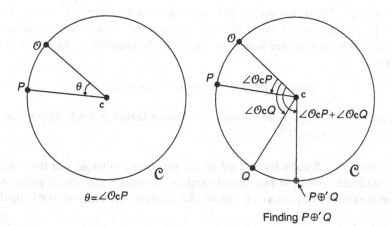

Finding $P \oplus' Q$

We now define a binary operation $\oplus'$ on $\mathcal{C}$ as follows: For two points on the circle P and Q, we simply sum the two angles $\angle OcP$ and $\angle OcQ$ modulo 2π, and define $P \oplus' Q$ be the unique point such that $\angle Oc(P \oplus' Q) = \angle OcP + \angle OcQ$. It is easy to verify that $\oplus'$ is a well-defined binary operation having $\mathcal{O}$ as its identity element. Moreover, every point has an inverse: The inverse of P is the point Q on the circle satisfying $\angle OcQ = 2\pi - \angle OcP$ modulo 2π. Since the "sum'" of three points is computed by adding their corresponding angles modulo 2π, and ordinary addition modulo 2π is associative, we conclude that $\oplus'$ is associative as well. It is this *periodic* feature of $\oplus'$ that hints at a possible connection with periodic functions.

Challenge 7.1   *Verify that for all $P, Q \in \mathcal{C}$, $P \oplus Q = P \oplus' Q$. (Hint: Use the fact that the chord L_{PQ} and the line $L_{O, P \oplus Q}$ are parallel.)*

We now wonder whether the group $\langle \mathcal{C}, \oplus \rangle$ is a disguised (a.k.a. an isomorphic) version of a more familiar infinite abelian group. Specifically, can we find a continuous group homomorphism φ from the familiar infinite abelian group $\langle \mathbb{R}, + \rangle$ into $\langle \mathcal{C}, \oplus \rangle$ such that

$$\varphi : \langle \mathbb{R}, + \rangle \longrightarrow \langle \mathcal{C}, \oplus \rangle$$

is both injective and surjective?

Challenge 7.2   *Show that there cannot exist a continuous isomorphism from $\langle \mathbb{R}, + \rangle$ to $\langle \mathcal{C}, \oplus \rangle$. (Hint: Show that the sequence $\{\varphi(n): n = 1, 2, \ldots\}$ must have a limit point in $\mathcal{C}$ but the sequence $\{1, 2, 3, \ldots\}$ does not converge in $\mathbb{R}$.)*

In view of the previous challenge, we now relax our standards by forgoing injectivity. If we allow the surjective homomorphism φ to have a nontrivial kernel, then it is possible to exploit the periodic nature of the group law $\oplus$ to find such a map; indeed, such a continuous homomorphism can be explicitly given in terms of the cosine and sine functions—powerful evidence that the familiar function e^z may be lurking just around the corner.

Challenge 7.3   *Let us view the circle $\mathcal{C}$ as lying in the xy-plane and as being defined by the equation $(x - a)^2 + (y - b)^2 = r^2$. Suppose that $\mathcal{O}$ has coordinates (x_0, y_0). Show that there exists a real number c such that the mapping φ: $\langle \mathbb{R}, + \rangle \to \langle \mathcal{C}, \oplus \rangle$ given by*

$$\varphi(x) = (r \cos(x + c) + a, r \sin(x + c) + b)$$

is a continuous surjective homomorphism whose kernel is $2\pi \mathbb{Z}$. (Hint: Let $c = \tan^{-1}((y_0 - b)/(x_0 - a))$.)

A critical new feature introduced in the previous challenge was the necessity for a coordinate system. In fact, coordinates are essential in the investigation of the arithmetic nature of the points on $\mathcal{C}$. To ask, for example, whether a point P is algebraic or transcendental assumes that P *has* coordinates—in which case we are actually asking whether one or all of the *coordinates* of P is algebraic or transcendental.

We have already established the transcendence of many points on $\mathcal{C}$ when all of the parameters a, b, c, x_0, y_0, r are algebraic.

Challenge 7.4   *Let* $\mathcal{C}$ *and* $\mathcal{O}$ *be as in Challenge 7.3. Deduce from Hermite's Theorem (Theorem 3.1) or from the weak version of the Schneider–Lang Theorem (Theorem 5.19) that if* a, b, c, x_0, y_0, r, *and* α *are real algebraic numbers,* $\alpha \neq -c$, *then* $P = \varphi(\alpha)$ *has transcendental coordinates. (Hint: If either coordinate of* $\varphi(\alpha)$ *is algebraic, then so is the other coordinate, and therefore so is* $e^{(c+\alpha)i}$.)

The values of the homomorphism $\varphi(x) = (r \cos(x + c) + a, r \sin(x + c) + a)$ are more easily studied if we first normalize $\mathcal{C}$ to be the unit circle; that is, $\mathcal{C} : x^2 + y^2 = 1$, and fix $\mathcal{O} = (1, 0)$. We then find that our continuous group homomorphism becomes $\varphi(z) = (\cos z, \sin z)$ or, viewing $\mathcal{C}$ as living in the complex plane, the more familiar

$$\varphi(z) = \cos z + i \sin z.$$

This last representation of $\varphi(z)$ indicates that we have recovered the result that we have repeatedly exploited throughout this book: The mapping $z \mapsto e^z$ is a continuous group homomorphism from $\langle \mathbb{R}, + \rangle$ onto $\langle \mathbb{R}^\times, \cdot \rangle$, or more generally from $\langle \mathbb{C}, + \rangle$ onto $\langle \mathbb{C}^\times, \cdot \rangle$. We close this brief discussion of the circle with a challenge that reveals a property of $\oplus$ that we will return to in our search for a group law on more exotic curves.

Challenge 7.5   *Use the geometric definition of* $\oplus$ *to show that any two points on the normalized unit circle that lie on the same vertical line sum to the identity element.*

Moving beyond the circle.   One of the most elegant of all observations in mathematics is that the function we seek, the Weierstrass $\wp$-function, plays a role analogous to that of e^z but with a different underlying geometric group structure. While it is not difficult to extend our previous work to show that the points on any conic section in the xy-plane form an abelian group with a geometrically defined binary operation, it is less obvious, but just as true, that the points on certain other curves have such a geometric group structure. The most basic geometric groups that are not isomorphic to ones arising from points on conic sections are the undeniably important elliptic curves.

We recall that a curve defined by a quadratic polynomial in x and y always yields a conic section. An *elliptic curve* is the set of all points satisfying the polynomial equation $f(x, y) = 0$, where $f(x, y)$ has degree 3 or, in some cases, degree 4. Since our eventual goal is to obtain the transcendence of some especially interesting numbers, we will develop just enough of the general theory to allow us to apply the methods that we have already developed. Consequently, rather than begin with the most general elliptic curve, and show that through a judicious change of variables its equation can be put into a certain standard form, we will now simply restrict ourselves to those equations that are already in the so-called *Weierstrass form:*

$$y^2 = 4x^3 - g_2 x - g_3, \tag{7.1}$$

where $g_2, g_3 \in \mathbb{C}$ and the polynomial $4x^3 - g_2 x - g_3$ has three *distinct* zeros. (The motivation for the subscripts on the coefficients arises from the theory of *modular forms*—an extremely important area of number theory that we will never mention again.)

A group law on the solutions to $y^2 = 4x^3 - g_2 x - g_3$. In our discussion of the geometric group law on the circle, we visualized the circle as lying in the xy-plane, effectively restricting our attention to the *real* solutions of $x^2 + y^2 = 1$. The advantage of working over $\mathbb{R}$ is that it is possible to provide an elementary geometric description for $\oplus$. Following a similar strategy, we now introduce a binary operation on the points of the elliptic curve defined by (7.1) in the special case in which both g_2 and g_3 are real numbers and we restrict our attention to only real solutions. We let $E(\mathbb{R})$ denote the set of *real* points on the curve, that is,

$$E(\mathbb{R}) = \left\{ (x, y) \in \mathbb{R}^2 : y^2 = 4x^3 - g_2 x - g_3 \right\}.$$

We will later write $E(\mathbb{C})$ for the complex points on the curve given in (7.1).

In defining the group $\langle \mathcal{C}, \oplus \rangle$, we first fixed a point $\mathcal{O}$, which later became the identity element. By Challenge 7.5 we saw that if $\mathcal{O}$ is selected to be the most eastern (or western) point on the circle, then $\oplus$ has the property that

> *Any two points that are on the same vertical*
> *line sum to the identity element.* (7.2)

We now take this important observation as the inspiration for determining a binary operation on the points of $E(\mathbb{R})$.

We proceed without specifying which point of the curve will act as the binary operation's identity element—that mysterious point will be revealed later. Instead, we begin by defining the inverse of a point. Given a point $P \in E(\mathbb{R})$, we draw the vertical line through P and call it L_P. This line is either tangent to $E(\mathbb{R})$ or intersects it at a second point. In the latter case, we call that second point of intersection the *inverse of* P, denoted by P^{-1}. If L_P is tangent to $E(\mathbb{R})$, then we declare that $P^{-1} = P$. We remark that P^{-1} is geometrically well-defined, since, as we will verify in Challenge 7.7, any vertical line will intersect $E(\mathbb{R})$ in at most two points.

Because a non-vertical line will intersect the curve $E(\mathbb{R})$ in three, not necessarily distinct, points, we can define the binary operation on $E(\mathbb{R})$ *indirectly* by extending observation (7.2) and declaring that:

> *Any three collinear points on* $E(\mathbb{R})$ *will "sum" to the identity element.*

The geometric rule for adding *any* two points on $E(\mathbb{R})$ is implicitly defined in the previous declaration. Specifically, given two distinct points $P, Q \in E(\mathbb{R})$, we let L_{PQ}

be the line that passes through P and Q. If L_{PQ} is *not* a vertical line, then since $E(\mathbb{R})$ is defined by a cubic equation, the line L_{PQ} intersects $E(\mathbb{R})$ in three, not necessarily distinct, points. If we let the third point of intersection be denoted by R, then $P \oplus Q \oplus R = \mathcal{O}$ and so $R = (P \oplus Q)^{-1}$. Therefore we now define $P \oplus Q$ to equal R^{-1}; that is, $P \oplus Q$ is on the vertical line through R. If P and Q are the same point, then the line L_{PQ} is taken to be the tangent line to $E(\mathbb{R})$ at P.

We illustrate this binary operation for the particular elliptic curve $y^2 = 4x(x-1)(x+1)$.

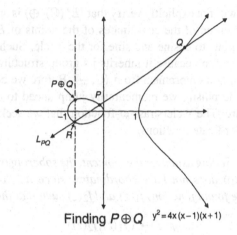

Finding $P \oplus Q$ $y^2 = 4x(x-1)(x+1)$

We have yet to describe the identity element $\mathcal{O}$ for this binary operation. In order to locate $\mathcal{O}$, we mandate that points lying on vertical lines should behave as points lying on non-vertical lines do with respect to the curve $E(\mathbb{R})$; that is, we now declare that all vertical lines that intersect $E(\mathbb{R})$ cross it at a third point "off at infinity." Since we have already proclaimed that any three collinear points on $E(\mathbb{R})$ sum to the identity element, and two *finite* points on any vertical are inverses of one another, it follows that this point "off at infinity" acts as an identity element for $\oplus$, which we denote by $\mathcal{O}$. We adjoin this "point" to $E(\mathbb{R})$ and denote this new set by $E^*(\mathbb{R})$.

Challenge 7.6 *Show that if* $P = (x_1, y_1)$ *and* $Q = (x_2, y_2)$ *are distinct points on the elliptic curve defined by equation* (7.1) *with* $x_1 \neq x_2$, *then the x-coordinate of* $P \oplus Q$ *is given by*

$$-x_1 - x_2 + \frac{1}{4}\left(\frac{y_2 - y_1}{x_2 - x_1}\right)^2.$$

(Hint: Let $y = mx + b$ *be the equation for the line containing P and Q. Then consider* $(mx + b)^2 = 4x^3 - g_2 x - g_3$ *and use the fact that the sum of the three solutions to this equation equals* $m^2/4$.)

Extending $\oplus$ *to the complex points on* $E(\mathbb{C})$. We note that the result of the previous challenge holds whether the coordinates of P and Q are real or complex. So we can immediately extend the operation $\oplus$ from $E^*(\mathbb{R})$ to $E^*(\mathbb{C})$ by applying the algebraic expression in Challenge 7.6 to define $P \oplus Q$, or by simply extending our mandate that any three collinear points sum to the identity element. In particular, to find the inverse of a point $P = (x_1, y_1) \in E(\mathbb{C})$, we seek a line that intersects $E(\mathbb{C})$ in only one other point. In fact, the complex line we seek is simply $x = x_1$.

Challenge 7.7 *Prove that a line intersects $E(\mathbb{C})$ in fewer than three points only if it is defined by an equation $x = \zeta$, for some complex number ζ.*

Even though we will not explicitly verify that $\langle E^*(\mathbb{C}), \oplus \rangle$ is an abelian group, we do seek a parameterization of the coordinates of the points of $E^*(\mathbb{C})$ by functions that play roles analogous to cosine and sine for the circle. Such efforts will reveal that $\langle E^*(\mathbb{C}), \oplus \rangle$ is a group because it inherits its group structure through a continuous surjective group homomorphism from $\langle \mathbb{C}, + \rangle$. Before we begin our search for such a group homomorphism, we momentarily jump ahead to offer a quick snapshot of the big picture: The Weierstrass $\wp$-function that we seek will be one of the homomorphism's coordinate functions.

Thinking out loud. *Setting aside, for the moment, the sobering issue that the identity element of $\langle E^*(\mathbb{C}), \oplus \rangle$ does not have coordinates, since it's "off at infinity," let's suppose that we have found functions $f_1(z)$ and $f_2(z)$ such that the mapping*

$$\varphi : z \mapsto (f_1(z), f_2(z))$$

is a continuous group homomorphism from $\langle \mathbb{C}, + \rangle$ onto $\langle E^*(\mathbb{C}), \oplus \rangle$.

If we let $W \subseteq \mathbb{C}$ denote the kernel of φ, then for any $w \in W$, $\varphi(z + w) = \varphi(z) \oplus \varphi(w) = \varphi(z) \oplus \mathcal{O} = \varphi(z)$. This relationship implies that $\varphi(z)$ is periodic with respect to $W = \ker(\varphi)$. Moreover, for any $w \in W$, $\varphi(w) = (f_1(w), f_2(w))$ must correspond to the identity element of $E^*(\mathbb{C})$. Just as with the cosine and sine functions used to parameterize the circle, here too we will see that the functions $f_1(z)$ and $f_2(z)$ must be related by an algebraic identity. Thus for any $z_0 \in \mathbb{C}$, either both $f_1(z_0)$ and $f_2(z_0)$ are finite or both are infinite. Given that $(f_1(w), f_2(w))$ is the identity element of $E^*(\mathbb{C})$, we conclude that both $f_1(w)$ and $f_2(w)$ must be infinite (otherwise the point $(f_1(w), f_2(w))$ would be on the curve $E(\mathbb{C})$ and would not be the identity element "off at infinity"). Since the coordinate functions for $\varphi(z)$ are assumed to be continuous, we know that for any $w \in W$, $\lim_{z \to w} |f_2(z)| = \infty$. For any given $w \in W$, the simplest such function is $f_2(z) = \frac{1}{z-w}$.

The previous brief informal daydream suggests that we might search among functions $f(z)$ that are periodic with respect to some set W and of the form $f(z) = \frac{g(z)}{h(z)}$, where $g(z)$ and $h(z)$ are entire functions with $h(z)$ having as its zeros precisely the elements of W. We begin this search with an exploration into which sets $W \subseteq \mathbb{C}$ can be a set of periods for a non-constant entire function, or in our case the *reciprocal* of a non-constant, entire function. Our intuitive discussion will lead us to the simplest and most natural candidate for one of the coordinate functions for the homomorphism φ. That candidate will, in fact, be the long-awaited Weierstrass $\wp$-function.

7.3 Entire periodic functions

In our proof of the Six Exponentials Theorem (Theorem 5.16) we employed the result that for any three $\mathbb{Q}$-linearly independent complex numbers w_1, w_2, w_3, the collection of points

$$W = \{n_1 w_1 + n_2 w_2 + n_3 w_3 : n_l \in \mathbb{Z}\}$$

must contain a convergent subsequence. We then applied this observation to conclude that a *nonzero* entire function cannot vanish at all the points of W.

In order to extend our previous work, we now turn our attention to "elliptic functions." To introduce the concept of an elliptic function, we pose a question motivated by our study of the periodic function e^z: Can there exist a non-constant entire function periodic with respect to a set W containing three $\mathbb{Q}$-linearly independent numbers? More generally, given $\mathbb{Q}$-linearly independent complex numbers $w_1, w_2, \ldots, w_L$ and the associated set $W = \{n_1 w_1 + \cdots + n_L w_L : n_l \in \mathbb{Z}\}$, can there exist a non-constant entire function $f(z)$ that is periodic modulo W? That is, can $f(z + w) = f(z)$ for all $w \in W$?

A simple argument shows that the answer to our question is no, a non-constant entire function cannot be periodic with respect to a set W containing more than two $\mathbb{Q}$-linearly independent numbers.

Challenge 7.8 *Prove that if $f(z)$ is an entire function with a convergent sequence of periods $\{w_n\}$, then $f(z)$ is a constant function. (Hint: Suppose $\lim_{n \to \infty} w_n = w$. Then the entire function $g(z) = f(z) - f(w)$ has a convergent sequence of zeros.) Conclude that if $L \geq 3$ and $f(z)$ is an entire function, periodic modulo W, then $f(z)$ is a constant function.*

In view of the previous challenge, we conclude that if we wish to find a non-constant entire function $f(z)$ that is periodic modulo $W = \{n_1 w_1 + \cdots + n_L w_L : n_l \in \mathbb{Z}\}$, then either $L = 1$ or $L = 2$. When $L = 1$, we already know that such a function exists; in particular, we have our perennial favorite function

$$f(z) = e^{\frac{2\pi i}{w_1} z}.$$

The case $L = 2$ is much more subtle and interesting. Indeed, as we claimed at the beginning of this chapter, we will now discover, it turns out that our requirement that $f(z)$ be *entire* is too stringent. We recall that if a function $f(z)$ is periodic with respect to the set $W = \mathbb{Z} w_1 + \mathbb{Z} w_2$, then it is *doubly periodic*. We note that if $f(z)$ is doubly periodic modulo $\mathbb{Z} w_1 + \mathbb{Z} w_2$, then the function defined by $g(z) = f(w_1 z)$ is doubly periodic modulo $\mathbb{Z} + \mathbb{Z}(w_2/w_1)$. The $\mathbb{Q}$-linear independence of 1 and w_2/w_1 is equivalent to the irrationality of w_2/w_1. If w_2/w_1 is real, then, as we saw in Section 5 of Chapter 5, W contains a convergent subsequence of points. Therefore by Challenge 7.8, we know that $g(z)$, and hence $f(z)$, must be constant.

The previous observations reveal an important insight regarding the set W: If $W = \mathbb{Z} w_1 + \mathbb{Z} w_2$ is to be a set of periods for a non-constant entire function, then

$w_2/w_1 \notin \mathbb{R}$. This restriction on the lattice of periods was foreshadowed in our opening remarks from Section 7.1. Therefore, our only hope for hunting down a non-constant entire function with $\mathbb{Q}$-linearly independent periods is to uncover one for which the periods are generated by exactly two numbers w_1, w_2, with $w_2/w_1 \notin \mathbb{R}$. Although this hope is not quite possible to realize, we will now see that it *is* possible to find a doubly periodic function of the form $g(z)/h(z)$, where $g(z)$ and $h(z)$ are non-constant entire functions satisfying the requirements we have imposed. Toward this end, we make two important observations.

Observation 1.   If $f(z)$ is doubly periodic with respect to the lattice $W = \mathbb{Z}w_1 + \mathbb{Z}w_2$, then $f(z)$ is completely determined by its values on the set

$$\mathcal{F} = \{r_1 w_1 + r_2 w_2 : 0 \leq r_1 < 1, \ 0 \leq r_2 < 1\}. \tag{7.3}$$

The set $\mathcal{F}$ is called a *fundamental domain* for the lattice W.

Challenge 7.9   Show that for any $z \in \mathbb{C}$ there exists a unique $w \in W$ such that $z + w \in \mathcal{F}$.

If a doubly periodic function $f(z)$ is entire (or merely continuous), then $|f(z)|$ attains a local maximum when restricted to any compact subset of its domain. Since the closure of $\mathcal{F}$ is compact and $f(z)$ is entire, there exists a real number M such that

$|f(z)| \leq M$ for all z in the closure of $\mathcal{F}$, and thus for all $z \in \mathcal{F}$.

By the previous challenge we now can conclude that for all $z \in \mathbb{C}$, $|f(z)| \leq M$. Therefore an entire doubly periodic function is bounded.

Observation 2.   Liouville, famous for his transcendental number from Chapter 1, also made important contributions to the theory of functions of a complex variable. Of special interest to us is the complete classification of all bounded entire functions.

THEOREM 7.1 (LIOUVILLE'S THEOREM)   *A bounded entire function is a constant function.*

Sketch of proof.   Suppose that $f(z)$ is an entire bounded function, say $|f(z)| \leq M$ for all $z \in \mathbb{C}$. For any two complex numbers z_1 and z_2, we let γ be a simple closed curve in $\mathbb{C}$ having both z_1 and z_2 within its interior. Then by Cauchy's Integral Formula (see Appendix),

$$f(z_1) - f(z_2) = \frac{1}{2\pi i} \int_\gamma \left(\frac{1}{\zeta - z_1} - \frac{1}{\zeta - z_2} \right) f(\zeta) d\zeta.$$

If we specialize γ to be the circle centered at z_1 whose radius r satisfies $r > 2|z_1 - z_2|$, then parameterizing γ by $\zeta = z_1 + re^{i\theta}$, for $0 \leq \theta < 2\pi$, we obtain the estimate

$$|f(z_1) - f(z_2)| < \frac{1}{2\pi} \int_0^{2\pi} \frac{|z_1 - z_2|M}{\frac{1}{2}r^2} d\theta = 2|z_1 - z_2|\frac{M}{r^2} < \frac{M}{r},$$

which, letting $r \to \infty$, reveals that $f(z)$ is constant and completes our sketch.   ∎

Putting the two previous observations together, we arrive at the following conclusion, which illustrates why there must exist points w at which our yet-to-be-identified non-constant doubly periodic function $\wp(z)$ cannot be defined:

An entire doubly periodic function is a constant function.

In light of this conclusion we are forced to extend our search for a doubly periodic function beyond the comfortable convergent realm of entire functions. Rather than consider the most general situation, we again focus on the simplest non-entire functions, namely, the collection of all functions that can be represented as a *ratio* of entire functions. Specifically, for any two nonzero entire functions $g(z)$ and $h(z)$ we consider the function

$$f(z) = \frac{g(z)}{h(z)}.$$

Such a function is called a *meromorphic function*. Notice that $f(z)$ is analytic except possibly at the zeros of $h(z)$. Moreover, since $h(z)$ is not identically zero, we know that its zeros are isolated, and so the points where $f(z)$ is not analytic form a discrete subset of $\mathbb{C}$.

Challenge 7.10 *Suppose that $g(z)$ and $h(z)$ have no zeros in common, and let $z_1, z_2, \ldots$ denote the zeros of $h(z)$. Then show that for each index n,*

$$\lim_{z \to z_n} |f(z)| = \infty.$$

The numbers z_n in Challenge 7.10 are called the *poles* of $f(z)$; and as we will see, the poles of $\wp(z)$ form its period lattice.

A series representation for a meromorphic function.   Suppose that $f(z) = g(z)/h(z)$ is a meromorphic function with a pole at the point $z = p$. Since p is a zero of $h(z)$, and $h(z)$ is not identically zero, there exists a smallest positive integer n_p such that $h^{(n_p)}(p) \neq 0$. This assertion implies, as we have seen throughout this book, that the function $h(z)/(z-p)^{n_p}$ is defined and nonzero at $z = p$. Thus the function $F(z) = (z-p)^{n_p} f(z)$ does *not* have a pole at $z = p$ and can be represented by a convergent power series centered at $z = p$. After finding this power series, and dividing by $(z-p)^{n_p}$, we find that in some neighborhood of p,

$$f(z) = \frac{a_{-n_p}}{(z-p)^{n_p}} + \frac{a_{-n_p+1}}{(z-p)^{n_p-1}} + \cdots + \frac{a_{-1}}{(z-p)^1} + \sum_{n=0}^{\infty} a_n(z-p)^n.$$

Such a series is called the *Laurent series for* $f(z)$ at $z = p$. The next challenge provides further insights into the structure of doubly periodic meromorphic functions but does require some background knowledge of complex analysis. The following challenge is designed only to inspire the function in (7.4) that follows.

Challenge 7.11 *Show that if we let γ be any positively oriented simple closed contour around p whose interior contains no other poles of the meromorphic function $f(z)$, then*

$$\int_\gamma f(\zeta)d\zeta = 2\pi i a_{-1}.$$

Next, if we further assume that $f(z)$ is doubly periodic with respect to the lattice W and that $\mathcal{F}$ is the fundamental domain for W defined by (7.3), then we can shift $\mathcal{F}$ slightly by some $\xi \in \mathbb{C}$, if necessary, so that no poles of $f(z)$ lie on the boundary of $\mathcal{F} + \xi$. Let γ denote the contour defined by the boundary of $\mathcal{F} + \xi$. Then using the periodicity, show that

$$\int_\gamma f(\zeta)\,d\zeta = 0.$$

Combining the two results of Challenge 7.11, we discover that if a doubly periodic meromorphic function is to have a pole at each element of W, then in its series expansion about the pole $z = w$, the coefficient a_{-1} must equal zero. The simplest such series is

$$f(z) = \frac{a_{-2}}{(z-w)^2} + \sum_{n=0}^{\infty} a_n(z-w)^n, \qquad (7.4)$$

where $a_{-2} \neq 0$. In the next section we will see that with an appropriate choice of coefficients $a_{-2}, a_0, a_1, \ldots$, this series will yield the long-sought-after $\wp(z)$.

7.4 Pinning down $\wp(z)$ and uncovering a group homomorphism

If we wish to construct a doubly periodic function having a double pole at each element of $W = \mathbb{Z}w_1 + \mathbb{Z}w_2$, where w_2/w_1 is not a real number, and possessing no other poles, then the function's series expansion centered at each $w \in W$ must be of the form (7.4). The simplest such function is

$$\sum_{w\in W} \frac{1}{(z-w)^2}. \qquad (7.5)$$

Unfortunately, this series does not converge to a meromorphic function. However, the following challenge reveals that the series

$$\sum_{w\in W} \frac{1}{(z-w)^3} \qquad (7.6)$$

is absolutely and uniformly convergent. Why would such an observation be relevant? The answer is that applying the next challenge, we can integrate the series in (7.6) term-wise and thus produce a meromorphic function with a structure similar to (7.5).

Challenge 7.12 *Show that the series in (7.6) converges absolutely (and uniformly) by first establishing the lower bound $|nw_1 + mw_2| \geq c(w_1, w_2)(|n| + |m|)$, where the positive constant $c(w_1, w_2)$ depends only on w_1 and w_2. Then show that there are $4N$ pairs of integers (n, m) with $|n| + |m| = N$, and thus*

$$\sum_{w \in W'} \frac{1}{|w|^3} \leq \frac{4}{c(w_1, w_2)^3} \sum_{N=1}^{\infty} N^{-2},$$

where W' denotes the nonzero lattice points of W.

The function defined by the series $\sum_{w \in W} \frac{1}{(z-w)^3}$ is periodic with respect to W, since replacing z by $z + w'$ for some $w' \in W$ simply corresponds to a rearrangement of the series. And although this function is not of the form we seek, its antiderivative is. Specifically, integrating this series term by term yields a series

$$g(z) = -\frac{1}{2} \sum_{w \in W} \left(\frac{1}{(z-w)^2} + c(w) \right),$$

where the numbers $c(w)$ are the constants of integration. Weierstrass showed that it is possible to choose those constants so that the series for $g(z)$ defines a meromorphic function.

Challenge 7.13 *Show that if we take $c(0) = 0$ and for $w \neq 0$, $c(w) = 1/w^2$, the above series converges absolutely for $z \notin W$. (Hint: Show that if $|w| > 2|z|$, then*

$$\left| \frac{1}{(z-w)^2} - \frac{1}{w^2} \right| \leq 10 \frac{|z|}{|w|^3}.)$$

Defining the constants of integration as suggested in the previous challenge and multiplying through by -2, we *finally* obtain the definition of the elusive *Weierstrass $\wp$-function*:

$$\wp(z) = \frac{1}{z^2} + \sum_{w \in W'} \left(\frac{1}{(z-w)^2} - \frac{1}{w^2} \right), \qquad (7.7)$$

where again W' denotes the nonzero elements of $W = \mathbb{Z}w_1 + \mathbb{Z}w_2$. Thus we remark that the $\wp$-function implicitly depends on the lattice W, just as the exponential function $e^{(2\pi i z)/w}$ depends on the lattice $\mathbb{Z}w$.

Although it is not immediately obvious from the series expression in (7.7), $\wp(z)$ *is* a periodic function with respect to the lattice W. This assertion can be established through a clever but simple observation. The trick is to define two new functions by

$$f_1(z) = \wp(z + w_1) - \wp(z) \quad \text{and} \quad f_2(z) = \wp(z + w_2) - \wp(z),$$

where again we recall that $W = \mathbb{Z}w_1 + \mathbb{Z}w_2$. We now observe that $f_1(z)$ and $f_2(z)$ are meromorphic functions whose derivatives are identically zero; thus, they are constant functions. Evaluating $f_n(-w_n/2)$ for $n = 1$ and 2, and noting that $\wp(z)$ is an even function reveals that these functions are identically zero.

Challenge 7.14 *Verify the claim of the previous sentence and then deduce that $\wp(z)$ is periodic with respect to the lattice $W = \mathbb{Z}w_1 + \mathbb{Z}w_2$.*

The long and winding road to the coordinate functions for $E^*(\mathbb{C})$.   Our goal remains to find a parameterization of the points on $E^*(\mathbb{C})$. One of the key reasons the cosine and sine functions parameterize the unit circle is that they are connected by the same relation that defines the circle: $\cos^2 x + \sin^2 x = 1$. If we hope to have $\wp(z)$ represent one of the two coordinate functions on $E^*(\mathbb{C})$, then we seek another function $h(z)$ such that $\wp(z)$ and $h(z)$ are connected by the same relation that defines $E(\mathbb{C})$, that is,

$$h(z)^2 = 4\wp(z)^3 - g_2\wp(z) - g_3.$$

Next we will discover that the function $h(z)$ is easily found; indeed, we have already alluded to it in (7.6)—it is precisely the function $\wp'(z)$.

Our first step is uncovering a relationship between $\wp(z)$ and $\wp'(z)$, which comes into focus by representing $\wp(z)$ as a series centered at $z = 0$. Of course, if $w \in W$, then $-w$ must also be in W. Hence replacing z by $-z$ simply leads to a rearrangement of the terms in the infinite series (7.7). Therefore we conclude that $\wp(z) = \wp(-z)$, and so the coefficients of the odd powers of z must all equal 0. Thus we may express the Laurent series for $\wp(z)$ about $z = 0$ as

$$\wp(z) = \frac{1}{z^2} + c_0 + c_2 z^2 + c_4 z^4 + \cdots.$$

In view of (7.7), we see that

$$\wp(z) - \frac{1}{z^2} = \sum_{w \in W'} \left(\frac{1}{(z-w)^2} - \frac{1}{w^2} \right),$$

and since the right-hand side of the above identity vanishes at $z = 0$, we conclude that the coefficient c_0 in the Laurent series for $\wp(z)$ must equal 0. That is, we now have

$$\wp(z) = \frac{1}{z^2} + c_2 z^2 + c_4 z^4 + \cdots. \tag{7.8}$$

Formal differentiation of the Laurent series (7.8) yields the convergent series

$$\wp'(z) = -\frac{2}{z^3} + 2c_2 z + 4c_4 z^3 + \cdots. \tag{7.9}$$

Challenge 7.15 *Using the Laurent series for $\wp(z)$ and $\wp'(z)$, convince yourself that there exists an entire function $h(z)$ satisfying*

$$\wp'(z)^2 - 4\wp(z)^3 + 20c_2\wp(z) = -28c_4 + z^2 h(z).$$

(Hint: Formally expand the Laurent series for $\wp(z)^3$ and $\wp'(z)^2$ and show that the series representation for $\wp'(z)^2 - 4\wp(z)^3 + 20c_2\wp(z) + 28c_4$ is a power series in z^2. Applying your findings, deduce that since $\wp'(z)^2 - 4\wp(z)^3 + 20c_2\wp(z)$ is a periodic function with respect to the lattice W, it follows that it is, in fact, a constant function. Finally, let $z \to 0$ and conclude that $\wp'(z)^2 - 4\wp(z)^3 + 20c_2\wp(z) + 28c_4$ is identically zero.)

Challenge 7.15 reveals that the values for the functions $\wp(z)$ and $\wp'(z)$ are connected by an relationship that defines the type of elliptic curve upon which we have defined a geometric group law. But what are the explicit values for the coefficients g_2 and g_3 of the elliptic curve? Those values can be found by looking back at the infinite series (7.7).

Challenge 7.16 *Let $g_2 = 60 \sum_{w \in W'} \frac{1}{w^4}$ and $g_3 = 140 \sum_{w \in W'} \frac{1}{w^6}$. Show that*

$$20c_2 = g_2 \quad and \quad 28c_4 = g_3.$$

(Remark: We introduced the coefficients 60 and 140 because they lead to a particularily simple relationship between $\wp(z)$ and $\wp'(z)$.)

We have shown that for all complex numbers $z \notin W$,

$$\wp'(z)^2 = 4\wp(z)^3 - g_2\wp(z) - g_3,$$

and so $\wp(z)$ satisfies a differential equation that is formally the same as the polynomial equation defining the elliptic curve $E(\mathbb{C})$. This observation implies that the function

$$z \mapsto (\wp(z), \wp'(z))$$

maps $\mathbb{C} \setminus W$ into $E(\mathbb{C})$. In fact, this mapping is surjective, although we will not verify this important claim. Instead, we discover how this mapping leads to a group homomorphism from $\langle \mathbb{C}, + \rangle$ onto $\langle E^*(\mathbb{C}), \oplus \rangle$.

Uncovering the fundamental group homomorphism. The mapping $z \mapsto (\wp(z), \wp'(z))$ is perfectly well-behaved as long as we restrict our attention to complex numbers $z \notin W$, and given our assertion that this mapping is surjective, it provides a parameterization of the points in $E(\mathbb{C})$. In order to construct the associated group homomorphism employing our mapping, we must accomplish three tasks. First, we must extend the domain of this mapping to all of $\mathbb{C}$. Next, we must extend the range of the mapping to all of $E^*(\mathbb{C})$. And finally, we must verify that this extended mapping preserves the algebraic structure, that is, that it is indeed a group homomorphism.

Extending our map to the appropriate domain and range is easily accomplished. We simply define the map $\Phi : \mathbb{C} \to E^*(\mathbb{C})$ by

$$\Phi(z) = \begin{cases} (\wp(z), \wp'(z)) & \text{if } z \notin W, \\ \mathcal{O} & \text{if } z \in W. \end{cases}$$

While this mapping clearly maps $\mathbb{C}$ *onto* $E^*(\mathbb{C})$, a fundamental question remains: Is it a group homomorphism? To establish that Φ is a homomorphism, we return to our mandate that any three collinear points on $E^*(\mathbb{C})$ should sum to the identity element.

Our goal is to verify that for all $z_1, z_2 \in \mathbb{C}$,

$$\Phi(z_1 + z_2) = \Phi(z_1) \oplus \Phi(z_2). \tag{7.10}$$

We first restrict our attention to complex numbers for which none of z_1, z_2, and $z_1 \pm z_2$ are in W. In this case the homomorphism relationship in (7.10) can be obtained from the explicit representation of the group law on $E(\mathbb{C})$: For distinct $P = (x_1, y_1)$ and $Q = (x_2, y_2)$ on $E(\mathbb{C})$, the x-coordinate of $P \oplus Q$ is given by

$$-x_1 - x_2 + \frac{1}{4}\left(\frac{y_2 - y_1}{x_2 - x_1}\right)^2.$$

Thus, we need only check that if $P = (\wp(z_1), \wp'(z_1))$ and $Q = (\wp(z_2), \wp'(z_2))$, then

$$\wp(z_1 + z_2) = -\wp(z_1) - \wp(z_2) + \frac{1}{4}\left(\frac{\wp'(z_2) - \wp'(z_1)}{\wp(z_2) - \wp(z_1)}\right)^2. \tag{7.11}$$

This verification is an application of Liouville's result (Theorem 7.1), which we leave as a challenge.

Challenge 7.17 *Fix a* $y \notin W$ *and define the function* $f(z)$ *by*

$$f(z) = \wp(z + y) + \wp(z) - \frac{1}{4}\left(\frac{\wp'(z) - \wp'(y)}{\wp(z) - \wp(y)}\right)^2.$$

First, show that the function

$$\frac{\wp'(z) - \wp'(y)}{\wp(z) - \wp(y)}$$

has a pole of order 1 at any element of the set

$$W_y = \{w, w + y, w - y \colon w \in W\}.$$

Then conclude that $f(z)$ *does not have a pole at any point in* W_y. *Second, show that* $f(z)$ *is a bounded entire function, and letting* $y \to 0$, *conclude that* $f(z) = -\wp(y)$.

By the previous challenge we have that when none of z_1, z_2, and $z_1 + z_2$ is a period, then the x-coordinate of $\Phi(z_1) \oplus \Phi(z_2)$ equals the x-coordinate of $\Phi(z_1 + z_2)$. A similar procedure allows us to conclude that the y-coordinates of $\Phi(z_1) \oplus \Phi(z_2)$ and of $\Phi(z_1 + z_2)$ also are equal.

We are now left with four cases to consider depending upon whether z_1, z_2, or $z_1 \pm z_2$ is in W. We begin with the easy case in which z_1 and z_2 are in W. Thus, since W is a lattice, we have that $z_1 + z_2 \in W$ and hence

$$\Phi(z_1 + z_2) = \mathcal{O} = \mathcal{O} \oplus \mathcal{O} = \Phi(z_1) \oplus \Phi(z_2).$$

Next we consider the case in which precisely one of the points, z_1 or z_2, is an element of W. If we assume that $z_1 \in W$ and $z_2 \notin W$, then we note that $z_1 + z_2 \notin W$. Since $\wp(z)$ is periodic modulo W, we know that $\wp'(z)$ is also periodic modulo W, and hence we have

$$\Phi(z_1 + z_2) = \big(\wp(z_1 + z_2), \wp'(z_1 + z_2)\big)$$
$$= \big(\wp(z_2), \wp'(z_2)\big)$$
$$= \mathcal{O} \oplus \big(\wp(z_2), \wp'(z_2)\big)$$
$$= \Phi(z_1) \oplus \Phi(z_2).$$

Challenge 7.18 *Verify that (7.10) holds in the final cases, in which neither z_1 nor z_2 belongs to W, but $z_1 \pm z_2$ does. (Hint: In the case $z_1 + z_2 \in W$, use the facts that $\wp(z)$ is periodic modulo W and that it is an even function to deduce that $\wp(z_1) = \wp(z_2)$. In the case $z_1 - z_2 \in W$, verify the result by taking the limit as $z_1 \to z_2$ in Challenge 7.6.)*

As gratifying as it may feel to have finally obtained a group homomorphism from $\langle \mathbb{C}, + \rangle$ to $\langle E^*(\mathbb{C}), \oplus \rangle$, it remains slightly unsettling that the identity element $\mathcal{O}$ is not represented by coordinates in the complex plane. We now offer a remedy for this queasiness by taming $\mathcal{O}$.

The informal description we gave for the identity element for $\langle E^*(\mathbb{C}), \oplus \rangle$ must be made formal in order to precisely define a group homomorphism $\Phi \colon \mathbb{C} \to E^*(\mathbb{C})$. Happily, our naïve, geometric description of $\mathcal{O}$ also has a firm algebraic definition. The key idea is to first introduce a third coordinate, z, into our picture and homogenize the equation $y^2 = 4x^3 - g_2 x - g_3$ via the substitution

$$y \mapsto \frac{y}{z} \quad \text{and} \quad x \mapsto \frac{x}{z}.$$

Clearing denominators we obtain

$$y^2 z = 4x^3 - g_2 x z^2 - g_3 z^3. \tag{7.12}$$

Although the set of all solutions to equation (7.12) lives in three-dimensional complex space, we can recover from it the points from $E(\mathbb{C})$ and more generally from $E^*(\mathbb{C})$. The critical observation is that any solution (x_0, y_0, z_0) to equation (7.12), with $z_0 \neq 0$, corresponds to a unique point on the original curve $E(\mathbb{C})$ through the following process: Since equation (7.12) is homogeneous, for any nonzero complex number λ, if the point (x_0, y_0, z_0) is on the curve, then so is $(\lambda x_0, \lambda y_0, \lambda z_0)$. In particular, if we take $\lambda = 1/z_0$, then a solution (x_0, y_0, z_0) to (7.12), with $z_0 \neq 0$, corresponds to a single point on $E(\mathbb{C})$, specifically, the point $(x_0/z_0, y_0/z_0)$.

This identification hints at where we might look for the identity element for $\oplus$, namely, among the *points* (x, y, z) whose coordinates satisfy equation (7.12) with $z = 0$. A little algebra shows that if $z = 0$, then x must equal 0, while y remains a free variable. If we again identify all solutions that are nonzero scalar multiples of each other with the same point on $E(\mathbb{C})$, then all of the points $(0, y, 0)$ may be identified with a single point $(0, 1, 0)$, which corresponds to the identity element $\mathcal{O}$ of the group $\langle E(\mathbb{C}), \oplus \rangle$.

We now identify $E^*(\mathbb{C})$ with the set $\{(x, y, 1): (x, y) \in E(\mathbb{C})\} \cup \{(0, 1, 0)\}$, and thus we can finally give a precise group homomorphism that is also analytic:

$$\Phi(z) = \begin{cases} (\wp(z), \wp'(z), 1) & \text{for } z \notin W, \\ (0, 1, 0) & \text{for } z \in W. \end{cases}$$

At long last we have realized our goal: We have found an algebraic analogue to e^z and an analytic group homomorphism from $\langle \mathbb{C}, + \rangle$ to a geometric group defined on an elliptic curve. In the next section we establish a transcendence result that generalizes the transcendence of π.

7.5 A new transcendence result

We begin by observing that the transcendence of π can be reformulated as a result involving the group homomorphism e^z.

THEOREM 7.2 *Every nonzero period of the function e^z is transcendental.*

There is a direct analogue of the previous result for the function $\wp(z)$ under a mild additional hypothesis.

THEOREM 7.3 *Suppose that the coefficients g_2 and g_3 of the differential equation*

$$\wp'(z)^2 = 4\wp(z)^3 - g_2\wp(z) - g_3,$$

are algebraic. Let W denote the lattice of periods for $\wp(z)$. Then every nonzero element of W is transcendental.

In Section 7.7 we present a proof of Theorem 7.3 under the slightly simplifying assumption that g_2 and g_3 are *integers*. Even with this additional assumption, Theorem 7.3 has the following spectacular corollary involving the classical gamma function, which we deduce in this section.

COROLLARY 7.4 *The real number*

$$\frac{\Gamma(1/4)^2}{\sqrt{\pi}}$$

is transcendental.

In the next section we will delve into the *gamma function* $\Gamma(z)$ and develop the necessary background. Obviously, we must exploit some properties of $\Gamma(z)$ in order to establish the corollary. In particular, we will verify the only two identities required in the proof of Corollary 7.4, which, for now, we simply state: For positive real numbers a and b,

$$\frac{\Gamma(a)\Gamma(b)}{\Gamma(a+b)} = \int_0^1 x^{a-1}(1-x)^{b-1}\,dx, \tag{7.13}$$

and for a complex number z for which neither z nor $1-z$ is a negative integer,

$$\Gamma(z)\Gamma(1-z) = \frac{\pi}{\sin(\pi z)}. \tag{7.14}$$

We now prove Corollary 7.4 by demonstrating that $\frac{\Gamma(1/4)^2}{\sqrt{\pi}}$ is algebraically dependent on an explicity computed period of our favorite elliptic curve

$$y^2 = 4x(x-1)(x+1) = 4x^3 - 4x.$$

We begin with the following general lemma.

LEMMA 7.5 *Suppose that we factor the polynomial differential equation of $\wp(z)$ over the complex numbers as*

$$\wp'(z)^2 = 4\wp^3(z) - g_2\wp(z) - g_3 = 4(\wp(z) - e_1)(\wp(z) - e_2)(\wp(z) - e_3). \tag{7.15}$$

If we write the period lattice for $\wp(z)$ as $W = \mathbb{Z}w_1 + \mathbb{Z}w_2$, then reordering $e_1, e_2,$ and e_3, if necessary, it follows that

$$\wp\left(\frac{w_1}{2}\right) = e_1, \quad \wp\left(\frac{w_2}{2}\right) = e_2, \quad and \quad \wp\left(\frac{w_1 + w_2}{2}\right) = e_3.$$

Proof. For notational convenience we write $w_3 = w_1 + w_2$. In view of (7.15), we need only show that for each $n = 1, 2, 3$, $\wp'(\frac{w_n}{2}) = 0$. But since $\wp(z)$ is even, it follows that $\wp'(z)$ is an odd function, and thus, in view of the fact that w_n is an element of the period lattice W, we have

$$\wp'\left(\frac{w_n}{2}\right) = \wp'\left(\frac{w_n}{2} - w_n\right) = \wp'\left(-\frac{w_n}{2}\right) = -\wp'\left(\frac{w_n}{2}\right).$$

The previous identity establishes the lemma. ∎

In order to connect $\frac{\Gamma(1/4)^2}{\sqrt{\pi}}$ with a period of $\wp(z)$, we study of the highly improper integral

$$I = \int_1^\infty \frac{dx}{\sqrt{4x^3 - 4x}}. \tag{7.16}$$

We first verify the amazing fact that $2I$ is a nonzero period of the elliptic curve $y^2 = 4x^3 - 4x$ and then show that the numbers I and $\frac{\Gamma(1/4)^2}{\sqrt{\pi}}$ are algebraically dependent.

LEMMA 7.6   *The number $2I$ is a nonzero period of the Weierstrass $\wp$-function that satisfies the differential equation $y' = 4y^3 - 4y$.*

Proof.   To establish this lemma, we first claim that the Weierstrass $\wp$-function associated with the differential equation

$$\left(\frac{dy}{dx}\right)^2 = 4y^3 - 4y$$

inverts the integral of (7.16) in the sense that

$$z = \int_{\wp(z)}^{\infty} \frac{dx}{\sqrt{4x^3 - 4x}}, \tag{7.17}$$

where the path of integration is any simple curve that does not contain a zero of the denominator, that is, which does not contain the numbers $-1, 0$, and 1.

We now sketch a proof of the claim, and instead of justifying each step, we refer the reader to any standard analysis text for further details. If we let ζ be a variable, then we can express z as a function of ζ through the integral

$$z = \int_{\zeta}^{\infty} \frac{dx}{\sqrt{4x^3 - 4x}},$$

with the same restriction on the path of integration. Upon differentiation of the inverse function we obtain

$$\frac{d\zeta}{dz} = \sqrt{4\zeta^3 - 4\zeta},$$

and therefore, after squaring both sides, it follows that ζ, as a function of z, satisfies the same differential equation as the Weierstrass elliptic function. Thus

$$\zeta = \wp(z + a) \quad \text{for some constant } a.$$

But a can be determined by examining the limit of the integral as ζ approaches ∞. As $\zeta \to \infty$, we have that $z \to 0$. Thus a must be pole w for $\wp(z)$. Recalling that $\wp(z)$ is periodic modulo its lattice of poles W, we see that $\zeta = \wp(z + w) = \wp(z)$, which implies the validity of (7.17).

Returning to our proof of the lemma, we note that the lower limit of integration, $x = 1$, in the integral I from (7.16), is a zero of the polynomial $4x^3 - 4x$, and thus by Lemma 7.5, we have that for some n, $\wp(w_n/2) = 1$. Hence

$$I = \int_{\wp(w_n/2)}^{\infty} \frac{dx}{\sqrt{4x^3 - 4x}} = \frac{w_n}{2},$$

and so $2I = w_n \in W \setminus \{0\}$, which completes our proof.   ∎

LEMMA 7.7   *The numbers I and $\frac{\Gamma(1/4)^2}{\sqrt{\pi}}$ are algebraically dependent.*

Proof. We begin with the simple change of variables $x = \frac{1}{\sqrt{u}}$ in (7.16) and find that

$$I = -\frac{1}{2}\int_1^0 \frac{u^{-\frac{3}{2}}\, du}{\sqrt{u^{-\frac{3}{2}} - u^{-\frac{1}{2}}}} = \frac{1}{2}\int_0^1 u^{-\frac{3}{4}}(1-u)^{-\frac{1}{2}}\, du.$$

We now observe that the last integral is precisely the one appearing in (7.13), and thus we conclude that

$$\frac{1}{2}\int_0^1 u^{-\frac{3}{4}}(1-u)^{-\frac{1}{2}}\, du = \frac{\Gamma(\frac{1}{4})\Gamma(\frac{1}{2})}{2\Gamma(\frac{3}{4})}.$$

Challenge 7.19 *Apply identity (7.14) to deduce that*

$$I = \frac{1}{2\sqrt{2}}\frac{\Gamma(\frac{1}{4})^2}{\sqrt{\pi}},$$

which establishes the lemma. Then apply Theorem 7.3 to prove Corollary 7.4. ∎

7.6 Exploring the gamma function and infinite products

In this section we briefly study the gamma function with the goal of sketching the proofs of identities (7.13) and (7.14).

The gamma function $\Gamma(z)$ plays an important role in probability theory, higher-dimensional geometry, and number theory. In probability theory $\Gamma(z)$ appears in one of the important moment-generating functions, in geometry $\Gamma(z)$ appears in the computations of volumes and surface areas of higher-dimensional spheres, and in number theory $\Gamma(z)$ is intrinsic in describing the distribution of prime numbers.

There are several equivalent definitions of the *gamma function*, the simplest is as an improper integral: For any complex number z with $\mathrm{Re}(z) > 0$, we define

$$\Gamma(z) = \int_0^\infty e^{-t}t^{z-1}\, dt. \tag{7.18}$$

While it is not at all transparent from this definition, one can show (although we will not) that $\Gamma(z)$ is not only defined, but is also analytic for all z with $\mathrm{Re}(z) > 0$.

LEMMA 7.8 (*a.k.a. Challenge 7.20*) *For positive real numbers a and b,*

$$\frac{\Gamma(a)\Gamma(b)}{\Gamma(a+b)} = \int_0^1 x^{a-1}(1-x)^{b-1}\, dx.$$

Sketch of the proof (a.k.a. Hints). First write the product $\Gamma(a)\Gamma(b)$ as a product of two improper integrals:

$$\Gamma(a)\Gamma(b) = \lim_{R\to\infty}\left(\int_0^R e^{-t}t^{a-1}\, dt\right)\left(\int_0^R e^{-u}u^{b-1}\, du\right).$$

Replacing t by t^2 and u by u^2 rewrite this product as

$$4 \lim_{R \to \infty} \int_0^R \int_0^R e^{-(t^2+u^2)} t^{2a-1} u^{2b-1} \, dt \, du.$$

Let $t = r \cos \theta$ and $u = r \sin \theta$, and assuming the limit of the double integral exists, show that the double integral equals

$$4 \left(\int_0^\infty e^{-r^2} r^{2(a+b)-1} dr \right) \left(\int_0^{\pi/2} \cos^{2a-1} \theta \sin^{2b-1} \theta \, d\theta \right),$$

which, applying the substitution $x = \cos^2 \theta$, yields

$$\Gamma(a)\Gamma(b) = \Gamma(a+b) \int_0^1 x^{a-1}(1-x)^{b-1} \, dx$$

and completes the outline of the proof.   ∎

Unfortunately, establishing identity (7.14) using the integral (7.13) is not such an easy matter, and thus we are forced to find a different expression for $\Gamma(z)$. To understand this alternative formulation of $\Gamma(z)$, we require the notion of an infinite product, which we will apply not only in our proof of Theorem 7.3, but also in the next chapter.

Finding an entire function with prescribed zeros.   Suppose we are presented with a collection of complex numbers, W, and are asked to find an entire function that vanishes at all $w \in W$. If W is a *finite* set, say $W = \{w_1, \ldots, w_N\}$, then the function $f(z) = (z - w_1) \cdots (z - w_N)$ vanishes on W and is entire since it is a polynomial function. However, if W is an infinite set, then such a product would contain infinitely many terms, and therefore it would not be so easy to determine whether the function is entire. Yet, depending on the set W it may or may not be possible to make sense of this infinite product and to apply it to define an entire function.

We must first decide what it means for the product of an infinite number of numbers to exist, and we take our clue from what it means for the sum of an infinite number of numbers to exist. An infinite product of nonzero numbers ρ_n, $\prod_{n=1}^{\infty} \rho_n$, *converges* precisely if the sequence of *partial products* $P_N = \prod_{n=1}^{N} \rho_n$ converges. The convergence of the sequence $\{P_N\}$ gives a necessary condition for the partial product to exist: Since $\frac{P_N}{P_{N-1}} = \rho_N$, we have $\lim_{n \to \infty} \rho_n = 1$. Thus it is more standard to consider infinite products of the form

$$\prod_{n=1}^{\infty} (1 - a_n), \tag{7.19}$$

where none of the terms $1 - a_n$ equals 0. Written in this form, a necessary condition for the convergence of this product is the more familiar $\lim_{n \to \infty} a_n = 0$.

If we write each term of the infinite product as an exponential of a logarithm, $1 - a_n = e^{\log(1-a_n)}$, where we have taken the principal branch of the log function, then it follows from the continuity of e^z that

$$\prod_{n=1}^{\infty}(1 - a_n) = \prod_{n=1}^{\infty} e^{\log(1-a_n)} = \exp\left(\sum_{n=1}^{\infty} \log(1 - a_n)\right).$$

The previous identities reveal that the product (7.19) converges absolutely if and only if the series $\sum_{n=1}^{\infty} \log(1 - a_n)$ converges absolutely, and as an aside we note that this series of logarithms converges absolutely if and only if the series $\sum_{n=1}^{\infty} a_n$ does.

Beginning with an infinite set of complex numbers $W = \{w_1, w_2, \ldots\}$, we return to the issue of finding an entire function that vanishes precisely at the elements of W. The naïve product

$$\prod_{n=1}^{\infty}(z - w_n)$$

cannot possibly converge for all $z \in \mathbb{C}$, because that conclusion would require having $\lim_{n\to\infty}(z - w_n) = 1$ for every choice of z. Yet if we restrict our attention to sets W with $\lim_{n\to\infty} |w_n| = \infty$, then we do know that for any complex number z, $\lim_{n\to\infty} |z/w_n| = 0$. This observation implies that there is hope that the infinite product

$$\prod_{n=1}^{\infty}\left(1 - \frac{z}{w_n}\right) \tag{7.20}$$

might converge. Notice, moreover, that if this product does converge to a function, then that function must vanish at each point $w_n \in W$.

Whether or not the product (7.20) converges depends entirely on the set W, but even without any specific information about W, we can make a few simple observations. Moving from a product of terms to the sum of their logarithms, we recall that when $|z/w_n| < 1$ we have the familiar series expansion

$$\log\left(1 - \frac{z}{w_n}\right) = -\frac{z}{w_n} - \frac{1}{2}\frac{z^2}{w_n^2} - \frac{1}{3}\frac{z^3}{w_n^3} - \cdots.$$

Heuristically, this observation implies that by focusing on the "dominating" terms, the series

$$\sum_{n=1}^{\infty} \log\left(1 - \frac{z}{w_n}\right)$$

is "farther from converging" than

$$\sum_{n=1}^{\infty}\left(\log\left(1 - \frac{z}{w_n}\right) + \frac{z}{w_n}\right),$$

which, in turn, is "farther from converging" than the series

$$\sum_{n=1}^{\infty} \left(\log\left(1 - \frac{z}{w_n}\right) + \frac{z}{w_n} + \frac{1}{2}\frac{z^2}{w_n^2} \right),$$

and so forth.

Challenge 7.20 *Convince yourself that the previous statements are somewhat meaningful and reasonable.*

Translating the previous observations back to the language of infinite products, we see that the product

$$\prod_{n=1}^{\infty} \left(1 - \frac{z}{w_n}\right)$$

is less likely to converge than the product

$$\prod_{n=1}^{\infty} \left(1 - \frac{z}{w_n}\right) e^{\frac{z}{w_n}},$$

which, in turn, is less likely to converge than the product

$$\prod_{n=1}^{\infty} \left(1 - \frac{z}{w_n}\right) e^{\frac{z}{w_n} + \frac{z^2}{2w_n^2}},$$

and so forth.

In fact, it is a delicate matter to determine which exponential factor, if any, will lead to the convergence of the associated infinite product for a given set W. For example, when W is the set of the positive integers, the convergence of the infinite product requires an exponential factor with only a linear term. That is,

$$\prod_{n=1}^{\infty} \left(1 - \frac{z}{n}\right) e^{z/n} \tag{7.21}$$

converges absolutely for all z; whereas the infinite product

$$\prod_{n=1}^{\infty} \left(1 - \frac{z}{n}\right)$$

does not converge.

If we want a function that vanishes only at the negative integers, then we simply replace n by $-n$ in the absolutely convergent series (7.21) and obtain

$$\prod_{n=1}^{\infty} \left(1 + \frac{z}{n}\right) e^{-z/n}, \tag{7.22}$$

which represents an entire function with a simple zero at each negative integer. Thus the reciprocal of this function,

$$\prod_{n=1}^{\infty} \left(1 + \frac{z}{n}\right)^{-1} e^{z/n},$$

represents a meromorphic function with a simple pole at each negative integer, and hence for any nonvanishing entire function $h(z)$,

$$\frac{h(z)}{z} \prod_{n=1}^{\infty} \left(1 + \frac{z}{n}\right)^{-1} e^{z/n} \qquad (7.23)$$

is a meromorphic function with a simple pole at each element of the set $\{0, -1, -2, \ldots\}$. Inspired by our informal observations, we now assert the following claim without proof.

UNSUBSTANTIATED CLAIM 7.9 *There exists a constant $\gamma \in \mathbb{C}$ such that if $h(z) = e^{-\gamma z}$, then the product (7.23) agrees with the gamma function. More precisely, for all $z \in \mathbb{C}$ with $\mathrm{Re}(z) > 0$,*

$$\Gamma(z) = \frac{e^{-\gamma z}}{z} \prod_{n=1}^{\infty} \left(1 + \frac{z}{n}\right)^{-1} e^{z/n}. \qquad (7.24)$$

Moreover the constant γ is Euler's constant *defined by*

$$\gamma = \lim_{n \to \infty} \left(1 + \frac{1}{2} + \cdots + \frac{1}{n} - \log n\right).$$

The infinite product (7.24) is an *analytic continuation*, or more precisely a *meromorphic continuation*, of the function $\Gamma(z)$ to the entire complex plane. Thus to evaluate $\Gamma(z)$ for some $z \in \mathbb{C}$ having a non-positive real part, we apply the infinite product from (7.24) rather than the integral in (7.13). We are now finally in a position to verify identity (7.14).

Challenge 7.21 *Use the infinite product representation for $\Gamma(z)$ to show that if neither z nor $1 - z$ is a negative integer, then*

$$\Gamma(z)\Gamma(1 - z) = \frac{\pi}{\sin(\pi z)}.$$

(Hint: Compare the infinite product on the left with the one on the right derived from the following infinite product expansion: $\sin(\pi z) = \pi z \prod_{n=-\infty, n \neq 0}^{\infty} (1 - \frac{z}{n}) e^{z/n}.$)

7.7 The proof of Theorem 7.3

The strategy of the proof of Theorem 7.3 perfectly parallels our earlier transcendence proofs (see, for example, the proof of the Gelfond–Schneider Theorem (Theorem 5.1)). The argument we now consider differs from the previous ones only in what might best be dubbed "technical points." We will offer an intuitive overview of the proof of Theorem 7.3 after we first address one technical difficulty.

As mentioned earlier, we assume that the differential equation for $\wp(z)$, or equivalently its associated elliptic curve, has integer coefficients. That is, we assume that both g_2 and g_3 are integers; indeed, for reasons that will be apparent momentarily, we will assume even more—that g_2 is an *even* integer. Although this assumption is not necessary for the validity of the theorem, it does simplify our estimates and hence allows us to appreciate the forest of ideas of the proof rather than focus on the individual trees of technical details.

Foremost among the technical differences between the present proof and earlier ones is that $\wp(z)$ is not an entire function, and so we are unable to directly apply the Maximum Modulus Principle—the crucial tool employed in our earlier arguments. Alas, there is no meromorphic version of this important principle, and thus we are forced to modify $\wp(z)$ appropriately.

Capping the poles of $\wp(z)$. Returning to the group homomorphism introduced at the close of Section 7.4,

$$z \mapsto (\wp(z), \wp'(z), 1), \tag{7.25}$$

we now wish to express it as an *equivalent* mapping

$$z \mapsto (p_1(z), p_2(z), p_3(z)),$$

where the coordinate functions $p_1(z), p_2(z)$, and $p_3(z)$ are entire functions. It is the analytic nature of the coordinate functions of this sought-after homomorphism that will enable us to apply the Maximum Modulus Principle.

Inspired by the notion that two triples (x, y, z) are equivalent if one is a nonzero scalar multiple of the other, we are led on a search for a nonzero entire function $d(z)$ such that both $d(z)\wp(z)$ and $d(z)\wp'(z)$ are entire functions. Such a function would then lead to the mapping

$$z \mapsto (d(z)\wp(z), d(z)\wp'(z), d(z)), \tag{7.26}$$

which is equivalent to the mapping (7.25) but would allow us to employ our previous method of proof. As we will now discover, such an entire function $d(z)$ does exist, and the mapping in (7.26) is the version of our group homomorphism that we seek.

The key to finding $d(z)$ is the observation that $\wp(z)$ has a *double* pole at each point $w \in W$, while $\wp'(z)$ has a *triple* pole at each $w \in W$. Thus, as a starting point, we wish to find a nonzero, entire function with a *triple* zero at each $w \in W$. Fortunately, we saw in Section 7.6 that using an appropriate infinite product, it is possible to construct just such a function.

The Weierstrass σ-function.   Although our eventual goal is to obtain a nonzero entire function with a triple zero at each pole of $\wp(z)$, we begin with the slightly more modest goal of finding a function $\sigma(z)$ with a *simple* zero at each point of W. Clearly, the function $d(z) = \sigma(z)^3$ would then have the desired triple zero at each simple zero of $\sigma(z)$. We recall that for a lattice $W = \mathbb{Z}w_1 + \mathbb{Z}w_2$, W' denotes the nonzero elements of W. Thus a natural choice for the function $\sigma(z)$ is the infinite product

$$z \prod_{w \in W'} \left(1 - \frac{z}{w}\right).$$

While this infinite product does not converge, we do happily report that, as was foreshadowed in our gamma function explorations, the infinite product

$$z \prod_{w \in W'} \left(1 - \frac{z}{w}\right) e^{\frac{z}{w} + \frac{z^2}{2w^2}}$$

does converge and thus defines an entire function. This infinite product is known as the *Weierstrass σ-function*, and it has a simple zero at each $w \in W$.

Challenge 7.22 *Show that the functions $\sigma(z)^2 \wp(z)$ and $\sigma(z)^3 \wp'(z)$ are entire functions.*

It is also clear from the infinite product expansion that $\sigma(z) = 0$ only for $w \in W$. Thus the mapping

$$z \mapsto \left(\sigma(z)^3 \wp(z), \sigma(z)^3 \wp'(z), \sigma(z)^3\right) \tag{7.27}$$

is the homomorphism we desire.

Challenge 7.23 *Using our identification of triples that are nonzero multiples of one another, convince yourself that the mapping defined in (7.27) is indeed a group homomorphism from $\langle\mathbb{C}, +\rangle$ to $\langle E^*(\mathbb{C}), \oplus\rangle$ equivalent to the original map $\Phi(z)$ defined in Section 7.4.*

The Weierstrass σ-function is precisely the tool that will enable us to apply the Maximum Modulus Principle in our proof of the transcendence of a nonzero period for an elliptic curve defined over $\mathbb{Z}$. Before facing the proof head-on, we first offer the following brief overview of the entire argument.

An intuitive overview of the Proof of Theorem 7.3

Following our well-worn outline, our first goal is to construct an auxiliary function with prescribed zeros that takes advantage of our assumption that a nonzero period of $\wp(z)$, say w, is algebraic. For simplicity, we assume that $w/2 \notin W$. In actuality

Continued

this assumption is no restriction at all, since for each nonzero w in the discrete lattice W, there must exist an integer n such that $w/2^n \notin W$. Since we wish to avoid the additional variable n floating around our proof, we simply assume that $n = 1$.

Step 1. We assume that the nonzero period $w \in W$ is algebraic and then construct an auxiliary function of the form

$$F(z) = \sum_{m=0}^{D_1-1} \sum_{n=0}^{D_2-1} a_{mn} z^m \wp(z)^n,$$

with relatively "small" integral coefficients a_{mn} that are not all zero, such that there exist parameters T and K so that

$$F^{(t)}\left(\frac{w}{2} + kw\right) = 0, \quad \text{for } 0 \le t < T \text{ and } 1 \le k \le K. \tag{7.28}$$

Note that under the hypothesis that w is algebraic, each of the values $F^{(t)}\left(\frac{w}{2} + kw\right)$, for $t \ge 0$ and $k \ge 1$, is an algebraic number.

Step 2. Since $F(z)$ is not identically zero, there exists a minimal index M such that for all integers t, $0 \le t < M$,

$$F^{(t)}\left(\frac{w}{2} + kw\right) = 0, \quad \text{for all } k, \ 1 \le k \le K, \tag{7.29}$$

yet there exists an integer k_0, $1 \le k_0 \le K$, for which

$$F^{(M)}\left(\frac{w}{2} + k_0 w\right) \ne 0.$$

Step 3. The function $H(z) = \sigma(z)^{2(D_2-1)} f(z)$ is an *entire* function satisfying

(i) $H^{(t)}\left(\frac{w}{2} + kw\right) = 0$, for $0 \le t < M$ and $1 \le k \le K$; and

(ii) $H^{(M)}\left(\frac{w}{2} + k_0 w\right) \ne 0$.

Applying the Maximum Modulus Principle to the function

$$G(z) = \frac{H(z)}{\prod_{1 \le k \le K} \left(z - \left(\frac{w}{2} + kw\right)\right)^{M-1}}$$

will reveal that $\left| H^{(M)}\left(\frac{w}{2} + k_0 w\right) \right|$ is small. On the other hand, in view of the product rule and our choice of M, we see that

$$H^{(M)}\left(\frac{w}{2} + k_0 w\right) = \sigma\left(\frac{w}{2} + k_0 w\right)^{2(D_2-1)} F^{(M)}\left(\frac{w}{2} + k_0 w\right).$$

Applying a lower bound for $\left| \sigma\left(\frac{w}{2} + k_0 w\right) \right|$, we conclude that the nonzero algebraic number $\left| F^{(M)}\left(\frac{w}{2} + k_0 w\right) \right|$ is small.

Step 4. We use the small nonzero algebraic number from Step 3 to obtain an integer that violates the Fundamental Principle of Number Theory.

The proof of Theorem 7.3.   We begin by assuming that the nonzero period $w \in W$ is algebraic, and, as we indicated earlier, that $w/2 \notin W$. We remark that as in our previous arguments, the integer parameters D_1, D_2, T, and K will be selected later.

Step 1.   Upon careful reflection on the previous outline, we discover that it does not address a hidden, additional complication that we must face: When we take the derivative of a function of the form

$$F(z) = \sum_{m=0}^{D_1-1} \sum_{n=0}^{D_2-1} a_{mn} z^m \wp(z)^n, \tag{7.30}$$

we obtain a polynomial expression in not only z and $\wp(z)$, but also in $\wp'(z)$. Moreover, it appears that higher derivatives of $F(z)$ will involve expressions containing higher derivatives of $\wp(z)$. For example, $F^{(k)}(z)$ could involve the functions $z, \wp(z), \wp'(z), \ldots, \wp^{(k)}(z)$. Thus we must account for the higher-order derivatives of the $\wp$-function when we construct a system of equations to be solved in order to obtain the integers a_{mn}, not all zero, satisfying

$$F^{(t)}\left(\frac{w}{2} + kw\right) = 0, \quad \text{for } 0 \le t < T \text{ and } 1 \le k \le K. \tag{7.31}$$

Our choice of zeros for the function $F(z)$ is guided by the observation that since $w/2 \notin W$, $F(z)$ is analytic at all points of the form $\frac{w}{2} + kw$, and if the coefficients a_{mn} are integers, then we hope to discover that the values $F^{(t)}\left(\frac{w}{2} + kw\right)$ are algebraic. In order to establish this algebraic hope, we now examine the role of the derivatives of $\wp(z)$ in each of the derivatives of $F(z)$. This task is a fairly routine matter, since differentiation of the differential equation for $\wp(z)$,

$$\wp'(z)^2 = 4\wp(z)^3 - g_2\wp(z) - g_3,$$

reveals the relationship

$$\wp''(z) = 6\wp(z)^2 - \frac{g_2}{2},$$

which allows us to express each derivative of $F(z)$ as a function of only z, $\wp(z)$, and $\wp'(z)$.

Challenge 7.24   *Show that for each pair of indices t and n, there exists an integral polynomial $P_{n,t}(x, y, X, Y)$ such that*

$$\left(\frac{d}{dz}\right)^t \left(\wp(z)^n\right) = P_{n,t}\left(\frac{g_2}{2}, g_3, \wp(z), \wp'(z)\right).$$

Moreover,

$$\deg_x(P_{n,t}) \le t, \ \deg_y(P_{n,t}) \le t, \ \deg_X(P_{n,t}) \le n + 2t, \ \deg_Y(P_{n,t}) \le 1,$$

and each coefficient of $P_{n,t}$ has absolute value at most $2^{n+5t}(2t)!$. (Hint: Establish this result by induction on t using $\wp''(z) = 6\wp(z)^2 - \frac{g_2}{2}$.)

If we apply the result of the previous challenge under the additional assumption that both $\frac{g_2}{2}$ and g_3 are integers, and if we further absorb the integral upper bound for the absolute values of the coefficients into one constant, then we obtain the following.

LEMMA 7.9 *For each pair of indices t and n, there exists a polynomial $p_{n,t}(X, Y) \in \mathbb{Z}[X, Y]$ satisfying*

$$\left(\frac{d}{dz}\right)^t \left(\wp(z)^n\right) = p_{n,t}(\wp(z), \wp'(z)),$$

where

$$\deg_X(p_{n,t}) \le n + 2t, \quad \text{and} \quad \deg_Y(p_{n,t}) \le 1,$$

and having coefficients of absolute value at most $c_1^{n+t \log t}$, where $c_1 = \{128, |g_2/2|, |g_3|\}$.

This lemma will play an important role in both our search for the integral coefficients of the auxiliary function $F(z)$ and later in demonstrating that under our assumption that w is algebraic, we can produce an *integer* between 0 and 1.

As we face the issue of generating our "small" coefficients a_{mn}, we recall that in our proof of the transcendence of one of $\xi, e^\xi, e^{\xi\zeta}$ in Chapter 5 (Theorem 5.2), we developed a system of equations analogous to the system (7.31). There we let K denote the field $\mathbb{Q}\left(\xi, e^\xi, e^{\xi\zeta}\right)$, and fixed a primitive element θ for K. Letting δ denote the degree of θ over $\mathbb{Q}$, in the proof of Theorem 5.2 we expressed each equation as $A_1 + A_2\theta + \cdots + A_\delta\theta^{\delta-1} = 0$, and then solved the system of linear equations having *integral* coefficients

$$A_1 = 0, \quad A_2 = 0, \quad \ldots, \quad A_\delta = 0. \tag{7.32}$$

We now wish to search for the analogous field K in the present context. Since we have assumed that $w/2 \notin W$, it follows from Lemma 7.5 that for any integer k,

$$\wp'\left(\frac{w}{2} + kw\right) = 0.$$

Thus we find that

$$\wp\left(\frac{w}{2} + kw\right) \in \{e_1, e_2, e_3\},$$

where we recall from Lemma 7.5 that e_1, e_2, and e_3 are the zeros of the polynomial $4x^3 - g_2x - g_3 = 0$. For notational simplicity we assume, without loss of generality, that $\wp(\frac{w}{2} + kw) = e_1$. Therefore we conclude that for each pair of integers $t \ge 0$ and $k \ge 1$,

$$F^{(t)}\left(\frac{w}{2} + kw\right)$$

is an algebraic number. Moreover, all such algebraic numbers lie in the field $K = \mathbb{Q}(w, e_1)$. If we let θ denote a fixed primitive element in K, then $K = \mathbb{Q}(\theta)$. We also can assume that θ is an algebraic integer, and, as before, we let δ denote the degree of θ.

We now wish to replace the system of equations (7.31) with a system of linear equations of the form (7.32). To carry out this equation conversion, we begin with the observation, whose verification we leave as a challenge, that for any triple of indices t, m, and n, there exists a polynomial $Q_{m,n,t}(X, Y, Z) \in \mathbb{Z}[X, Y, Z]$ such that

$$\left(\frac{d}{dz}\right)^t \left(z^m \wp(z)^n\right) = \sum_{s=0}^t \binom{t}{s} \left(\frac{d}{dz}\right)^{t-s} (z^m) \left(\frac{d}{dz}\right)^s (\wp(z)^n)$$

$$= Q_{m,n,t}(z, \wp(z), \wp'(z)), \qquad (7.33)$$

where each coefficient of $Q_{m,n,t}$ has absolute value at most $c_2^{n+t\log m}$, and

$$\deg_X(Q_{m,n,t}) \le m, \quad \deg_Y(Q_{m,n,t}) \le n + 2t, \quad \text{and} \quad \deg_Z(Q_{m,n,t}) \le 1.$$

Challenge 7.25 *Verify the previous bounds for the absolute values of the coefficients and the degrees of the polynomial $Q_{m,n,t}$.*

Applying identity (7.33) together with Lemma 7.5, we can express the values of the derivatives of the auxiliary function more explicitly as

$$F^{(t)}\left(\frac{w}{2} + kw\right) = \sum_{m=1}^{D_1-1} \sum_{n=0}^{D_2-1} a_{mn} Q_{m,n,t}\left(\left(\frac{1+2k}{2}\right)w, \wp\left(\frac{w}{2}\right), \wp'\left(\frac{w}{2}\right)\right)$$

$$= \sum_{m=1}^{D_1-1} \sum_{n=0}^{D_2-1} a_{mn} Q_{m,n,t}\left(\left(\frac{1+2k}{2}\right)w, e_1, 0\right). \qquad (7.34)$$

Challenge 7.26 *With the aid of Lemma 4.4 and Lemma 7.9, show that for any t, $0 \le t < T$, and any k, $1 \le k \le K$, there exist rational numbers $A_1, A_2, \ldots, A_\delta$, depending on t and k, with numerators of absolute value at most $K^{D_1} D_1^T c_3^{\delta(D_1+D_2+2T)}$ and denominators of absolute values at most $c_4^{D_1}$ such that identity (7.34) can be expressed as*

$$F^{(t)}\left(\frac{w}{2} + kw\right) = A_1 + A_2\theta + \cdots + A_\delta\theta^{\delta-1}. \qquad (7.35)$$

A now-standard application of Siegel's Lemma (Theorem 4.3) implies that if we have $D_1 D_2 > \delta KT$, then there exists a collection of integers a_{mn}, not all zero, satisfying

$$\max\{|a_{mn}|\} \le c_5^{\left(\log(D_1D_2)+D_1\log K+T\log D_1+\delta(D_1+D_2+2T)\right)\frac{\delta KT}{D_1D_2-\delta KT}}, \qquad (7.36)$$

such that $F(z)$ has our desired prescribed zeros.

Of course, the bounds on $|a_{mn}|$ from (7.36) and the bounds on $|F(z)|$ to come all depend on the parameters D_1, D_2, T, and K. To simplify these estimates, we now fix the relationship between these parameters. We remark that the parameters D_1 and D_2 will depend on the *free* parameters T and K. Specifically, we select D_1 and D_2 so that $D_1 D_2 > \delta KT$ and

$$\frac{\delta KT}{D_1 D_2 - \delta KT} \le \frac{1}{\log T}.$$

These inequalities can be realized if we define D_1 and D_2 by

$$D_1 = \left[\sqrt{\delta K}\, T + 1\right] \quad \text{and} \quad D_2 = \left[\sqrt{\delta K}\,(1 + \log T) + 1\right], \qquad (7.37)$$

where we recall that $[x]$ denotes the integer part of x.

Given our choice of parameters, for sufficiently large T the bound from (7.36) can be expressed as

$$0 < \max\{|a_{mn}|\} \le e^{T\sqrt{K}\log K}. \qquad (7.38)$$

Step 2. In order to find a nonzero value for the function $F(z)$, we first must demonstrate that the function is not identically zero (even though we know that the coefficients are not all zero). This observation is the content of the next challenge.

Challenge 7.27 *Show that the functions z and $\wp(z)$ are algebraically independent. (Hint: Use the obvious fact that the function z is entire and vanishes only at $z = 0$.)*

In view of Challenge 7.27, we see that $F(z)$ is not identically zero. Thus there exists a minimal integer M such that for all $0 \le t < M$,

$$F^{(t)}\left(\frac{w}{2} + kw\right) = 0 \quad \text{for all} \quad k, \; 1 \le k \le K, \qquad (7.39)$$

and a $k_0, 1 \le k_0 \le K$, such that

$$F^{(M)}\left(\frac{w}{2} + k_0 w\right) \ne 0.$$

This nonzero value is the nonzero algebraic number that will eventually lead to our violation of the Fundamental Principle of Number Theory.

Step 3. Of course, the function $F(z)$ is not analytic, but the function $\sigma(z)^{2(D_2-1)}F(z)$ is. Moreover, by (7.39) we have that the very useful function defined by

$$G(z) = \frac{\sigma(z)^{2(D_2-1)}F(z)}{\prod_{1 \le k \le K}\left(z - \left(\frac{w}{2} + kw\right)\right)^{M-1}}$$

is entire.

In order to provide an upper bound for the modulus of $G(z)$, we require the following fact, which we state here without proof: For any $\varepsilon > 0$, the function $\sigma(z)^2 \wp(z)$

has order of growth less than $2 + \varepsilon$. That is, given $\varepsilon > 0$, there exists an R_ε such that for every $R > R_\varepsilon$,

$$\left| \sigma(z)^2 \wp(z) \right|_R < e^{|z|^{2+\varepsilon}}. \tag{7.40}$$

We now estimate the modulus of $G(z)$ through an application of the Maximum Modulus Principle along a circle of radius $R = M^{1/4}$. In view of (7.40), we have that if we take R sufficiently large, which is equivalent to taking T (and thus implicitly M) sufficiently large, then

$$\left| \sigma(z)^2 \wp(z) \right|_R < e^{R^{\left(2+\frac{1}{4}\right)}}.$$

If $|w| \left(K + \frac{1}{2} \right) < \frac{1}{2}R$, which we will ensure later, then for all sufficiently large T, and all $z \in \mathbb{C}$ satisfying $|z| \leq R$, the Maximum Modulus Principle allows us to obtain

$$|G(z)|_R \leq \frac{D_1 D_2 e^{T\sqrt{K}} \log K R^{D_1} e^{D_2 R^{9/4}}}{\left(R - |\frac{w}{2}|(K+1) \right)^{K(M-1)}} \leq e^{-\frac{1}{8}KM \log M}.$$

Thus in view of our definition of $G(z)$, we have the following upper bound for our nonzero value of our auxiliary function,

$$\left| \sigma \left(\frac{w}{2} + k_0 w \right)^{2(D_2-1)} F^{(M)} \left(\frac{w}{2} + k_0 w \right) \right| \leq M! |G(z)|_R \prod_{k \neq k_0} |(k_0 - k)w|^{M-1}$$

$$< e^{M \log M} e^{-\frac{1}{8}KM \log M} (K|w|)^{KM}$$

$$< e^{-2\delta M \log M},$$

provided we take K to be a sufficiently large multiple of δ.

We now observe that since K does not depend on T, the number $\left| \sigma \left(\frac{w}{2} + k_0 w \right) \right|$ can be viewed as a constant. Thus for all sufficiently large T, dividing both sides of the previous inequality by $\left| \sigma \left(\frac{w}{2} + k_0 w \right) \right|^{2(D_2-1)}$ yields

$$\left| F^{(M)} \left(\frac{w}{2} + k_0 w \right) \right| < e^{-\delta M \log M}.$$

Step 4. We conclude our argument by appealing to the result of Challenge 7.26; namely, there exist rational numbers $A_1, A_2, \ldots, A_\delta$ having numerators of absolute value at most $K^{D_1} D_1^M c_3^{\delta(D_1+D_2+2M)} \leq c_6^{M \log T}$, where c_6 depends on our choice of K, and denominators of absolute values at most $c_4^{D_1}$ such that

$$F^{(M)} \left(\frac{w}{2} + k_0 w \right) = A_1 + A_2 \theta + \cdots + A_\delta \theta^{\delta-1}.$$

Given that we can multiply this identity through by a suitable integer to clear away all the denominators, we may as well assume that $A_1, A_2, \ldots, A_\delta$ are integers. Thus if we define

$$\mathcal{N} = (A_1 + A_2 \theta + \cdots + A_\delta \theta^{\delta-1}) \prod \left(A_1 + A_2 \theta' + \cdots + A_\delta \theta'^{\delta-1} \right),$$

where the product is over all the conjugates θ' of θ, $\theta' \neq \theta$, then by the symmetry of the conjugates we conclude that $\mathcal{N}$ is a nonzero integer. As in the proof of so many previous results (see, for example, the proof of Theorem 5.2), we have

$$0 < |\mathcal{N}| < e^{-\delta M \log M} \left(e^{M \log T} \right)^{\delta - 1} < e^{-M \log M} < 1,$$

provided that T, and thus M, is sufficiently large. Therefore we discover that $0 < |\mathcal{N}| < 1$, which *finally* contradicts the Fundamental Principle of Number Theory and, at long long last, completes the proof of Theorem 7.3. *Phew.*   ∎

7.8 The transcendence of gamma values and the Schneider–Lang Theorem revisited

The elliptic curve $y^2 = 4x^3 - 4x$, which gave rise to the transcendence of $\Gamma(\frac{1}{4})^2/\sqrt{\pi}$, possesses a special property not shared by all elliptic curves. To understand this property and see its consequences within the context of transcendence, we begin not with the elliptic curve but instead with the lattice. Given the effort that was required in Section 7.5 to compute a particular period of the $\wp$-function associated with this elliptic curve, it is not likely that we could easily display the lattice associated with this curve. Rather than take on such a challenge so late in the chapter and in the text, we instead offer a short, general overview.

Algebraic Excursion: An introduction to the theory of complex multiplication

We begin with $\mathbb{Q}$-linearly independent complex numbers w_1 and w_2 whose ratio is not real and let $W = \mathbb{Z}w_1 + \mathbb{Z}w_2$. It is clear that for any integer n, $nW \subseteq W$. More generally, we define the *endomorphism ring of W*, denoted by $\text{End}(W)$, by

$$\text{End}(W) = \{\zeta \in \mathbb{C} : \zeta W \subseteq W\}.$$

Thus we see that $\mathbb{Z} \subseteq \text{End}(W)$. In fact, as an aside, we remark that in "most" cases, $\text{End}(W) = \mathbb{Z}$. Viewing the multiplication of elements in W by an integer n as an "endomorphism" might sound a bit highbrow, but the reason for this language is that any element of $\text{End}(W)$ gives rise to an endomorphism of the points on the associated elliptic curve.

We have not yet made explicit the relationship between lattices and elliptic curves, but it was implicit throughout the entire development of this chapter. Given a lattice W as above, W has an associated Weierstrass $\wp$-function. That $\wp$-function satisfies a differential equation of the form $(\frac{df}{dz})^2 = 4f(z)^3 - g_2 f(z) - g_3$, where the coefficients g_2 and g_3 are determined by the lattice via the formulas from

Continued

Challenge 7.16. Through this connection, the multiplication of the lattice by n is connected with the iteration of addition on the associated elliptic curve.

Specifically, if P is a point in $E^*(\mathbb{C})$ and n is a positive integer, then the mapping $P \mapsto P \oplus P \oplus \cdots \oplus P$, where P is added to itself n times using the group law on $E^*(\mathbb{C})$, is in fact an endomorphism of $E^*(\mathbb{C})$. If n is negative, then P^{-1} is summed $|n|$ times. Since we have parameterized the *finite points* on $E^*(\mathbb{C})$ by $z \mapsto (\wp(z), \wp'(z))$, and this mapping is a homomorphism of groups, we have another way to represent how an integer n gives rise to an endomorphism of $E^*(\mathbb{C})$: For a nonzero integer n, there is the mapping $(\wp(z), \wp'(z)) \mapsto (\wp(nz), \wp'(nz))$. Since this operation is simply iterated addition, it is possible, although not necesssary, to give explicit formulas giving the coordinates of $(\wp(nz), \wp'(nz))$ in terms of the coordinates of $(\wp(z), \wp'(z))$.

For some lattices, $\mathrm{End}(W)$ contains $\mathbb{Z}$ as a *proper* subset; for example, if $W = \mathbb{Z} + \mathbb{Z}i$, then $iW \subseteq W$ and hence $i \in \mathrm{End}(W)$. A lattice whose endomorphism ring contains $\mathbb{Z}$ as a proper subset is said to have *complex multiplication*. The reason for this language is that, as the next challenge shows, if there exists a number $\zeta \in \mathrm{End}(W)$, with $\zeta \notin \mathbb{Z}$, then ζ cannot be a real number. Moreover, such a complex number must be an algebraic number whose degree equals two.

Challenge 7.28 *Let W be a lattice as above. First show that $\mathbb{Z} \subseteq \mathrm{End}(W)$ and then prove that if there exists a complex number $\zeta \in \mathrm{End}(W), \zeta \notin \mathbb{Z}$, then ζ cannot be a real number. Moreover, show that such a number ζ is an algebraic number whose degree equals two. (Hint: Use that fact that if w_1 and w_2 are generators of the lattice W, then ζw_1 and ζw_2 are elements of W and hence are of the form $aw_1 + bw_2$ for integers a and b.)*

In the case that W has complex multiplication by ζ, the associated elliptic curve has a corresponding endomorphism induced by ζ. It is easy to represent this endomorphism abstractly as $(\wp(z), \wp'(z)) \mapsto (\wp(\zeta z), \wp'(\zeta z))$. Moreover, just as there are formulas expressing the coordinates of the point obtained by summing a point n times, there are analogous formulas for the coordinates of the point $(\wp(\zeta z), \wp'(\zeta z))$ involving the values $\wp(z)$ and $\wp'(z)$.

Fortunately, in the case of the curve $y^2 = 4x^3 - 4x$, it is very easy to represent an endomorphism on the points $E(\mathbb{C})$ without resorting to complicated formulas. In particular the mapping $(x, y) \mapsto (-x, iy)$ is an endomorphism of the elliptic curve corresponding to the associated lattice that has i in its endomorphism ring.

Challenge 7.29 *Verify that if W is a lattice with complex multiplication by i, then $g_3 = 0$.*

For elliptic curves defined by an algebraic equation *and* having complex multiplication, Gregory Chudnovsky established the following better-than-spectacular theorem in late 1970s.

THEOREM 7.10   *Let $E(\mathbb{C})$ be an elliptic curve defined by the equation $y^3 = 4x^3 - g_2 x - g_3$, where g_2 and g_3 are algebraic numbers. If $E(\mathbb{C})$ has complex multiplication, then for any nonzero w in the lattice of periods W associated with $E(\mathbb{C})$,*

$$w \quad and \quad \pi$$

are algebraically independent.

The proof of Chudnovsky's Theorem involves technical details that we cannot include in such a short overview, but the argument does follow the same outline as all of our proofs and, of course, relies on the Fundamental Principle of Number Theory.

COROLLARY 7.11 $\Gamma(1/4)$ *is transcendental.*

Proof.   In Section 7.5 we established that $\Gamma(\frac{1}{4})^2/\sqrt{2\pi}$ is a nonzero period associated with the elliptic curve $y^3 = 4x^3 - 4x$. Thus by Theorem 7.10 we see that

$$\frac{\Gamma(1/4)^2}{\sqrt{2\pi}} \quad and \quad \pi$$

are algebraically independent. If $\Gamma(1/4)$ were algebraic, then

$$F(X, Y) = \Gamma(1/4)^4 - 2X^2 Y$$

would be a nonzero polynomial having algebraic coefficients with $F\big(\Gamma(\frac{1}{4})^2/\sqrt{2\pi}, \pi\big) = 0$. Thus we could conclude that

$$\frac{\Gamma(1/4)^2}{\sqrt{2\pi}} \quad and \quad \pi$$

are algebraically dependent, which contradicts Chudnovsky's Theorem. Hence $\Gamma(1/4)$ is transcendental.   ∎

The curve $y^2 = 4x^3 - 4x$ is not the only elliptic curve that satisfies the hypotheses of Chudnovsky's Theorem and has a mathematically attractive period. The curve defined by $y^2 = 4x^3 - 4$ has algebraic coefficients and, as can be verified, has complex multiplication by the primitive third root of unity, $\rho = e^{\frac{2\pi i}{3}}$, which is a zero of the polynomial $x^2 + x + 1$.

Challenge 7.30   *Verify that the mapping $(x, y) \mapsto (\rho x, -y)$ maps the elliptic curve defined by $y^2 = 4x^3 - 4$ into itself. In addition, show that if a lattice W has complex multiplication by ρ, then $g_2 = 0$.*

The previous challenge gives rise to the following result.

THEOREM 7.12   *The number*

$$\frac{\Gamma(\tfrac{1}{3})^3}{\pi}$$

is transcendental.

Proof.   The integral

$$2 \int_1^\infty \frac{dx}{\sqrt{4x^3 - 4}} \tag{7.41}$$

represents a nonzero period of the elliptic curve defined by $y^2 = 4x^3 - 4$, and so by Theorem 7.3, that quantity is transcendental. We leave it as a challenge to evaluate this integral and discover that it equals $\Gamma(\tfrac{1}{3})^3/(\sqrt[3]{16}\,\pi)$.   ∎

Challenge 7.31   *Show that the integral in (7.41) equals $\Gamma(\tfrac{1}{3})^3/(\sqrt[3]{16}\,\pi)$. (Hint: Make the change of variables $u = x^{-3}$ to write*

$$2 \int_1^\infty \frac{dx}{\sqrt{4x^3 - 4}} = \frac{1}{3} \int_0^1 u^{-5/6}(1 - u)^{-1/2} du,$$

and then show that this last integral equals $\frac{1}{3}\frac{\Gamma(1/6)\Gamma(1/2)}{\Gamma(2/3)}$. Now simplify the previous expression using (7.14) and the so-called "duplication formula":

$$\Gamma(2z) = \frac{1}{\sqrt{2\pi}} 2^{2z-\frac{1}{2}}\Gamma(z)\Gamma\left(z + \frac{1}{2}\right).)$$

With the result of the previous challenge in hand we can argue as we did in the proof of Corollary 7.11 to deduce the transcendence of yet another gamma value.

COROLLARY 7.13 $\Gamma(1/3)$ *is transcendental.*

We conclude this discussion with a challenge to prove that several values of the gamma function are transcendental.

Challenge 7.32   *From the transcendence of $\Gamma(1/4)$ and $\Gamma(1/3)$ deduce the transcendence of $\Gamma(2/3)$ and $\Gamma(3/4)$. Then prove that all of the numbers $\Gamma(1/2), \Gamma(1/3), \Gamma(1/4), \Gamma(2/3),$ and $\Gamma(3/4)$ are transcendental. (Hint: Use identity (7.14) together with the well-known identity $\Gamma(1/2) = \sqrt{\pi}$.)*

A brief return to the Schneider–Lang Theorem.   In the last section of Chapter 5, we sketched a proof of what we called the Weak Version of the Schneider–Lang Theorem (Theorem 5.19). Given that we have briefly discussed meromorphic functions in this chapter, we can now state the full version of the Schneider–Lang Theorem.

THEOREM 7.14 (THE SCHNEIDER–LANG THEOREM)   *Let F be a number field. Suppose that $f_1(z), f_2(z), \ldots, f_N(z)$ are meromorphic functions having orders of growth bounded above by ρ and satisfying the following two conditions:*

(i)   *$f_1(z)$ and $f_2(z)$ are algebraically independent.*
(ii) *The operator $\frac{d}{dz}$ maps the ring $F[f_1(z), f_2(z), \ldots, f_N(z)]$ into itself.*

Then for any finite extension E of F,

$$\mathrm{card}\Big\{z \in \mathbb{C} : f_1(z) \in E,\ f_2(z) \in E, \ldots, f_N(z) \in E\Big\} \le 2\rho[E : \mathbb{Q}].$$

The order of growth of a meromorphic function $f(z)$ can be defined in one of two ways, and both revolve around the observation that the poles of a meromorphic function must be isolated; for if the poles were not isolated, then the reciprocal function, which would be meromorphic, would have a convergent set of zeros and thus must be identically zero. Hence we can find an entire function $g(z)$ whose zeros are precisely the poles of $f(z)$, and then define the order of growth of $f(z)$ to be the order of growth of the *entire* function $g(z)f(z)$.

Equivalently, a meromorphic function has a *finite order of growth* if there is an exponent κ satisfying

$$\max\{|f(z)| : |z| = R\} \le e^{R^{\kappa}},$$

for all values of R that avoid the poles of $f(z)$. The infimum of all such κ is the order of growth of the function. The previous imprecise definition can be made precise by saying that the order of growth of a meromorphic function $f(z)$ equals

$$\limsup_{R \to \infty} \frac{\log\log\max\{|f(z)| : |z| = R\}}{\log R}.$$

While we do not prove the Schneider–Lang Theorem, we do remark that its proof is essentially an elaboration of our proof of the transcendence of a nonzero period of an elliptic curve defined over the algebraic closure $\bar{\mathbb{Q}}$. In fact, as the next challenge suggests, Theorem 7.3 follows immediately from the Schneider–Lang Theorem.

Challenge 7.33 *Deduce Theorem 7.3 from the Schneider–Lang Theorem. (Hint: Let $f_1(z) = z$ and $f_2(z) = \wp(z)$ and assume that the nonzero period w is algebraic (and as before, $w/2 \notin W$). Now show that each element of the infinite sequence of points $z_n = \frac{w}{2} + nw$, for $n \in \mathbb{Z}^+$, simultaneously makes $f_1(z)$ and $f_2(z)$ algebraic and moreover, all of those algebraic values lie in the number field $\mathbb{Q}(w, \wp(w/2))$.)*

We close this section by offering some wonderful consequences of the Schneider–Lang Theorem.

COROLLARY 7.15 *Suppose that $\wp(z)$ is a Weierstrass elliptic function defined over the algebraic numbers. Suppose that $\alpha, \beta, u,$ and w are complex numbers satisfying the following conditions*

(*i*) *α and β are both algebraic, $\alpha \neq 0, 1$, and β is not in the endomorphism ring of the lattice associated with $\wp(z)$;*

(*ii*) *$\wp(u)$ is algebraic, and;*

(*iii*) *w is a nonzero period for $\wp(z)$.*

Then each of the ten numbers

$$u, \quad w, \quad e^u, \quad e^w, \quad \log \alpha, \quad \wp(\alpha), \quad \wp(\log \alpha), \quad \wp(\beta u) \quad \wp(\beta \log \alpha), \quad \wp(\beta \alpha)$$

is transcendental.

We leave the verification of Corollary 7.15 as the final meta-challenge of this chapter. Is it possible to produce a result regarding transcendental numbers that, in some sense, "tops" a corollary that asserts that ten numbers are all at once transcendental? While it certainly *is* possible, we will not attempt such a feat here. Instead, we will close our journey through classical transcendental number theory by transcending numbers themselves and considering transcendence issues in a far more formal setting.

Number 8

$$1 + \frac{1}{T^2 - T} + \frac{1}{(T^4 - T)(T^2 - T)^2} + \frac{1}{(T^8 - T)(T^4 - T)^2(T^2 - T)^4} + \cdots$$

Transcending Numbers and Discovering a More Formal e:
Function fields and the transcendence of $e_C(1)$

8.1 Moving beyond numbers

We conclude our tour of classical transcendental number theory by transcending the world of numbers themselves and ascending to the realm of formal power series. Specifically, we consider transcendence issues within the setting of function fields in a single variable over a finite field. While this theory has important implications in many different areas of mathematics, our goal here is to discover an object in this context that is analogous to the all-important exponential function e^z.

Before introducing function fields, we first pause to reflect back. While the various arguments from the preceding chapters have been elaborate and technically complicated, they all possessed a basic common trait: Central to each argument was the fundamental fact that e^z is both a group homomorphism and a periodic function. Through our study of formal power series in this chapter, we will revisit these fundamental features of e^z, see reflections of these properties in the new setting at hand, and discover connections with the Weierstrass $\wp$-function. These paths will lead us to find analogues of the celebrated numbers e and π and will allow us to prove that these new objects are indeed *transcendental*.

We now fix a prime number p, let q be a fixed positive integral power of p, and let F_q denote the finite field with q elements. Recall that F_q has characteristic p. The polynomial ring $F_q[T]$ will play the role of the integers $\mathbb{Z}$. Just as the smallest field containing $\mathbb{Z}$ is the field of all rational numbers, $\mathbb{Q}$, the smallest field containing $F_q[T]$ is the field of all rational functions $F_q(T)$, which we will denote by k.

Within this framework, the indeterminate T is transcendental over F_q, since all of its powers are F_q-linearly independent. However, we wish to study elements that are transcendental over k, which clearly T itself is not, since it is a zero of $f(z) = z - T \in k[z]$. Yet we will show that the quantity represented by the convergent series in this chapter's title is *transcendental* over $F_2(T)$. We begin by constructing a natural field extension of k where transcendental objects reside. In order to motivate the field extension construction to come, we first offer a whirlwind overview of the making of a real number.

An intuitive look at the construction of the real numbers

We begin with the field of rational numbers $\mathbb{Q}$ and immediately bring the absolute value $|\ |$ onto the scene. The absolute value is a function from $\mathbb{Q}$ into the non-negative rational numbers that possesses three defining traits: For any rational numbers r and s,

(i) $|r| = 0$ if and only if $r = 0$;
(ii) $|rs| = |r||s|$; and
(iii) $|r + s| \leq |r| + |s|$, the so-called *triangle inequality*.

The absolute value empowers us to measure distances between elements of $\mathbb{Q}$ and thus determine closeness. We recall that with respect to the absolute value $|\ |$, $\mathbb{Q}$ is not *complete*; that is, there are Cauchy sequences of rational numbers that do not converge to a rational number. For example, the sequence of rational numbers

$$0.1, \ 0.11, \ 0.110001, \ 0.110001000000000000000001, \ \ldots \qquad (8.1)$$

converges to Liouville's number $\sum_{n=1}^{\infty} 10^{-n!}$, which we saw in Chapter 1 is not in $\mathbb{Q}$. Intuitively, if we expand the field $\mathbb{Q}$ to include all limit points of all Cauchy sequences of rational numbers, then we have constructed an extension field of $\mathbb{Q}$, sometimes denoted by $\mathbb{Q}_\infty$ and more often referred to as the *real numbers*.

Equivalently and more precisely, we can express the complete field $\mathbb{Q}_\infty$ (a.k.a. $\mathbb{R}$) formally as the collection of all Laurent series in $1/10$ with coefficients from the finite set $\{0, 1, 2, \ldots, 9\}$:

$$\mathbb{Q}_\infty = \left\{ \sum_{n=-N}^{\infty} d_n \left(\frac{1}{10} \right)^n : N \in \mathbb{Z} \text{ and } d_n \in \{0, 1, 2, \ldots, 9\} \right\}.$$

Moreover, we can dismiss the formalism if we introduce the notion of a convergent infinite series. Given $a_1, a_2, \ldots \in \mathbb{Q}$, we say the infinite series $\sum_{n=1}^{\infty} a_n$ *converges* if its sequence of partial sums $\left\{ \sum_{n=1}^{N} a_n : N = 1, 2, \ldots \right\}$ is a Cauchy sequence. We then define the *sum of the series* to be the limit of this sequence of partial sums in $\mathbb{Q}_\infty$.

Challenge 8.1 *Show that the sequence of partial sums of any element from $\mathbb{Q}_\infty$ is a Cauchy sequence.*

Using this challenge we can identify each element of $\mathbb{Q}_\infty$ with a decimal expansion. Thus, for example, since the sequence of partial sums of $\sum_{n=0}^{\infty} 1/n!$ can

Continued

---

be shown to be a Cauchy sequence, we conclude that the infinite series converges to an element of $\mathbb{Q}_\infty$, one we have come to know as e. We note that the Laurent expansion of e—more familiarly known as the decimal expansion—is given by

$$e = 2\left(\frac{1}{10}\right)^0 + 7\left(\frac{1}{10}\right)^1 + 1\left(\frac{1}{10}\right)^2 + 8\left(\frac{1}{10}\right)^3 + 2\left(\frac{1}{10}\right)^4 + 8\left(\frac{1}{10}\right)^5 + \cdots.$$

We are now ready to develop a parallel theory over function fields by introducing an absolute value on the field $k = F_q(T)$ and then constructing the completion of the field k with respect to this new notion of closeness.

8.2 An intimate interlude—How to get close in function fields

For $\alpha \in k$, we know that $\alpha = \frac{f(T)}{g(T)}$ for some polynomials $f(T), g(T) \in F_q[T]$, with $g(T)$ not identically zero. Since the coefficients of the numerator and denominator of α all reside in the *finite* ground field F_q, they do not offer us a robust enough measure of relative sizes as heights did for polynomials in $\mathbb{Z}[z]$. Therefore in order to define an absolute value on k, we are led to consider the *degrees* of the polynomials in the numerators and denominators of the elements of k rather than their *coefficients*.

We define the map $|\ \ |_\infty : k \to [0, +\infty)$ as follows: We first declare $|0|_\infty = 0$; that is, the function maps the zero polynomial to 0. If we now suppose that $\alpha \in k \setminus \{0\}$ is expressed as $\alpha = \frac{f(T)}{g(T)}$, where $f(T), g(T) \in F_q[T]$, then we define

$$|\alpha|_\infty = \left|\frac{f(T)}{g(T)}\right|_\infty = q^{\deg(f) - \deg(g)}.$$

We remark that for any polynomial $f(T) \in F_q[T]$, we have $|f(T)|_\infty = q^{\deg(f)}$, where we recall that the degree of the zero polynomial is $-\infty$. This map exhibits several important properties. In fact, we have the following.

PROPOSITION 8.1 *The map $|\ \ |_\infty : k \to [0, +\infty)$ as defined above is an absolute value on k.*

Proof. In order to establish the proposition, we must show that the following three conditions hold: $|\alpha|_\infty = 0$ if and only if $\alpha = 0$; for all $\alpha_1, \alpha_2 \in k$, $|\alpha_1 \alpha_2|_\infty = |\alpha_1|_\infty |\alpha_2|_\infty$; and finally the triangle inequality: $|\alpha_1 + \alpha_2|_\infty \leq |\alpha_1|_\infty + |\alpha_2|_\infty$.

From the definition of $|\ \ |_\infty$, it immediately follows that $|\alpha|_\infty = 0$ if and only if $\alpha = 0$, that is, if and only if α is the zero polynomial. Next, if $\alpha_1, \alpha_2 \in k$, with

$\alpha_1 = \frac{f_1(T)}{g_1(T)}$ and $\alpha_2 = \frac{f_2(T)}{g_2(T)}$, then we see that

$$|\alpha_1 \alpha_2|_\infty = \left| \frac{f_1(T) f_2(T)}{g_1(T) g_2(T)} \right|_\infty = q^{\deg(f_1 f_2) - \deg(g_1 g_2)} = q^{\deg(f_1) - \deg(g_1)} q^{\deg(f_2) - \deg(g_2)}$$

$$= \left| \frac{f_1(T)}{g_1(T)} \right|_\infty \left| \frac{f_2(T)}{g_2(T)} \right|_\infty = |\alpha_1|_\infty |\alpha_2|_\infty.$$

Thus it only remains for us to verify the triangle inequality. Applying the identity we have just established, it follows that

$$|\alpha_1 + \alpha_2|_\infty = \left| \frac{f_1(T)}{g_1(T)} + \frac{f_2(T)}{g_2(T)} \right|_\infty = \left| \frac{f_1(T) g_2(T) + f_2(T) g_1(T)}{g_1(T) g_2(T)} \right|_\infty$$

$$= |f_1(T) g_2(T) + f_2(T) g_1(T)|_\infty \left| \frac{1}{g_1(T) g_2(T)} \right|_\infty$$

$$= \frac{|f_1(T) g_2(T) + f_2(T) g_1(T)|_\infty}{|g_1(T) g_2(T)|_\infty}$$

$$\leq \frac{q^{\max\{\deg(f_1) + \deg(g_2), \deg(f_2) + \deg(g_1)\}}}{q^{\deg(g_1) + \deg(g_2)}}$$

$$= \max \left\{ q^{\deg(f_1) - \deg(g_1)}, q^{\deg(f_2) - \deg(g_2)} \right\},$$

and hence we deduce that

$$|\alpha_1 + \alpha_2|_\infty \leq \max\{|\alpha_1|_\infty, |\alpha_2|_\infty\}. \tag{8.2}$$

Given that both $|\alpha_1|_\infty$ and $|\alpha_2|_\infty$ are non-negative, we trivially conclude that

$$|\alpha_1 + \alpha_2|_\infty \leq |\alpha_1|_\infty + |\alpha_2|_\infty,$$

which completes our proof. ∎

A strong postscript to our proof. While we have established that the map $|\ |_\infty$ is an absolute value on the function field k, we unintentionally proved much more. In fact, inequality (8.2) is much stronger than the triangle inequality and is, in fact, referred to as the *strong triangle inequality*. An absolute value that satisfies the strong triangle inequality is said to be a *nonarchimedean* absolute value. If we momentarily return to the usual absolute value $|\ |$ on $\mathbb{Q}$, we have that, for example,

$$2 = |1 + 1| \nleq \max\{|1|, |1|\} = 1,$$

and hence the usual absolute value on $\mathbb{Q}$ does not satisfy the strong triangle inequality and thus is merely *archimedean*. As we will soon discover, the additional strength afforded to us by the strong triangle inequality will empower us to turn traditionally treacherous analysis into smooth sailing.

We now turn our attention to the basic question of whether or not k is complete with respect to the absolute value $|\ |_\infty$. Let us consider the following sequence in k:

$$\frac{1}{T}, \quad \frac{T+1}{T^2}, \quad \frac{T^5+T^4+1}{T^6}, \quad \frac{T^{23}+T^{22}+T^{18}+1}{T^{24}}, \dots,$$

which when rewritten as

$$\frac{1}{T}, \quad \frac{1}{T}+\frac{1}{T^2}, \quad \frac{1}{T}+\frac{1}{T^2}+\frac{1}{T^6}, \quad \frac{1}{T}+\frac{1}{T^2}+\frac{1}{T^6}+\frac{1}{T^{24}}, \dots$$

begins to resemble the sequence in (8.1) with $\frac{1}{10}$ replaced by $\frac{1}{T}$. More precisely, our sequence is given by

$$\left\{ \sum_{n=1}^{N} \left(\frac{1}{T}\right)^{n!} : N = 1, 2, \dots \right\}.$$

We now assert that this sequence is a Cauchy sequence in k. To verify this claim, we observe that for any given $\varepsilon > 0$, there exists an index N such that $|(\frac{1}{T})^{N!}|_\infty = |\frac{1}{T^{N!}}|_\infty = q^{-N!} < \varepsilon$. Hence for any indices $s \geq t \geq N$, the strong triangle inequality implies that

$$\left| \sum_{n=1}^{s} \left(\frac{1}{T}\right)^{n!} - \sum_{n=1}^{t} \left(\frac{1}{T}\right)^{n!} \right|_\infty$$

$$= \left| \sum_{n=t+1}^{s} \left(\frac{1}{T}\right)^{n!} \right|_\infty \leq \max\left\{ \left|\left(\frac{1}{T}\right)^{(t+1)!}\right|_\infty, \left|\left(\frac{1}{T}\right)^{(t+2)!}\right|_\infty, \dots, \left|\left(\frac{1}{T}\right)^{s!}\right|_\infty \right\}$$

$$= \max\left\{ q^{-(t+1)!}, q^{-(t+2)!}, \dots, q^{-s!} \right\} = q^{-(t+1)!} \leq q^{-N!} < \varepsilon,$$

which establishes that we indeed have a Cauchy sequence.

The same reasoning used in Chapter 1 to show that the non-repeating decimal expansion $\sum_{n=1}^{\infty} 10^{-n!}$ is not in $\mathbb{Q}$ can be employed to show that $\sum_{n=1}^{\infty} \left(\frac{1}{T}\right)^{n!}$ is not in k. Thus we can confirm any hunches we may have had that k is *not* complete with respect to $|\ |_\infty$. We can complete k to obtain a complete field denoted by k_∞ just as we completed $\mathbb{Q}$ to obtain the field of real numbers $\mathbb{Q}_\infty$. Specifically, we let k_∞ be the field of *formal Laurent series* over F_q, that is,

$$k_\infty = \left\{ \sum_{n=-N}^{\infty} a_n \left(\frac{1}{T}\right)^n : N \in \mathbb{Z} \text{ and } a_n \in F_q \right\}. \tag{8.3}$$

Therefore we discover that the elements of k_∞ formally resemble the elements of $\mathbb{Q}_\infty$. There is one major difference between the complete fields $\mathbb{Q}_\infty$ and k_∞, and this important distinction revolves around the strong triangle inequality. In k_∞ we have the following wonderful theorem, which is a calculus student's fantasy when working over $\mathbb{R}$:

THEOREM 8.2 *Let* $\{\alpha_n\}$ *be a sequence in* k. *Then the infinite series* $\sum_{n=1}^{\infty} \alpha_n$ *converges in* k_∞ *if and only if*

$$\lim_{n \to \infty} |\alpha_n|_\infty = 0.$$

Challenge 8.2   *Prove Theorem 8.2   (Hint: Consider the argument that was employed to verify that the sequence* $\{\sum_{n=1}^{N} (1/T)^{n!}\}$ *is Cauchy in* k.)

In view of the definition in (8.3) it easily follows that k_∞ is uncountable and hence must contain elements that are *transcendental* over k. It is in the complete field k_∞ that we begin our search for an analogue of e.

8.3 A formal search for e

In 1936 Leonard Carlitz introduced a function $e_C(z) : k_\infty \to k_\infty$ that is the natural analogue to the usual exponential function e^z, but in the end is more closely related to the Weierstrass $\wp$-function. The Carlitz exponential function is a periodic entire function whose power series representation resembles that of e^z. We can describe this function either through its power series and then show that it is periodic, or through a study of periodic functions and then show that there is one such function whose power series is formally the same as that of e^z. Here we will pursue the first approach and discover the beautiful exponential function found by Carlitz.

In order to uncover an analogue of the power series $e^z = \sum_{n=0}^{\infty} \frac{z^n}{n!}$ over k_∞, we begin by uncovering the basic shape such a power series should possess. We will then move to the interesting quest of finding an analogue for $n!$ in $F_q[T]$.

The function $e_C(z)$ we wish to construct must not only have a power series expansion analogous to that of e^z, but should also share several other important properties of e^z. Foremost among the important properties of the exponential function is that it preserves algebraic structure as a group homomorphism from the additive group of $\mathbb{R}$ onto the multiplicative group of the positive real numbers; that is,

$$e^{a+b} = e^a e^b \quad \text{for all } a, b \in \mathbb{R}.$$

In fact, more structure is preserved: For all positive integers n and m, and real numbers a and b, we have

$$e^{na+mb} = \left(e^a\right)^n \left(e^b\right)^m. \tag{8.4}$$

Inspired by this simple observation, we now view k_∞ as an F_q-module, and thus the analogue of (8.4) involves F_q-linearity; that is, we wish the function $e_C(z)$ to satisfy the ring-structure-preserving relationship

$$e_C(a\xi + b\zeta) = a\,e_C(\xi) + b\,e_C(\zeta) \quad \text{for all } a, b \in F_q \quad \text{and all } \xi, \zeta \in k_\infty.$$

As the next proposition reveals, the previous identity implies that the power series for $e_C(z)$ must be of the form

$$e_C(z) = \sum_{r=0}^{\infty} d_r z^{q^r}. \tag{8.5}$$

PROPOSITION 8.3 *Let $L(z)$ denote the formal power series given by $L(z) = \sum_{n=0}^{\infty} a_n z^n$, where $a_n \in k$. Then $L(z)$ is F_q-linear if and only if $a_n = 0$ for all n that are not a power of q. That is, a power series over k is F_q-linear if and only if it has the form*

$$\sum_{r=0}^{\infty} a_r z^{q^r}.$$

Challenge 8.3 Prove the proposition. (Hints and suggestions: First establish the result for polynomial functions $f(z) \in k[z]$ by induction. Specifically, show that since the characteristic of F_q is p, $f(z_1 + z_2) = f(z_1) + f(z_2)$ for all $z_1, z_2 \in k_\infty$ if and only if all the exponents in $f(z)$ are powers of p. Next apply the fact that q is the smallest integer such that $a^q = a$ for all $a \in F_q$ to conclude that we have F_q-linearity if and only if the exponents of $f(z)$ are, in fact, powers of q. Now extend this result to power series.)

Thus in view of (8.5), it only remains to determine the appropriate coefficients $d_r \in k$ so that $e_C(z)$ formally resembles the power series for e^z. It therefore seems natural to select the coefficients to be reciprocals of *factorials*. Unfortunately, in this context, it is not clear what *factorial* even means. As we will discover in the next section, there is an intuitive definition for the factorial function in $F_q[T]$ that beautifully mirrors $n!$ for positive integers n and leads to the exponential function $e_C(z)$.

8.4 Finding the factorial in $F_q[T]$

We inspire the new notion of factorial in our present setting by carefully reconsidering the classical factorial over the positive integers. One way to view $n!$ is as $n! = \prod_{0 < m \leq n} m$. However, since we have no natural ordering on the elements of $F_q[T]$, we require a different point of view. Specifically, for a positive integer n, we define $n!'$ by

$$n!' = \prod_{0 < |m| \leq n} m, \tag{8.6}$$

which clearly satisfies $n!' = (-1)^n (n!)^2$. While this new "factorial" may appear silly, it does inspire our first ill-fated attempt at a generalization. For a polynomial $P(T) \in F_q[T]$, it now seems reasonable to consider

$$P(T)!' = \prod_{\substack{\deg(Q) \leq \deg(P), \\ Q(T) \text{ nonzero}}} Q(T).$$

We remark that since $|Q|_\infty = q^{\deg Q}$, the previous product can be rewritten so as to perfectly parallel the product in (8.6). In particular, we have that

$$P(T)!' = \prod_{0 < |Q|_\infty \leq |P|_\infty} Q(T).$$

Just as our definition of $n!'$ contained unwelcomed redundancies, so does the choice of $P(T)!'$. Specifically, any polynomial $Q(T) \in F_q[T]$ having $\deg(Q) \leq \deg(P)$ is represented in this product $(q-1)$ times, since if $\deg(Q) \leq \deg(P)$, then for any nonzero element $a \in F_q$, $\deg(aQ) = \deg(Q) \leq \deg(P)$. This observation implies that for any nonzero polynomial $Q(T)$ with $\deg(Q) \leq \deg(P)$, the product defining $P(T)!'$ contains the product $\prod_{a \in F_q^*} aQ(T) = Q(T)^{q-1} \prod_{a \in F_q^*} a$. Since F_q^*, the set nonzero elements of F_q, is a multiplicative group wherein -1 is the only element other than 1 that is its own inverse, we can simplify the last expression to $-Q(T)^{q-1}$. In view of this repetitious shortcoming, we can take another a step closer to finding the factorial by counting each polynomial $Q(T)$ in the product of $P(T)!'$ only once. In other words, we are led to consider $P(T)!''$ defined by

$$P(T)!'' = \prod_{\substack{\deg(Q) \leq \deg(P), \text{ with} \\ Q(T) \text{ nonzero and } monic}} Q(T).$$

We can simplify our definition of $P(T)!''$ by noting that in fact, it depends only on the *degree* of $P(T)$ and not on the polynomial $P(T)$ itself. To emphasize this important observation, we introduce some notation that both reflects this fact and leads to Carlitz's fruitful choice. For a positive integer n, we define $D_n \in F_q[T]$ by

$$D_n = D_n(T) = \prod_{\substack{Q(T) \in F_q[T], \, Q(T) \\ \text{monic, with } \deg(Q) = n}} Q(T).$$

Thus for any $P(T) \in F_q[T]$ with $\deg(P) = d$, we have the "factorial-esque" expression

$$P(T)!'' = D_d D_{d-1} \cdots D_1.$$

In a perhaps surprising turn of events, we will now discover that the individual D_n's, rather than the product of D_n's, will play the role of factorials within $F_q[T]$.

A more primal view of n!. The redundancy issues we observed earlier can also be resolved by examining the role of irreducible polynomials in these products. To inspire this important connection, we begin by recasting the usual factorial function $n!$ in terms of the $\mathbb{Z}$-analogue of irreducible polynomials: Prime numbers.

While it is clear that any positive prime number $p \leq n$ is a factor in $n!$; it is perhaps only slightly less clear that the number of integers less than or equal to n that p divides equals $\left[\frac{n}{p}\right]$, where we again write $[x]$ for the integer part of x. A few moments' reflection reveals that the power of p that appears in the prime factorization of $n!$ is $\left[\frac{n}{p}\right] + \left[\frac{n}{p^2}\right] + \cdots$. Thus for any positive integer n we have

$$n! = \prod_{p \text{ prime}} p^{s_p}, \quad \text{where } s_p = \left[\frac{n}{p}\right] + \left[\frac{n}{p^2}\right] + \cdots .$$

If we consider the analogous product in $F_q[T]$, then we finally arrive at our definition of the factorial function over $F_q[T]$. Specifically, for any polynomial $P(T) \in F_q[T]$, we define the *factorial of $P(T)$, $P(T)!$,* by

$$P(T)! = \prod_{\substack{Q(T) \in F_q[T],\ Q(T) \\ \text{monic and irreducible}}} Q(T)^{s_Q}, \quad \text{where } s_Q = \left[\frac{|P|_\infty}{|Q|_\infty}\right] + \left[\frac{|P|_\infty}{|Q|_\infty^2}\right] + \cdots .$$

Just as we noted with $P(T)!''$, here we remark that $P(T)!$ depends only on $\deg(P)$. Moreover, if $\deg(P) = t$, then the next challenge asserts the key factorial fact that $P(T)! = D_t$.

Challenge 8.4 *Verify the identity*

$$D_t = \prod_{\substack{Q(T) \in F_q[T],\ Q(T) \\ \text{monic and irreducible}}} Q(T)^{s_Q}, \quad \text{where } s_Q = \left[\frac{q^t}{q^{\deg(Q)}}\right] + \left[\frac{q^t}{q^{2\deg(Q)}}\right] + \cdots .$$

(Hint: Let $Q(T)$ be a fixed monic irreducible polynomial of degree $d \leq t$. Count the number of monic polynomials of degree t that are divisible by $Q(T)$, $Q(T)^2, \ldots, Q(T)^{[t/d]}$ and then verify that $Q(T)$ appears precisely s_Q times in the factorization of D_t into irreducible polynomials over F_q.)

These observations certainly confirm that D_n is an amazing analogue of $n!$.

Factoring factorials. To further develop the parallel between D_n and $n!$, we recall that in our proof of the transcendence of e we utilized the basic fact that for positive integers m and n,

$$m! \text{ divides } n! \text{ if and only if } m \leq n.$$

Of course, this elementary identity holds because if $m < n$, then $n! = n \times (n - 1) \times \cdots \times (n - (n - m - 1)) \times m!$. We now produce the analogue of this divisibility property for our new factorial D_n. In particular, for $m < n$, we wish to study how D_m relates to D_n. Not surprisingly, such a divisibility property in this context will enter into the proof that the analogue of e in k_∞ is transcendent. We uncover this relationship between D_m and D_n through an alternative albeit equivalent representation of D_n. This new formulation of D_n follows from an important result concerning the factorization of polynomials into irreducible polynomials in the ring $F_q[T][z]$.

Algebraic Excursion: Irreducible polynomials over F_q

The factorial formula that we seek requires a fairly deep result concerning zeros of polynomials with coefficients in F_q. We begin with the following proposition.

PROPOSITION 8.4 *The polynomial $T^{q^t} - T$ equals the product of all monic irreducible polynomials $P(T) \in F_q[T]$ with $\deg(P)$ dividing t.*

The truth of Proposition 8.4 depends on the following perhaps surprising claim: *The elements of the field F_{q^t} are precisely all the zeros of the polynomial $f(T) = T^{q^t} - T$.*

Challenge 8.5 *Employ Lagrange's Theorem, which asserts that if G is a finite group, then the order of an element of G divides the order of the group G, to verify the previous claim.*

Proof of Proposition 8.4. To establish the proposition, we first remark that if we let $F_{q^t} = \{\alpha_1, \alpha_2, \ldots, \alpha_{q^t}\}$, then the previous claim reveals that

$$T^{q^t} - T = \prod_{n=1}^{q^t} (T - \alpha_n). \tag{8.7}$$

Now let $P(T) \in F_q[T]$ be an irreducible factor of $T^{q^t} - T$ over F_q. In view of (8.7), we have that $P(T)$ is monic and has, for some n, α_n as a zero. We now write $d = \deg(P)$. Since the field extension $F_q(\alpha_n)$ is a subfield of F_{q^t}, $F_q(\alpha_n)$ is a vector space of dimension d over F_q, and since F_{q^t} is a vector space of dimension t over F_q, we conclude that d divides t. Therefore we discover that *every* irreducible factor $P(T)$ of $T^{q^t} - T$ over F_q satisfies $\deg(P)|t$.

Next, suppose that $P(T) \in F_q[T]$ is a monic irreducible polynomial of degree d with $d|t$. If α is a zero of $P(T)$, then the field $F_q(\alpha)$ is isomorphic to the field F_{q^d}, since they are both vector spaces of dimension d over F_q. However, from basic field theory it follows that in any fixed algebraic closure of F_q there is only *one* field of any given degree. Thus $F_q(\alpha) = F_{q^d}$ and hence $\alpha \in F_{q^d}$. By the previous claim we now conclude that α is a zero of $T^{q^d} - T$. Since $d|t$, $T^{q^d} - T$ divides $T^{q^t} - T$, and so $P(T)$ is a factor of $T^{q^t} - T$, which completes our proof. ∎

As an immediate consequence of Proposition 8.4 we have:

COROLLARY 8.5 *A monic irreducible polynomial of degree d divides $T^{q^t} - T$ if and only if d divides t.*

In fact, we remark that the product in (8.7) together with Proposition 8.4 implies the following more general observation, which is the main result of this excursion.

Continued

PROPOSITION 8.6 *If $P(T)$ is a monic irreducible polynomial whose degree divides t, then $P(T)$ appears in the factorization of $T^{q^t} - T$ precisely once. That is,*

$$T^{q^t} - T = \prod_{\substack{Q(T) \in F_q[T],\ Q(T)\ \text{monic},\\ \text{irreducible with } \deg(Q)|t}} Q(T).$$

Challenge 8.6   *Give a proof of Proposition 8.6.*

Proposition 8.6 provides us with a clue as to how to rewrite the expression

$$D_t = \prod_{\substack{Q(T) \in F_q[T],\ Q(T)\\ \text{monic and irreducible}}} Q(T)^{s_Q}, \quad \text{where } s_Q = \left[\frac{q^t}{q^{\deg(Q)}}\right] + \left[\frac{q^t}{q^{2\deg(Q)}}\right] + \cdots,$$

(8.8)

in a form that will allow us to uncover the relationship between D_n and D_m for $m < n$. The basic idea is to regroup the irreducible factors $Q(T)$ from the previous product so that we can apply Proposition 8.6. In particular, the following challenge allows us to conclude that for $t \geq 1$,

$$D_t = \left(T^{q^t} - T\right)\left(T^{q^{t-1}} - T\right)^q \left(T^{q^{t-2}} - T\right)^{q^2} \cdots \left(T^q - T\right)^{q^{t-1}} \quad (8.9)$$

(we recall that by definition, $D_0 = 1$).

Challenge 8.7   *Verify identity (8.9) by applying Proposition 8.6 to rewrite the product as*

$$\prod_{n=0}^{t-1} \prod_{\substack{Q(T) \in F_q[T],\ Q(T)\ \text{monic},\\ \text{irreducible with } \deg(Q)|t-n}} Q(T)^{q^n}.$$

Then show that the total exponent for a given monic irreducible polynomial $Q(T)$ appearing in the above product is

$$\sum_{m=1}^{[t/\deg(Q)]} q^{t-m\deg(Q)},$$

which equals the exponent s_Q.

For an integer $t \geq 0$, Carlitz introduced the simplifying notational convention of writing

$$[t] = T^{q^t} - T.$$

In view of this new notation, we can express identity (8.9) in the more suggestive "factorial-evoking" form

$$D_t = [t][t-1]^q [t-2]^{q^2} \cdots [1]^{q^{t-1}}.$$

Thus for $m \leq n$ we have

$$D_n = [n][n-1]^q [n-2]^{q^2} \cdots [m+1]^{q^{n-m-1}} [m]^{q^{n-m}} [m-1]^{q^{n-m+1}} \cdots [1]^{q^{n-1}},$$

which, if we write the exponent on [1] as $q^{m-1+(n-m)}$, yields

$$D_n = [n][n-1]^q [n-2]^{q^2} \cdots [n-(n-m-1)]^{q^{n-m-1}}$$

$$\times \left([m][m-1]^q \cdots [1]^{q^{m-1}} \right)^{q^{n-m}}$$

$$= [n][n-1]^q [n-2]^{q^2} \cdots [n-(n-m-1)]^{q^{n-m-1}} D_m^{q^{n-m}}.$$

Therefore we see that for any positive integer r and any integer $j \geq 0$, it follows that

$$D_{r+j} = [r+j][r+j-1]^q [r+j-2]^{q^2} \cdots [r+1]^{q^{j-1}} D_r^{q^j},$$

and we at long last uncover the stronger than anticipated fact that

$$D_r^{q^j} \text{ divides } D_{r+j}. \tag{8.10}$$

Given our extended intuitive development of the factorial in $F_q[T]$, hopefully we are not surprised to discover that Carlitz adopted D_n as the analogue of $n!$ and thus defined an analogue of e^z over k_∞ by

$$e_C(z) = \sum_{r=0}^{\infty} \frac{z^{q^r}}{D_r}.$$

It is a testament to Carlitz's insight that $e_C(z)$ not only *resembles* the function e^z but also shares many of its analytic and algebraic properties. Even more astonishing is that $e_C(z)$ also exhibits some fundamental properties enjoyed by the elliptic functions introduced in Chapter 7. Thus this one basic function provides a bridge between the F_q-analogues of both e^z and $\wp(z)$, thereby solidifying its rightful place on the list of the most important functions in number theory.

8.5 The transcendence of $e_C(1)$

Before we consider the analytic and algebraic properties of the function $e_C(z)$, we first establish the transcendence of the value corresponding to e, namely $e_C(1)$.

Several years after Carlitz introduced his exponential function, Luther Wade began to examine the transcendence of special values of $e_C(z)$. The proof we develop here for the transcendence of $e_C(1)$ is due to Wade and perfectly parallels Hermite's proof of the transcendence of e from Chapter 2.

Every transcendence proof we have seen has ultimately invoked the Fundamental Principle of Number Theory. Thus it should not be very shocking to learn that the proof of the transcendence of $e_C(1)$ will depend on an $F_q[T]$-analogue of that basic principle. Since the ring $F_q[T]$ does not come equipped with a natural ordering, we first restate the Fundamental Principle of Number Theory without reference to the far-too-familiar notion of *betweenness*:

THE FUNDAMENTAL PRINCIPLE OF NUMBER THEORY REVISITED. *If $\mathcal{N}$ is a nonzero integer, then $|\mathcal{N}| \geq 1$.*

This statement is clearly equivalent to our original principle but subtly shifts the emphasis from the *ordering* of integers to the *absolute values* of integers. Using the absolute value $|\ |_\infty$ on $F_q[T]$, we can now state a natural $F_q[T]$-analogue of this principle:

THE FUNDAMENTAL PRINCIPLE OF NUMBER THEORY REVISITED IN $F_q[T]$. *If $P(T)$ is a nonzero element of $F_q[T]$, then $|P(T)|_\infty \geq 1$.*

We are now ready to establish the transcendence of $e_C(1)$.

THEOREM 8.7 *The formal power series*

$$e_C(1) = \sum_{r=0}^{\infty} \frac{1}{D_r} = 1 + \frac{1}{T^q - T} + \frac{1}{(T^{q^2} - T)(T^q - T)^q}$$

$$+ \frac{1}{(T^{q^3} - T)(T^{q^2} - T)^q(T^q - T)^{q^2}} + \cdots$$

is transcendental over $F_q(T)$.

An intuitive overview of the proof of the transcendence of $e_C(1)$.

We hope that by now, the following outline will have a familiar feel.

Step 1. We assume that $e_C(1)$ is algebraic over $F_q(T)$. Thus there exists a nonzero polynomial $Q(z) \in F_q[T][z]$ satisfying $Q(e_C(1)) = 0$.

Step 2. We substitute the power series representation of $e_C(1)$ into the assumed identity $Q(e_C(1)) = 0$ and obtain an infinite series that sums to 0.

Step 3. We split the infinite series from Step 2 into two pieces: A polynomial main term $\mathcal{P}$, and its tail $\mathcal{T}$, both elements of $F_q(T)$. We then multiply our identity through by an appropriate denominator $\delta \in F_q[T]$, so that $\delta\mathcal{P}$ and $\delta\mathcal{T}$ are in $F_q[T]$.

Continued

Step 4. We show that it is possible to complete Step 3 in such a manner that $\delta \mathcal{T}$ is a nonzero element of $F_q[T]$ that violates the Fundamental Principle of Number Theory in $F_q[T]$.

Just as in the proof of Theorem 2.3, this proof depends on a judicious choice of $\mathcal{P}$ and (thus indirectly) $\mathcal{T}$. Happily, we have already developed much of the detailed analysis of polynomials with coefficients in $F_q[T]$ that is required to make such a difficult choice.

Proof of Theorem 8.7.   We begin by assuming that $e_C(1)$ is algebraic and let $Q(z) \in F_q[T][z]$ be a nonzero polynomial satisfying $Q(e_C(1)) = 0$. We now show that $Q(z)$ can be selected so as to be in sympathy with the power series of $e_C(z)$. In particular, the following lemma allows us to assume that $Q(z)$ has an especially attractive form; namely, the degree of each term of $Q(z)$ is a power of q. Once the lemma is established we may apply Proposition 8.3 to deduce the extremely valuable property that $Q(\xi + \zeta) = Q(\xi) + Q(\zeta)$ for all $\xi, \zeta \in k_\infty$.

LEMMA 8.8   *Every nonzero polynomial in $F_q[T][z]$ divides a polynomial of the form $\sum_{j=l}^{m} A_j z^{q^j}$ where the coefficients $A_j = A_j(T)$ lie in $F_q[T]$ and neither A_l nor A_m is the zero polynomial.*

*Proof of Lemma 8.8 (a.k.a. **Challenge 8.8**).   Suppose that the polynomial $Q(z)$ has degree d. Show that if we divide each monomial z^{q^j}, for $j = 0, 1, \ldots, d$, by $Q(z)$, we obtain $d + 1$ identities of the form*

$$z^{q^j} = Q_j(z)Q(z) + R_j(z),$$

where $Q_j(z)$ and $R_j(z)$ are polynomials in $F_q[T][z]$ with $\deg(R_j) < d$. Thus for each j, if we write

$$R_j(z) = b_{j,d-1}z^{d-1} + b_{j,d-2}z^{d-2} + \cdots + b_{j,0},$$

then show that

$$B = \begin{pmatrix} b_{0,d-1} & b_{1,d-1} & \cdots & b_{d,d-1} \\ b_{0,d-2} & b_{1,d-2} & \cdots & b_{d,d-2} \\ \vdots & \vdots & & \vdots \\ b_{0,0} & b_{1,0} & \cdots & b_{d,0} \end{pmatrix}$$

is a $d \times (d+1)$ matrix over $F_q[T]$. Argue that there must exist a nonzero vector $\vec{V} \in k^{d+1} = (F_q(T))^{d+1}$ such that $B\vec{V} = \vec{0}$. Thus prove that there exist rational functions $V_j \in k$, not all identically zero, satisfying

$$\sum_{j=0}^{d} V_j z^{q^j} \equiv 0 \quad mod\ Q(z).$$

Finally, verify that we can then multiply this congruence by an element of $F_q[T]$ to establish the lemma. ∎

In view of Lemma 8.8, we can assume, without loss of generality, that the polynomial $Q(z)$ that vanishes at $e_C(1)$ is of the form $Q(z) = \sum_{j=l}^{m} A_j z^{q^j}$, with A_l and A_m nonzero elements of $F_q[T]$. Thus given our assumption that $Q(e_C(1)) = 0$, together with Proposition 8.3, we see that

$$Q(e_C(1)) = \sum_{j=l}^{m} A_j \left(\sum_{r=0}^{\infty} \frac{1}{D_r} \right)^{q^j} = \sum_{j=l}^{m} A_j \left(\sum_{r=0}^{\infty} \frac{1}{D_r^{q^j}} \right) = 0.$$

We now rewrite this infinite series by reordering the terms in the summation so as to simplify the denominators. (For the analytically fastidious reader, we note that by the strong triangle inequality, all convergence is absolute, and thus we may rearrange the order of the terms of the series without changing the value of the series.) In particular, for $t \geq l$, if we define the expression

$$B_t = \sum_{\substack{i+j=t \\ l \leq j \leq m}} \frac{A_j D_t}{D_i^{q^j}},$$

then

$$\sum_{j=l}^{m} A_j \left(\sum_{r=0}^{\infty} \frac{1}{D_r^{q^j}} \right) = \sum_{t=l}^{\infty} \frac{B_t}{D_t} = 0. \tag{8.11}$$

Challenge 8.9   *Given the definition of the B_t's, verify identity* (8.11) *and then prove the spectacular fact that $B_t \in F_q[T]$. (Hint: Recall from* (8.10) *that if $r + j = t$, $D_r^{q^j}$ divides D_t. Now use the relationship*

$$\frac{D_t}{D_r^{q^j}} = [t][t-1]^q[t-2]^{q^2} \cdots [r+1]^{q^{t-r+1}}$$

to establish that each B_t is in $F_q[T]$.)

As in our proof of the transcendence of e, we now fix an index $b > l$, and break the series of (8.11) into two sums:

$$\sum_{t=l}^{b} \frac{B_t}{D_t} + \sum_{t=b+1}^{\infty} \frac{B_t}{D_t} = 0. \tag{8.12}$$

Multiplying (8.12) by D_b results in two terms $\mathcal{P}_b$ and $\mathcal{T}_b$ defined, respectively, by

$$D_b \sum_{t=l}^{b} \frac{B_t}{D_t} + D_b \sum_{t=b+1}^{\infty} \frac{B_t}{D_t} = \mathcal{P}_b + \mathcal{T}_b = 0. \tag{8.13}$$

Challenge 8.10 *Show that $\mathscr{P}_b \in F_q[T]$ by verifying that for $t \leq b$,*

$$\frac{B_t D_b}{D_t} = \sum_{j=l}^{m} A_j[b][b-1]^q[b-2]^{q^2} \cdots [b-t+1]^{q^{t-1}} \Delta_j,$$

where

$$\Delta_j = \frac{[b-t]^{q^t}[b-t-1]^{q^{t+1}} \cdots [1]^{q^b}}{[t-j]^{q^j}[t-j-1]^{q^{j+1}} \cdots [1]^{q^t}}$$

is in $F_q[T]$.

Since (8.13) is equivalent to $\mathscr{T}_p = -\mathscr{P}_b$, the previous challenge allows us to conclude that $\mathscr{T}_p \in F_q[T]$. It is the polynomial $\mathscr{T}_p$ that will eventually contradict the Fundamental Principle of Number Theory in $F_q[T]$. Thus as a first step, we must verify that $\mathscr{T}_b$ is a *nonzero* polynomial. This assertion is equivalent to the far easier task of showing that the *finite* sum $\mathscr{P}_b$ is nonzero.

From divisibility to $\mathscr{P}_b \neq 0$. As with every transcendence proof, the most difficult step is establishing that the element we carefully constructed is nonzero. We recall that in our proof of the transcendence of e, we employed divisibility properties of the integers (and hence indirectly used unique factorization) to verify that the integer we constructed was nonzero. In the present proof it is divisibility properties of polynomials over a finite field that are required to establish the corresponding result. Specifically, we will show that $\mathscr{P}_b$ is nonzero by demonstrating that there exists a nonzero polynomial $G(T) \in F_q[T]$ that divides all but one of the summands in $\mathscr{P}_b$. In effect, we show that $\mathscr{P}_b \neq 0$ by showing that $\mathscr{P}_b \not\equiv 0 \mod G(T)$.

Without any fanfare we reveal that $G(T) = [b-l]$, that is, $G(T) = T^{q^{b-l}} - T$; and we henceforth assume that b is taken to be so large that

$$\max\{\deg(A_l), \deg(A_{l+1}), \ldots, \deg(A_m)\} < q^{b-l}, \tag{8.14}$$

where we recall that the A_j's are the coefficients of the polynomial $Q(z)$. The next challenge reveals that $G(T)$ divides $B_t D_b / D_t$ for all t satisfying $l < t \leq b$ but does *not* divide $B_l D_b / D_l$.

Challenge 8.11 *Show that $\mathscr{P}_b = \sum_{t=l}^{b} \frac{B_t D_b}{D_t}$ is not the zero polynomial by first writing each of the terms implicit in the sum $B_t D_b / D_t$ as*

$$\frac{A_j D_b}{D_{t-j}^{q^j}} = A_j[b][b-1]^q[b-2]^{q^2} \cdots [b-t+1]^{q^{t-1}}$$

$$\times \left(\frac{[b-t]^{q^t}[b-t-1]^{q^{t+1}} \cdots [1]^{q^b}}{[t-j]^{q^j}[t-j-1]^{q^{j+1}} \cdots [1]^{q^t}} \right),$$

and then noting that the right-hand side is a product of two factors both from $F_q[T]$:

$$A_j[b][b-1]^q[b-2]^{q^2}\cdots[b-t+1]^{q^{t-1}} \quad \text{and} \quad \frac{[b-t]^{q^t}[b-t-1]^{q^{t+1}}\cdots[1]^{q^b}}{[t-j]^{q^j}[t-j-1]^{q^{j+1}}\cdots[1]^{q^t}}.$$

Now show that for all $t \geq l+1$, $G(T)$ divides the first factor, while for $t = l$, $G(T)$ does not divide either factor. (Hint: It might be useful to apply the observations that $[m]$ divides $[n]$ if and only if $T^{q^m} - T$ divides $T^{q^n} - T$; which holds if and only if every irreducible factor of $T^{q^m} - T$ divides $T^{q^n} - T$, which holds if and only if every divisor of m is a divisor of n, which holds if and only if m divides n.)

As $\mathcal{T}_b = -\mathcal{P}_b$, the previous challenge implies that $\mathcal{T}_p$ is a *nonzero* polynomial in $F_q[T]$. We now are ready to run head-on into our contradiction.

The fundamentally impossible upper bound: $|\mathcal{T}_b|_\infty < 1$. We begin by recalling that

$$\mathcal{T}_b = D_b \sum_{t=b+1}^{\infty} \frac{B_t}{D_t} = \sum_{t=b+1}^{\infty} \sum_{j=l}^{m} \frac{A_j D_b}{D_{t-j}^{q^j}}.$$

Since we wish to produce an upper bound for the quantity $|\mathcal{T}_b|_\infty$, we require bounds for the degrees of the various polynomials in the numerators and denominators of the terms in the series for $\mathcal{T}_b$. By (8.9), we see that D_b is a product of b factors each of degree q^b; hence it easily follows that $\deg(D_b) = bq^b$. Similarly, we have that $\deg\left(D_{t-j}^{q^j}\right) = q^j\left((t-j)q^{t-j}\right) = (t-j)q^t$. Given that $l \leq j \leq m$ and $b+1 \leq t$, we conclude that

$$(b+1-m)q^{b+1} \leq \deg\left(D_{t-j}^{q^j}\right).$$

Pulling all these observations together, we see that

$$\left|\frac{A_j D_b}{D_{t-j}^{q^j}}\right|_\infty \leq q^{\max_{l \leq j \leq m}\{\deg(A_j)\}+bq^b-(b+1-m)q^{b+1}}.$$

Challenge 8.12 *Show that for all sufficiently large b, $\left|A_j D_b/D_{t-j}^{q^j}\right|_\infty < 1$ for each $j = l, \ldots, m$. Thus $\mathcal{T}_b$ is an infinite sum of terms each having absolute value less than 1. Apply the strong triangle inequality to deduce that $|\mathcal{T}_b|_\infty < 1$.*

Therefore, by the previous challenge, our assumption that $e_C(1)$ is algebraic has led us to a nonzero element of $F_q[T]$ that violates the Fundamental Principle of Number Theory in $F_q[T]$, and thus we have established the transcendence of $e_C(1)$. ∎

Closing remarks regarding the transcendence of $e_C(1)$. We conclude this section with several observations. First, we note that the transcendence of $e_C(1) = \sum_{r=0}^{\infty} \frac{1}{D_r}$ depends only upon the divisibility properties of the terms in the series and not on any other special properties $e_C(z)$ might possess. Realizing this fact, Wade also established the transcendence of any series of the form

$$\sum_{r=0}^{\infty} \frac{C_r}{D_r},$$

where infinitely many of the polynomials $C_r = C_r(T) \in F_q[T]$ are nonzero and satisfy the additional condition that the sequence $\{\deg(C_r)\}$ does not grow too quickly as a function of r. Specifically, for infinitely many r,

$$\deg(C_r) \leq (r - (c_r + 1))q^r - (r - 1)q^{(r-1)},$$

where $\lim_{r \to \infty} c_r = \infty$.

Next we remark that Carlitz's first analogue of the exponential function in this context was actually the *alternating* series

$$e(z) = \sum_{k=0}^{\infty} (-1)^k \frac{z^{q^k}}{D_k}.$$

While the function $e_C(z)$ more closely parallels the function e^z and now is viewed as more standard, we do note that with hardly any modifications, the previous argument establishes the transcendence of $e(1)$.

Finally, we note that Wade also demonstrated the transcendence of $f(1)$, where $f(z) = \sum_{r=0}^{\infty} \frac{z^{q^r}}{[r]}$, which is another possible analogue to e^z in this context. The issue of why $e_C(z)$ is a closer and thus better analogue to e^z is addressed in the next section.

8.6 A repeated look at $e_C(z)$ through periodicity

In order to set the scene for our analysis ahead, we first return to the more familiar exponential function e^z, and recall that this function is periodic modulo $2\pi i\mathbb{Z}$; that is, for any $n \in \mathbb{Z}$, $e^{z+2\pi in} = e^z$. We remark that the periods are integer multiples of $2\pi i$, a number that we can view as the product of $2i$, an algebraic number, and π, an important constant that we now understand is transcendental. Here we find the analogues of these observations within k_∞. We will discover that our journey into periodicity will naturally lead us to an element in k_∞ that corresponds to π in $\mathbb{R}$.

Our goal here is to construct an entire function that captures the periodicity of e^z. We recall that in the present setting, the polynomial ring $F_q[T]$ plays the same role in the complete field k_∞ as $\mathbb{Z}$ plays in $\mathbb{R}$. Therefore we initially seek an analogue to the function $e^{2\pi iz}$, which is periodic with respect to $\mathbb{Z}$. In other words, we first seek an entire function $f: k_\infty \to k_\infty$ that is periodic modulo $F_q[T]$; that is, $f(z + A) = f(z)$ for all $A \in F_q[T]$. Our hope is to then connect this function with $e_C(z)$, just as classically the function $e^{2\pi i z}$ is connected by a change of variables to e^z.

A wonderfully simple insight.   Suppose we have a function $f(z)$ with two very special properties: First, $f(A) = 0$ for all $A \in F_q[T]$; and second, $f(z)$ is F_q-linear. We now claim that this special function $f(z)$ would be periodic with respect to $F_q[T]$: For any $A \in F_q[T]$, we have $f(z+A) = f(z) + f(A) = f(z)$. This simple insight offers us a path to follow: Construct an F_q-linear function that vanishes on all of $F_q[T]$.

We saw in the previous chapter that it is possible to construct a function with prescribed zeros by looking at an appropriately modified infinite product. But we want more in the present setting than simply an entire function vanishing at all elements of $F_q[T]$. We want an $F_q[T]$-linear function that vanishes at all elements of $F_q[T]$. Moreover, we also want this function to lead to an F_q-analogue to e^z. We accomplish this feat not by simply writing down an infinite product and examinig its convergence, but by obtaining this function as a *limit* of F_q-linear polynomials that vanish on larger and larger nested subsets of $F_q[T]$.

We first specify the subsets of $F_q[T]$ that will be the zeros of a sequence of polynomials. For a positive integer d, we let $\mathcal{F}(d) = \{A \in F_q[T] : \deg(A) < d\}$, and as a first attempt to construct a function that vanishes at all points of $F_q[T]$, we define the polynomial

$$e_d(z) = \prod_{A \in \mathcal{F}(d)} (z - A),$$

which does indeed vanish for all $A \in F_q[T]$ of degrees less than d. Before getting too attached to this function, we caution that we will eventually have to modify the function $e_d(z)$ before we let d approach infinity, since the limit of these polynomials as $d \to \infty$ is not an entire function. As an illustration, for $z = T^{-1}$, the limit

$$\lim_{d\to\infty} e_d\left(T^{-1}\right) = \lim_{d\to\infty} \prod_{A \in \mathcal{F}(d)} \left(\frac{1}{T} - A\right)$$

does not exist, since $\left|\prod_{A \in \mathcal{F}(d)} \left(\frac{1}{T} - A\right)\right|_\infty$ increases without bound as $d \to \infty$.

Before we adjust our polynomial $e_d(z)$ so that the limit converges to an entire function, we first discover a spectacular property about our natural function $e_d(z)$: It is F_q-linear, and hence periodic with respect to $\mathcal{F}(d)$. To establish this surprising claim, we first represent $e_d(z)$ explicitly as a polynomial whose coefficients, in the limit, converge to quantities that begin to resemble the coefficients of the power series for $e_C(z)$.

Determining $e_d(z)$ via determinants.   One of the most insightful ways of analyzing the product in $e_d(z)$ follows from the realization that $e_d(z)$ is connected with determinants of very special matrices. In particular, we will discover that $e_d(z)$ is actually a ratio of two such determinants.

The connection between determinants and the function $e_d(z)$ becomes clearer if we view $e_d(z)$ as a multiple product:

$$e_d(z) = \prod_{a_0 \in F_q} \prod_{a_1 \in F_q} \cdots \prod_{a_{d-1} \in F_q} \left(z - \left(a_0 + a_1 T + \cdots + a_{d-1}T^{d-1}\right)\right), \quad (8.15)$$

which slightly resembles the well-known *Vandermonde determinant*

$$\det \begin{pmatrix} 1 & 1 & \cdots & 1 \\ x_1 & x_2 & \cdots & x_n \\ x_1^2 & x_2^2 & \cdots & x_n^2 \\ \vdots & \vdots & & \vdots \\ x_1^{n-1} & x_2^{n-1} & \cdots & x_n^{n-1} \end{pmatrix} = \prod_{1 \le t < s \le n} (x_s - x_t).$$

We now introduce the F_q-analogue to the Vandermonde determinant, the less well known Moore determinant.

For $w_1, w_2, \ldots, w_n$ in k_∞, we define the *Moore determinant* by

$$\det \begin{pmatrix} w_1 & w_2 & \cdots & w_n \\ w_1^q & w_2^q & \cdots & w_n^q \\ w_1^{q^2} & w_2^{q^2} & \cdots & w_n^{q^2} \\ \vdots & \vdots & & \vdots \\ w_1^{q^{n-1}} & w_2^{q^{n-1}} & \cdots & w_n^{q^{n-1}} \end{pmatrix},$$

which we denote by $\Delta(w_1, w_2, \ldots, w_n)$. We remark that one can verify the following identity by induction, which more closely resembles (8.15):

$$\Delta(w_1, w_2, \ldots, w_n) = \prod_{i=1}^{d} \left(\prod_{a_{i-1} \in F_q} \prod_{a_{i-2} \in F_q} \cdots \prod_{a_1 \in F_q} (w_i + a_{i-1}w_{i-1} + \cdots + a_1 w_1) \right).$$

The following result highlights one of the important properties of Moore determinants, which is also the key to connecting them with $e_C(z)$.

PROPOSITION 8.9 *Let $w_1, w_2, \ldots, w_n$ be elements of k_∞. Then $w_1, w_2, \ldots, w_n$ are linearly independent over F_q if and only if $\Delta(w_1, w_2, \ldots, w_n) \neq 0$.*

Challenge 8.13 *Prove Proposition 8.9. (Hint: First assume that $\Delta(w_1, w_2, \ldots, w_n) \neq 0$, and that there exist $a_1, a_2, \ldots, a_n$ all in F_q satisfying $a_1 w_1 + a_2 w_2 + \cdots + a_n w_n = 0$. Exponentiating the identity to the power q^r, for $r = 0, 1, \ldots, n-1$,* *we obtain the linear system*

$$
\begin{pmatrix}
w_1 & w_2 & \cdots & w_n \\
w_1^q & w_2^q & \cdots & w_n^q \\
w_1^{q^2} & w_2^{q^2} & \cdots & w_n^{q^2} \\
\vdots & \vdots & & \vdots \\
w_1^{q^{n-1}} & w_2^{q^{n-1}} & \cdots & w_n^{q^{n-1}}
\end{pmatrix}
\begin{pmatrix}
a_1 \\ a_2 \\ \vdots \\ a_n
\end{pmatrix}
=
\begin{pmatrix}
0 \\ 0 \\ 0 \\ \vdots \\ 0
\end{pmatrix}.
\tag{8.16}
$$

Since $\Delta(w_1, w_2, \ldots, w_n) \neq 0$, it follows that $a_1 = a_2 = \cdots = a_n = 0$.

On the other hand, if $w_1, w_2, \ldots, w_n$ are linearly independent over F_q, then we establish $\Delta(w_1, w_2, \ldots, w_n) \neq 0$ by induction. The base case $n = 1$ is immediate. We now assume that the result holds for any index $m < n$. If $\Delta(w_1, w_2, \ldots, w_n) = 0$, then there exist $a_1, a_2, \ldots, a_n$ in k_∞, not all zero, such that (8.16) holds. Show that we lose no generality in assuming that $a_1 = 1$. Viewing the linear system (8.16) as n equations, for each $r = 1, 2, \ldots, n - 1$, we raise the rth equation to the qth power and subtract it from the $(r + 1)$th equation. This reduction leads to the following system of linear equations:

$$
(a_2 - a_2^q)w_2^q + \cdots + (a_n - a_n^q)w_n^q = 0
$$
$$
\vdots \qquad \vdots \qquad \vdots
$$
$$
(a_2 - a_2^q)w_2^{q^{n-1}} + \cdots + (a_n - a_n^q)w_n^{q^{n-1}} = 0.
$$

Show that since $w_2, w_3, \ldots, w_n$ are F_q-linearly independent, $w_2^q, w_3^q, \ldots, w_n^q$ are also F_q-linearly independent. Thus by induction we have that $\Delta(w_2^q, \ldots, w_n^q) \neq 0$. Hence we have that $a_2 = a_2^q, \ldots, a_n = a_n^q$, and therefore all of $a_1 = 1, a_2, \ldots a_n$ must lie in F_q, which contradicts the F_q-linear independence of $w_1, w_2, \ldots, w_n$.)

Armed with Proposition 8.9, it is not too difficult to relate $e_d(z)$ to a ratio of two Moore determinants. Indeed, we have the following attractive result.

THEOREM 8.10

$$
e_d(z) = \frac{\Delta(1, T, T^2, \ldots, T^{d-1}, z)}{\Delta(1, T, T^2, \ldots, T^{d-1})}.
\tag{8.17}
$$

Proof. By Proposition 8.9 we see that z_0 is a solution to $\Delta(1, T, \ldots, T^{d-1}, z) = 0$ if and only if the quantities $1, T, \ldots, T^{d-1}, z_0$ are F_q-linearly dependent, a circumstance that holds if and only if $z_0 \in \mathcal{F}(d)$. Thus the polynomial $\Delta(1, T, \ldots, T^{d-1}, z)$ has precisely the same zeros as $e_d(z)$. Since both $\Delta(1, T, \ldots, T^{d-1}, z)$ and $e_d(z)$ are polynomials, we conclude that there exists a number $\eta \in k_\infty$ satisfying

$$
e_d(z) = \eta \Delta(1, T, \ldots, T^{d-1}, z).
$$

Thus we will have established the theorem if we verify that

$$\eta = \frac{1}{\Delta(1, T, \dots, T^{d-1})}.$$

We compute η by comparing the leading coefficients of $e_d(z)$ and $\Delta(1, T, \dots, T^{d-1}, z)$. The leading coefficient of $e_d(z)$ is 1, since $e_d(z)$ is a product of expressions of the form $z - A$ with $A \in \mathcal{F}(d)$. Next we calculate the leading coefficient of $\Delta(1, T, \dots, T^{d-1}, z)$ by recalling that

$$\Delta(1, T, \dots, T^{d-1}, z) = \det \begin{pmatrix} 1 & T & \cdots & T^{d-1} & z \\ 1 & T^q & \cdots & T^{(d-1)q} & z^q \\ 1 & T^{q^2} & \cdots & T^{(d-1)q^2} & z^{q^2} \\ \vdots & \vdots & & \vdots & \vdots \\ 1 & T^{q^d} & \cdots & T^{(d-1)q^d} & z^{q^d} \end{pmatrix}. \tag{8.18}$$

Thus the leading coefficient of $\Delta(1, T, T^2, \dots, T^{d-1}, z)$ is the minor obtained by deleting the last row and the last column. This determinant is precisely $\Delta(1, T, \dots, T^{d-1})$, which completes the proof. ∎

We can now apply Theorem 8.10 to obtain an explicit polynomial representation for $e_d(z)$. We calculate the numerator of (8.17) by expanding the determinant in (8.18) along the last column to deduce that

$$\Delta(1, T, T^2, \dots, T^{d-1}, z) = \sum_{j=0}^{d} (-1)^{d-j} M_j z^{q^j},$$

where M_j is the minor obtained by deleting the jth row and last column of the matrix. However, we note that each minor M_j is the transpose of a Vandermonde determinant, so each is easily evaluated as

$$M_j = \prod_{\substack{0 \le t < s \le d \\ s \ne j, t \ne j}} \left(T^{q^s} - T^{q^t} \right). \tag{8.19}$$

We now discover that the M_j terms can be related to the now more-familiar factorials D_r that appear in the function $e_C(z)$.

Challenge 8.14 *First show that we can rewrite the product representation of M_j in (8.19) as*

$$M_j = \frac{\prod_{0 \le t < s \le d}(T^{q^s} - T^{q^t})}{\prod_{0 \le t < j}(T^{q^j} - T^{q^t}) \prod_{j < s \le d}(T^{q^s} - T^{q^j})}. \tag{8.20}$$

Next verify that the numerator in (8.20) equals the product $D_d D_{d-1} \cdots D_1$, by first writing

$$\prod_{0 \le t < s \le d} \left(T^{q^s} - T^{q^t} \right) = \prod_{s=1}^{d} \prod_{t < s} \left(T^{q^s} - T^{q^t} \right).$$

Finally, establish that the first product in the denominator of (8.20) equals D_j and that the second product in the denominator equals

$$\left([d - j][d - j - 1] \cdots [1] \right)^{q^j}.$$

For a positive integer j, if we introduce the standard notation,

$$L_j = [j][j - 1] \cdots [1],$$

then the previous challenge allows us to conclude that

$$M_j = \frac{\prod_{i=1}^{d} D_i}{D_j L_{d-j}^{q^j}}.$$

Thus we can express the numerator of $e_d(z)$ from (8.17) as

$$\sum_{j=0}^{d} (-1)^{d-j} \frac{\prod_{i=1}^{d} D_i}{D_j L_{d-j}^{q^j}} z^{q^j}.$$

We can apply the same ideas used in Challenge 8.14 to deduce that the denominator $\Delta(1, T, \ldots, T^{d-1})$ of $e_d(z)$ from (8.17) can be rewritten as $\prod_{i=1}^{d-1} D_i$. Putting all these identities together, we discover that

$$e_d(z) = \sum_{j=0}^{d} (-1)^{d-j} \frac{\prod_{i=1}^{d} D_i}{D_j \, L_{d-j}^{q^j} \prod_{i=1}^{d-1} D_i} z^{q^j} = \sum_{j=0}^{d} (-1)^{d-j} \frac{D_d}{D_j L_{d-j}^{q^j}} z^{q^j}. \qquad (8.21)$$

Basking in the glow of the big picture.   The formula we have just found for $e_d(z)$ is beautiful for several reasons. First, since the only exponents on z are powers of q, we conclude from Proposition 8.3 that $e_d(z)$ is F_q-linear. Second, we note by (8.21) that $e_d(z)$ has coefficients involving the factorials D_t, and thus perhaps we have stumbled upon a connection between $e_d(z)$ and the exponential function $e_C(z)$ as we optimistically had hoped.

Recall that our desire is to construct an entire F_q-linear function that vanishes for *all* $A \in F_q[T]$. Unfortunately, as we have already seen, we cannot simply take the limit of the polynomials $e_d(z)$ as $d \to \infty$, since that limit, $\prod_{A \in F_q[T]}(z - A)$, diverges for all $z \notin F_q[T]$, and hence is not an entire function. Thus we need to modify the polynomials $e_d(z)$ so that the new limit as $d \to \infty$ exists. We proceed in a manner that parallels our development of the Weierstrass σ-function from the previous chapter.

An intuitive rationale for the modification to our polynomials

Recall that for a lattice $W = \mathbb{Z}w_1 + \mathbb{Z}w_2$ the Weierstrass σ-function is defined by the product

$$\sigma(z) = z \prod_{w \in W, w \neq 0} \left(1 - \frac{z}{w}\right) e^{\frac{z}{w} + \frac{1}{2}\frac{z^2}{w^2}}. \qquad (8.22)$$

Ignoring for the moment any pesky convergence issues (that is, ignoring the exponential factor in (8.22)), we might conjecture that our best hope for convergence should come from an analogous product of the form

$$z \prod_{A \in F_q[T], A \neq 0} \left(1 - \frac{z}{A}\right).$$

This product inspires us to define an analogous polynomial $\bar{e}_d(z)$ by

$$\bar{e}_d(z) = z \prod_{A \in \mathcal{F}(d), A \neq 0} \left(1 - \frac{z}{A}\right).$$

We remark that $A \in \mathcal{F}(d)$ if and only if $-A \in \mathcal{F}(d)$. Thus we have

$$\bar{e}_d(z) = z \prod_{A \in \mathcal{F}(d), A \neq 0} \left(\frac{1}{A}(z + A)\right) = \frac{e_d(z)}{\prod_{A \in \mathcal{F}(d), A \neq 0} A}.$$

Thus it appears that dividing the polynomial $e_d(z)$ by $\prod_{A \in \mathcal{F}(d), A \neq 0} A$ is the natural modification necessary to produce our desired function. With these intuitive insights as our guide, we now move toward our periodic function.

Challenge 8.15 *Verify the identity*

$$\prod_{A \in \mathcal{F}(d), A \neq 0} A = (-1)^d \frac{D_d}{L_d}.$$

(Hint: Recall that if $Q(T) \in \mathcal{F}(d)$, then for each nonzero $a \in F_q$, $aQ(T)$ is also in $\mathcal{F}(d)$. Using the fact that $\prod_{a \in F_q^*} a = -1$, conclude that

$$\prod_{\substack{Q(T) \in F_q[T], \\ \deg(Q) = i}} Q(T) = -\left(\prod_{\substack{Q(T) \in F_q[T], Q(T) \\ \text{monic with } \deg(Q) = i}} Q(T)^{q-1}\right) = -D_i^{q-1},$$

and then establish the identity.)

We note that $\bar{e}_d(z)$ vanishes for all $A \in \mathcal{F}(d)$. Also, by the previous challenge and (8.21) we see that

$$\bar{e}_d(z) = z \prod_{A \in \mathcal{F}(d), A \neq 0} \left(1 + \frac{z}{A}\right) = \sum_{j=0}^{d} (-1)^j \left(\frac{L_d}{L_{d-j}^{q^j}}\right) \frac{z^{q^j}}{D_j}. \qquad (8.23)$$

We now demonstrate that $\lim_{d \to \infty} \bar{e}_d(z)$ exists and is closely related to $e_C(z)$. This task is not easily accomplished, since the coefficients of $\bar{e}_d(z)$ involve the ratios $L_d / L_{d-j}^{q^j}$, which depend on the parameter d that is approaching infinity. Thus our goal is to replace the coefficients

$$\frac{L_d}{L_{d-j}^{q^j}} \qquad (8.24)$$

by quantities whose dependence on d is more manageable.

We recall that $L_d = [d][d-1] \cdots [1]$ and hence is a product of d polynomials in $F_q[T]$ and has degree

$$q^d + q^{d-1} + \cdots + q = \frac{q^{d+1} - q}{q - 1}.$$

Thus in order to simplify the quotient in (8.24), we compare L_d with a simpler product of d polynomials from $F_q[T]$ of degree $\frac{q^{d+1} - q}{q-1}$. In particular, we consider

$$[1]^{q^{d-1}} [1]^{q^{d-2}} \cdots [1] = [1]^{\frac{q^d - 1}{q - 1}}.$$

In an attempt to connect these two polynomials, we define the quantity

$$\zeta_d = \frac{[1]^{\frac{q^d - 1}{q - 1}}}{L_d}.$$

It follows that

$$\frac{L_d}{L_{d-j}^{q^j}} = [1]^{\frac{q^d - 1}{q-1} - \frac{q^d - q^j}{q-1}} \frac{\zeta_{d-j}^{q^j}}{\zeta_d} = [1]^{\frac{q^j - 1}{q-1}} \frac{\zeta_{d-j}^{q^j}}{\zeta_d},$$

and so we can express $\bar{e}_d(z)$ from (8.23) as

$$\bar{e}_d(z) = \frac{1}{\zeta_d} \sum_{j=0}^{d} (-1)^j \left([1]^{\frac{q^j - 1}{q-1}} \zeta_{d-j}^{q^j}\right) \frac{z^{q^j}}{D_j}.$$

We now proceed to evaluate $\lim_{d \to \infty} \bar{e}_d(z)$ by first showing that $\lim_{d \to \infty} \zeta_d$ exists in k_∞.

Challenge 8.16 *Verify the identity*

$$\zeta_d = \prod_{j=1}^{d-1} \left(1 - \frac{[j]}{[j+1]}\right) \tag{8.25}$$

by showing that each factor of the product can be rewritten as

$$1 - \frac{[j]}{[j+1]} = \frac{[1]^{q^j}}{[j+1]}.$$

Now using (8.25), *prove that* $\lim_{d\to\infty} \zeta_d$ *exists in* k_∞. *(Hint: First show that* $|\zeta_d - \zeta_{d-1}|_\infty \to 0$ *as* $d \to \infty$.)

We let $\zeta_* \in k_\infty$ denote $\lim_{d\to\infty} \zeta_d$ and for each index d, define $\delta_d \in k_\infty$ to be the offset given by $\zeta_d = \delta_d + \zeta_*$. Therefore by F_q-linearity, we have

$$
\begin{aligned}
\bar{e}_d(z) &= \frac{1}{\zeta_d} \sum_{j=0}^{d} (-1)^j \left([1]^{\frac{q^j-1}{q-1}} (\delta_{d-j} + \zeta_*)^{q^j}\right) \frac{z^{q^j}}{D_j} \\
&= \frac{1}{\zeta_d} \sum_{j=0}^{d} (-1)^j \left([1]^{\frac{q^j-1}{q-1}} \delta_{d-j}^{q^j}\right) \frac{z^{q^j}}{D_j} + \frac{1}{\zeta_d} \sum_{j=0}^{d} (-1)^j \left([1]^{\frac{q^j-1}{q-1}} \zeta_*^{q^j}\right) \frac{z^{q^j}}{D_j},
\end{aligned}
\tag{8.26}
$$

where, we are delighted to report, the terms of the second summation are *independent* of d. Next we claim that the first sum approaches 0 as $d \to \infty$. That is, for any $z_0 \in k_\infty$,

$$\lim_{d\to\infty} \frac{1}{\zeta_d} \sum_{j=0}^{d} (-1)^j \left([1]^{\frac{q^j-1}{q-1}} \delta_{d-j}^{q^j}\right) \frac{z_0^{q^j}}{D_j} = 0.$$

Challenge 8.17 *Verify the previous claim by first establishing that*

$$|D_j|_\infty = q^{jq^j}, \quad \left|[1]^{\frac{q^j-1}{q-1}}\right|_\infty = q^{q\frac{q^j-1}{q-1}}, \quad and \quad \left|\delta_{d-j}^{q^j}\right|_\infty = q^{q^{d-j}(q-1)}.$$

Now conclude that for any $z_0 \in k_\infty$,

$$\left|(-1)^j \left([1]^{\frac{q^j-1}{q-1}} \delta_{d-j}^{q^j}\right) \frac{z_0^{q^j}}{D_j}\right|_\infty \to 0 \quad as \ d \to \infty.$$

Thus (8.26) together with the previous challenge allows us to conclude that $\lim_{d\to\infty} \bar{e}_d(z)$, which we now denote by $\bar{e}(z)$, exists. Moreover,

$$\bar{e}(z) = \frac{1}{\zeta_*} \sum_{j=0}^{\infty} (-1)^j \left([1]^{\frac{q^j-1}{q-1}} \zeta_*^{q^j}\right) \frac{z^{q^j}}{D_j}.$$

The dramatic conclusion. We have produced a power series $\bar{e}(z)$ that vanishes at all elements of $F_q[T]$ and is F_q-linear. Thus we conclude that $\bar{e}(z)$ is periodic modulo $F_q[T]$. Moreover, we remark that the denominators of the coefficients in $\bar{e}(z)$ agree with those of $e_C(z)$. In fact, just as in the case of the functions $e^{2\pi i z}$ and e^z, here we see that a simple change of variables reveals the relationship between the functions $\bar{e}(z)$ and $e_C(z)$. As the challenge below suggests, there exists a number π_C in the algebraic closure of k_∞ such that

$$\bar{e}(z) = \frac{1}{\pi_C} e_C(\pi_C z).$$

Challenge 8.18 *Let λ be any $(q-1)$st root of $-[1]$. If we define $\pi_C = \lambda \zeta_*$, then verify the identity*

$$\bar{e}(z) = \frac{1}{\pi_C} e_C(\pi_C z).$$

If we now replace z by $\pi_C^{-1} z$, then the previous identity is transformed into

$$e_C(z) = \pi_C \bar{e}(\pi_C^{-1} z).$$

Since $\bar{e}(z)$ is periodic modulo $F_q[T]$, the above identity reveals that $e_C(z)$ is periodic modulo $\pi_C F_q[T]$, which is analogous to the fact that e^z is periodic modulo $2\pi i \mathbb{Z}$. Thus through this elaborate discussion of periodicity we have discovered the Carlitz analogue of π, namely π_C.

In the next section we present an overview of the proof of the transcendence of π_C. In that sketch, and again in the final section of this chapter, we will be required to reach beyond the complete field k_∞. The necessity for this extended point of view follows from the observation that when the function $e_C(z)$ is evaluated at *algebraic* values, the resulting quantities might not be elements of k_∞. To rectify this situation, we introduce a new field, $\bar{k}_\infty$, that is both complete and algebraically closed, and thus is the analogue of $\mathbb{C}$. In contrast to the construction of $\mathbb{C}$ from the complete field $\mathbb{R}$, where the algebraic closure turns out also to be a complete field, here the algebraic closure of k_∞ is not complete. The field $\bar{k}_\infty$ is obtained by completing the algebraic closure of k_∞. We note, without proof, that the complete field $\bar{k}_\infty$ remains algebraically closed.

8.7 The transcendence of π_C

We have shown that the Carlitz function $e_C(z)$ has a power series expansion resembling that of e^z; and just as e^z is periodic modulo $2\pi i \mathbb{Z}$, $e_C(z)$ is periodic modulo $\pi_C F_q[T]$. The analogy between e^z and $e_C(z)$ becomes deeper still, since we now discover that the analogue of $2\pi i$ in this context, π_C, is transcendental over $k = F_q(T)$.

THEOREM 8.11 *The element π_C is transcendental over k.*

An intuitive sketch of the proof. We begin by assuming that π_C is algebraic and use this assumption to construct a power series that is amenable to the analysis we employed to establish the transcendence of $e_C(1)$. Unfortunately, there is no readily available series representation for π_C, and hence we find our desired series by using the fact that $e_C(\pi_C) = 0$; that is, we apply the power series for $e_C(z)$ evaluated at $z = \pi_C$:

$$\sum_{r=0}^{\infty} \frac{(\pi_C)^{q^r}}{D_r} = 0. \tag{8.27}$$

Since we have assumed that π_C is algebraic, we can use its alleged conjugates to play a role in our eventual contradiction of the Fundamental Principle of Number Theory in $F_q[T]$. To bring these conjugates into play, we multiply the identity in (8.27) by the power series expression for $e_C(z)$ evaluated at each of the conjugates of π_C. This product can then be rewritten as a somewhat more easily analyzed series. It is possible to view this new series as a sum of a polynomial term and an infinite series tail-term. After multiplying by a suitable denominator, we can show that the tail-term of the series is a nonzero element of $F_q[T]$ whose absolute value is strictly less than 1. This violates the Fundamental Principle of Number Theory and thus establishes the transcendence of π_C.   ∎

8.8 Revisiting $\wp(z)$ and moving beyond Carlitz

We remarked earlier that the Carlitz function $e_C(z)$ is analogous to both the usual exponential function e^z and the Weierstrass $\wp$-function. The parallels between $e_C(z)$ and $\wp(z)$ begin where the parallels between $e_C(z)$ and e^z end.

The primary difference between $e_C(z)$ and e^z arises from an almost invisible difference between $\mathbb{Z}$ and $A = F_q[T]$: In $\mathbb{Z}$, multiplication corresponds to repeated addition, but no such interpretation holds over A. More precisely, for any positive integer n and any $z \in \mathbb{C}$, $nz = z + \cdots + z$, where z is added n times. Hence the fact that e^z is a homomorphism from an additive group to a multiplicative group yields the identity

$$e^{nz} = (e^z)^n.$$

There is no such simple identity connecting the product of elements in k_∞ by elements from A, because multiplication, even by the indeterminate $T \in A$, does not correspond to repeated addition. Therefore we are unable to deduce from the homomorphism property of $e_C(z)$ a connection between $e_C(Tz)$ and $e_C(z)$. Yet those two quantities are indeed related, and it is here that $e_C(z)$ begins to emulate the Weierstrass $\wp$-function.

A new action of T on $\bar{k}_\infty$. We recall from Chapter 7 that for certain elliptic curves the endomorphism ring is larger than $\mathbb{Z}$, and those elliptic curves are precisely the ones that have so-called complex multiplication. For an elliptic curve with complex multiplication by some complex number $\rho \notin \mathbb{Z}$, ρ acts on the points of the elliptic curve through the associated $\wp$-function. Specifically, for a point $P \in E$ that is the image of some $z_P \in \mathbb{C}$ under the group homomorphism $z \mapsto (\wp(z), \wp'(z))$, we define the action $\rho \circ P = (\wp(\rho z_P), \wp'(\rho z_P))$, provided that ρz_P is not in the lattice of periods for the $\wp$-function. Implicit in this analysis is the fact that the functions $\wp(\rho z)$ and $\wp(z)$ are related by certain explicit relations.

To understand how T can act on $\bar{k}_\infty$, we first observe that there is a *trivial T-action*: For $v \in \bar{k}_\infty$, we let $T \circ v = T \times v$. This T-action is not a particularly enlightening, since it corresponds to the usual multiplication by the indeterminant T. However, through the Carlitz exponential function, we can define a *nontrivial* action of T on $\bar{k}_\infty$. The central component of this new T-action is a relationship between the two functions $e_C(Tz)$ and $e_C(z)$.

Challenge 8.19 *Use the series expansion $e_C(z) = \sum_{r=0}^{\infty} \frac{z^{q^r}}{D_r}$ and the definition of the denominators D_r to show that*

$$e_C(Tz) = Te_C(z) + e_C(z)^q. \tag{8.28}$$

A fruitful interpretation of the relationship in (8.28) is that there exists an operator ϕ_T acting on the values of $e_C(z)$ satisfying

$$\phi_T \circ e_C(z) = e_C(Tz).$$

In order to exploit this relationship between $e_C(z)$ and $e_C(Tz)$ to define a nontrivial T-module structure on $\bar{k}_\infty$, we now report, without proof, the fact that $e_C(z)$ is surjective. Thus for any point $v \in \bar{k}_\infty$, there exists a $z_v \in \bar{k}_\infty$ such that $e_C(z_v) = v$. We now define the action of T on v by $T \circ v = e_C(Tz_v)$.

Just as an elliptic curve with complex multiplication by ρ has an endomorphism ring containing $\mathbb{Z}$ as a proper subset, here ϕ_T demonstrates that the Carlitz exponential function imposes a structure onto $\bar{k}_\infty$—which is the analogue of $E^*(\mathbb{C})$—possessing a set of endomorphisms properly containing $\mathbb{F}_q$. Moreover, through iterations of ϕ_T, it is possible to describe a T^n-action for every $n \geq 1$. Thus since $e_C(z)$ is $\mathbb{F}_q$-linear, we obtain an action for every element of $\mathbb{F}_q[T]$.

The action of the polynomial $\phi_T(X) = TX + X^q$ on $\bar{k}_\infty$ implicitly given in (8.28) via the Carlitz exponential function $e_C(z)$ is a special case of a much more general phenomenon that was uncovered by Vladimir Drinfeld late in the last century. Drinfeld showed that there was a correspondence between polynomials of the form

$$\phi_T(X) = TX + a_1 X^q + a_2 X^{q^2} + \cdots + a_d X^{q^d} \tag{8.29}$$

and entire $\mathbb{F}_q$-linear functions $e_{\phi_T}(z)$ with a lattice of periods W in $\bar{k}_\infty$. Indeed, any one of these three objects, a T-action as in (8.29), an entire $\mathbb{F}_q$-linear function whose first coefficient equals 1, or a lattice of periods in $\bar{k}_\infty$, uniquely determines the other two. Drinfeld's seminal work was recognized in 1990 when he was awarded the Fields Medal for his important contributions to mathematics.

What is especially appealing about Drinfeld's correspondence is that the *degree* of ϕ_T, defined to be d when ϕ_T is expressed as in (8.29), equals the A-rank of the lattice of periods of the associated entire, $\mathbb{F}_q$-linear function, $e_{\phi_T}(z)$. In fact, T-actions of different degrees give rise to algebraically independent "Drinfeld exponential functions." Moreover, two "sufficiently different" T-actions of a given degree will have algebraically independent exponential functions.

The previous phenomenon does not occur in the case $d = 1$, the situation considered by Carlitz, essentially because the exponential function associated with a T-action of the form

$$\phi_T(X) = TX + aX^q, \quad \text{with } a \neq 0,$$

is entirely determined by the value of a. This assertion is easy to see once we explicitly have the relationship between a general polynomial of the form (8.29) and its associated function $e_{\phi_T}(z)$.

THEOREM 8.12   *Suppose $\phi_T(X)$ is any polynomial of the form*

$$\phi_T(X) = TX + a_1X^q + \cdots + a_dX^{q^d},$$

with $d > 0$, and $a_i \in \bar{k}_\infty$ for all $i = 1, 2, \ldots, d$. Then there exists a unique F_q-linear power series $e_{\phi_T}(z) = \sum_{r=0}^{\infty} c_r z^{q^r}$, with $c_0 = 1$, such that

$$\phi_T \circ e_{\phi_T}(z) = e_{\phi_T}(Tz). \tag{8.30}$$

Sketch of the proof.   If we begin with the desired relationship $\phi_T \circ e_{\phi_T}(z) = e_{\phi_T}(Tz)$ and replace z by $T^{-1}z$, then we have

$$\phi_T \circ e_{\phi_T}(T^{-1}z) = e_{\phi_T}(z).$$

Substituting the series representation for $e_{\phi_T}(z)$ into this expression and then rewriting the left-hand side by grouping together all terms of the form z^{q^r}, we obtain the following recurrence relation. For each index r,

$$\left(T^{q^r} - T\right)c_r = \sum_{i=1}^{r} a_i c_{r-i}^{q^i}.$$

Therefore each c_r is completely determined by the coefficients of $\phi_T(X)$ and the previous $c_0, c_1, \ldots, c_{r-1}$. Declaring $c_0 = 1$ thus uniquely determines c_r for all $r > 0$, which completes our sketch.   ∎

Challenge 8.20   Let $\phi_T(X) = TX + aX^q$, with $a \neq 0$, and let $e_a(z) = e_{\phi_T}(z)$ as in (8.30) be its associated exponential function. Verify that

$$a^{1/(q-1)}e_a(z) = e_C\left(a^{1/(q-1)}z\right).$$

We note that the proof of Theorem 8.12 implies that the power series coefficients of $e_{\phi_T}(z)$ all reside in any field containing all of the coefficients of $\phi_T(X)$, for example, in $k(a_1, \ldots, a_d)$. In particular, of considerable interest in examining the transcendence of numbers associated with a given T-action is the situation in which the coefficients of $\phi_T(X)$ are all algebraic over k. In this case we see that the coefficients of the power series for the associated exponential function $e_{\phi_T}(z)$ are also algebraic over k. This observation lies behind the following extension, established by Jing Yu in 1983, of the transcendence of π_C.

THEOREM 8.13   *Suppose that $\phi_T(X) = TX + a_1 X^q + \cdots + a_d X^{q^d}$, where the coefficients $a_1, a_2, \ldots, a_d$ are algebraic over k. Then any nonzero element in the lattice of periods of the function associated with $\phi_T(X)$ is transcendental.*

While we do not even hint at the proof of this modern result, we do note that the proof of this theorem, as well as the Drinfeld analogues of the theorems of Hermite, Lindemann–Weierstrass, Gelfond–Schneider, and Schneider–Lang, all follow the same outline that we employed throughout this short journey through classical transcendental number theory. As expected, these increasingly sophisticated arguments ultimately rely upon the most basic result we have—the Fundamental Principle of Number Theory.

Appendix

Selected highlights from complex analysis

A.1 Analytic and entire functions

In this short appendix we offer some highlights of various fundamental notions from complex function theory. The remarks that follow are by no means designed to be complete, and the justifications are certainly not rigorous. If anything, these observations are concise. There are many excellent complex analysis texts that readers can consult for additional background or details.

Let $U \subseteq \mathbb{C}$ be an open subset of the complex plane and let $f : U \to \mathbb{C}$ be a complex-valued function. Given a point $z_0 \in U$, we say that the *derivative of f at $z = z_0$ exists* if the limit

$$\lim_{z \to z_0} \frac{f(z) - f(z_0)}{z - z_0}$$

exists. If this limit does exist, then we define the *derivative of f at $z = z_0$*, denoted by $f'(z_0)$, to be the value of the limit. The function f is said to be *analytic on the open set U* if its derivative exists for every $z \in U$. A function that is analytic on the whole complex plane is called an *entire function*.

In a dramatic departure from real analysis, we record two surprising results from complex function theory. First, suppose we decompose the analytic function $f(z)$ into its real and imaginary parts as $f(z) = u(z) + iv(z)$. Then writing $z = x + iy$, the real-valued functions $u(z) = u(x + iy)$ and $v(z) = v(x + iy)$ satisfy the following two partial differential equations:

$$u_x(x + iy) = v_y(x + iy) \quad \text{and} \quad u_y(x + iy) = -v_x(x + iy). \tag{A.1}$$

These famous differential equations are known as the *Cauchy–Riemann equations*.

The second surprising result is that if f is analytic on U, then so is its derivative, and thus f can be differentiated an arbitrary number of times. One consequence of this incredible fact is that if a function $f(z)$ is entire, then given any point $z_0 \in \mathbb{C}$, f can be expressed as a power series centered about z_0 as

$$f(z) = \sum_{n=0}^{\infty} \frac{f^{(n)}(z_0)}{n!} (z - z_0)^n.$$

A.2 Contour integrals

Let $a < b$ be real numbers. A function $\gamma : [a, b] \to \mathbb{C}$ is called a *smooth path* if $\gamma'(t)$ exists for all $t \in [a, b]$ and $\gamma'(t)$ is a continuous function on $[a, b]$. Given a function $f : U \to \mathbb{C}$ and a smooth path $\gamma([a, b]) \subseteq U$, we define the *contour integral of f along γ*, denoted as $\int_\gamma f(z)\, dz$, by

$$\int_\gamma f(z)\, dz = \int_a^b f(\gamma(t))\gamma'(t)\, dt.$$

To illustrate the definition of the contour integral, we consider the following very simple example. Let $\gamma : [0, 1] \to \mathbb{C}$ be the smooth path defined by $\gamma(t) = e^{2\pi i t}$. Thus γ is the path around the unit circle traversed once in a counter-clockwise direction. We now compute the contour integral of the entire function $f(z) = z^2$ along γ:

$$\int_\gamma z^2\, dz = \int_0^1 \left(e^{2\pi i t}\right)^2 2\pi i e^{2\pi i t}\, dt = 2\pi i \int_0^1 e^{6\pi i t}\, dt$$

$$= \frac{1}{3}\left(e^{6\pi i t}\right)\Big|_0^1 = \frac{1}{3}(1 - 1) = 0.$$

A.3 Cauchy's integral formula

The previous example is actually a very special case of a much more general phenomenon that is codified in a famous result due to Augustin-Louis Cauchy. We say that a path $\gamma : [a, b] \to \mathbb{C}$ is a *simple closed path* if $\gamma(a) = \gamma(b)$, and the curve $\gamma([a, b])$ partitions its complement $\mathbb{C} \setminus \gamma([a, b])$ into exactly two pieces (the "inside" of γ and the "outside" of γ).

THEOREM A.1 (CAUCHY'S THEOREM) *Let $g : U \to \mathbb{C}$ be an analytic function and γ a simple closed smooth path in U. Then*

$$\int_\gamma g(z)\, dz = 0.$$

In fact, Cauchy's result is a special case of the following beautiful theorem.

THEOREM A.2 (CAUCHY'S INTEGRAL FORMULA) *Let $f : U \to \mathbb{C}$ be an analytic function and γ a simple closed smooth path in U traversed once in a counter-clockwise direction. Then for any integer $n \geq 0$ and any complex number z_0 contained in the inside of γ,*

$$f^{(n)}(z_0) = \frac{n!}{2\pi i} \int_\gamma \frac{f(z)}{(z - z_0)^{n+1}}\, dz.$$

We remark that if we apply Theorem A.2 with $n = 0$ and $f(z) = (z - z_0)g(z)$, then we immediately deduce the validity of Theorem A.1.

A.4 The Maximum Modulus Principle

In this section we provide an overivew of the proof of the key result from complex analysis that we exploited throughout our journey through transcendence. We begin with a *local version* of the Maximum Modulus Principle.

LEMMA A.3   *Let $f : U \to \mathbb{C}$ be an analytic function. Suppose that $z_0 \in U$, and R is a real number such that the open disk $D = D(z_0, R) = \{z \in \mathbb{C} : |z - z_0| < R\}$ is contained in U. If there exists a $z^* \in D$ such that $|f(z)| \leq |f(z^*)|$ for all $z \in D$, then f is constant on D.*

A sketch of the proof.   We will prove the lemma in the special case $z^* = z_0$. Let r be a fixed real number satisfying $0 < r < R$ and let $\gamma(t) = z_0 + re^{2\pi i t}$ for $0 \leq t \leq 1$. Thus γ is the path around the circle centered at z_0 of radius r, traversed once in a counter-clockwise direction. We also note that $\gamma(t) \in D$ for all $0 \leq t \leq 1$. Applying Cauchy's Integral Formula with $n = 0$, we see that

$$f(z_0) = \frac{1}{2\pi i} \int_\gamma \frac{f(z)}{(z - z_0)} \, dz = \frac{1}{2\pi i} \int_0^1 \frac{f\left(z_0 + re^{2\pi i t}\right) 2\pi i r e^{2\pi i t}}{re^{2\pi i t}} \, dt.$$

The previous identity together with our hypotheses implies that

$$|f(z_0)| = \left| \int_0^1 f\left(z_0 + re^{2\pi i t}\right) dt \right|$$

$$\leq \int_0^1 \left| f\left(z_0 + re^{2\pi i t}\right) \right| dt$$

$$\leq \int_0^1 |f(z_0)| \, dt = |f(z_0)|,$$

and hence

$$|f(z_0)| = \int_0^1 \left| f\left(z_0 + re^{2\pi i t}\right) \right| dt.$$

Thus it follows that

$$\int_0^1 \left(|f(z_0)| - \left| f\left(z_0 + re^{2\pi i t}\right) \right| \right) dt = 0, \tag{A.2}$$

where, by our hypothesis, $|f(z_0)| - \left| f\left(z_0 + re^{2\pi i t}\right) \right| \geq 0$. Since the function $G(r, t) = |f(z_0)| - \left| f\left(z_0 + re^{2\pi i t}\right) \right|$ is continuous, in view of (A.2) we conclude that $G(r, t)$ is identically zero, and hence $|f(z)| = |f(z_0)|$ for all $z \in D$. That is, we see that the function $|f(z)|$ is constant on D, say $|f(z)| = c$ for some $c \in \mathbb{R}$. Now if we write $f(z) = u(z) + iv(z)$, where $u(z)$ and $v(z)$ are real-valued functions, then we see that

$$u(z)^2 + v(z)^2 = c^2.$$

Taking the partial derivatives of the previous identity with respect x and then y, where $z = x + iy$, we have

$$2u(z)u_x(z) + 2v(z)v_x(z) = 0 \quad \text{and} \quad 2u(z)u_y(z) + 2v(z)v_y(z) = 0,$$

which, in view of the Cauchy–Riemann equations (A.1), implies that

$$u(z)u_x(z) - v(z)u_y(z) = 0 \quad \text{and} \quad u(z)u_y(z) + v(z)u_x(z) = 0.$$

Multiplying the first identity by $u(z)$ and the second by $v(z)$ and adding the resulting identities reveals that

$$u(z)^2 u_x(z) + v(z)^2 u_x(z) = \left(u(z)^2 + v(z)^2 \right) u_x(z) = c^2 u_x(z) = 0.$$

Thus we can conclude that $u_x(z) = 0$ for all $z \in D$. A similar argument shows that $u_y(z) = 0$ for all $z \in D$, and hence we establish that the function $u(z)$ is constant on D. An analogous line of reasoning can be applied to deduce that $v(z)$ is constant on D and therefore we find that $f(z)$ is constant on D, which completes our proof. ∎

THEOREM A.4 (THE MAXIMUM MODULUS PRINCIPLE) *Let $D \subseteq \mathbb{C}$ be an open disk and let $\overline{D}$ denote the union of D and its boundary. If $f : \overline{D} \to \mathbb{C}$ is a continuous function that is analytic on D, then $|f(z)|$ attains its maximum value at some point along the boundary of $\overline{D}$.*

Proof. Since the set $\overline{D}$ is compact, we know that the continuous function $|f(z)|$ attains its maximum value somewhere in $\overline{D}$. We now assume that its maximum is realized at some point in D. Thus by Lemma A.3 we see that $f(z)$ is constant on D, and hence by continuity, constant on all of $\overline{D}$, and therefore the maximum value of $|f(z)|$ does occur at some points along the boundary of $\overline{D}$. ∎

Acknowledgments

Our treatment was influenced by many sources. Of particular note, we wish to explicitly acknowledge four texts:

- *Basic Structures of Function Field Arithmetic*, by David Goss, Springer-Verlag, 1996.
- *Introduction to Transcendental Numbers*, by Serge Lang, Addison-Wesley, 1966.
- *Nombres transcendants*, by Michel Waldschmidt, Lecture Notes in Mathematics, 402, Springer-Verlag, 1974.
- *Transcendence Methods*, by Michel Waldschmidt, Queen's Papers in Pure and Applied Mathematics, 52. Queen's University, 1979.

We wish to thank David Masser, Joseph Silverman, Michel Waldschmidt, and Darren Creutz for their comments on an early draft of this book. At Springer, we would also like to acknowledge the tireless and creative efforts of MaryAnn Brickner, David Kramer, Joseph Piliero, and Mark Spencer. Finally, we wish to express our gratitude to our exquisite editor, Ina Lindemann, for her infectious enthusiasm and for making the publication process so pleasant.

Index

A-number 159
Algebraic
  Approximation 175
  Integer 121
  Numbers 4, 7
  Polynomial differential equation 144
Algebraically closed 6
Algebraically independent
  Functions 88
  Numbers 73
Analytic continuation 207
Analytic function 255
Analytically fastidious reader 237
Aping 19
Archimedean 226
Baker, A. 115
Bananas 19
Binomial coefficient 30
Burger's number 261
Cantor, G. 6, 13
Cardinality 93, 133
Carlitz, L. 228
Carlitz's exponential function 234
Catapult 45
Cauchy, A. 256
Cauchy Criterion 5
Cauchy's Integral Formula 256
Cauchy's Theorem 256
Cauchy-Riemann equations 255
Champernowne's number 20
Chudnovsky, G. 217
Chudnovsky's Theorem 218
Complete field 6, 224
Complex multiplication 217, 250
Complex numbers 6
Conjugate
  Algebraic numbers 7, 57
  Complete set 66
  Complete exponential sum 67

Conspiracies 117
Constants, evolving 125
Contour integral 256
Convergence
  Of a series 224
  Of a product 204
Degree
  Of an algebraic number 7
  Of a monomial 55
  Of a polynomial 7
Diagonalization 19
Difference, vive la 87
Diophantine approximation 133, 149
Diophantine equations 115
Dirichlet's Theorem 150
  Extension 154
Division algorithm 14
Doubly periodic 183, 191
Drinfeld, V. 251
Dyson, F. 24
e
  Irrationality 27
  Power series 26, 29
  Transcendence 38
Elementary symmetric functions 53
Elliptic curve 187
Elliptic function 191
Endomorphism ring 216
Entire function 77, 255
Euler, L. 27
Euler's constant 207
F_q-linearity 228
F_q-module 228
Factorial in $F_q[T]$ 231
Fantasy, calculus student's 228
Fields Medal 24, 115, 251
Finite order of growth 142, 220
Formal Laurent series 227
Fourier, J. 27

Fourier's approach, intuitive idea   27
Fractional part   128
Fundamental domain   192
Fundamental Principle of Number
      Theory   9, 253
   In $F_q[T]$ 235
   Revisited   235
Gamma function   200, 203
Gauss, C.F.   6
Gelfond, A.O.   113
Gelfond-Schneider Theorem   113
Group homomorphism   146, 228
Hadamard's inequality   180
Hata, M.   25
Height
   Of a polynomial   86
   Of a vector   94
   Of an algebraic number   103
Hermite, C.   43
Hilbert, D.   113
Imaginary part   7
Integer   1
   Infamous   129
Irrational numbers   5
Koksma, J.   181
Lagrange's Theorem   232
Lang, S.   144
Laurent series   193
Leading coefficient   7
Lindemann, F.   43
Lindemann-Weierstrass Theorem   44
Linear forms in logarithms   115
Linearly independent   44
Liouville, J.   6, 10
Liouville number   18
Liouville's method, intuitive idea   10
Liouville's Theorem   11, 14, 192
   Restated   18
   Restated Yet Again   18
   Generalized version   152
   Watered-down version   149
Mahler, K.   20, 147
Mahler's classification   159
Mahler's number   20
Maximum Modulus Principle   78, 258
   Local version   257
Mean Value Theorem   14, 16
Measure zero   19
Meromorphic continuation   207
Meromorphic function   193
Metamorphosis, coefficient   118
Modular forms   188
Modulus   7, 78

Monkey   18
Monomial   54
   Degree   55
Moore determinant   242
Natural numbers   1
Nonarchimedean   226
Number   6
   Enigmatic   75
Number field   120
Order of growth of a function   142
Paradox, intriguing   ix
Partial products   204
Phew   216
π
   Irrationality   40
   Transcendence   43
Pigeonhole Principle   91, 93
Point at infinity   189
Poles   193
Polynomial   7
   Limbo   147, 151
   Homogeneous   55
   Irreducible   7, 14
   Minimal   7, 153
   Monic   7
   Reducible   14
   Zero   7
Primitive element   119, 120
Primitive Element Theorem   121
Punch line   105
Rational numbers   2
Real numbers   5, 224
Real part   7
Resultant of polynomials   179
Rocky road   130
Roth, K.   24
Roth's Theorem   24
Rumor   159
S-number   159
Sailing, smooth   226
Schmidt, W.   182
Schneider, T.   113
Schneider-Lang Theorem   144, 220
   Weak Version   144
Siegel, C.   24, 159
Siegel's Lemma   94
Simple closed path   256
Six Exponentials Theorem   137
Smooth path   256
Spaghetti   62
Strong triangle inequality   226
Symmetric function   52
Symmetric polynomial   53

T-action, trivial   251
  Nontrivial   251
T-module   251
T-number   159
Thue, A.   24
Transcendence measure   182
Transcendence trivia   156
Transcendental
  Near-integer   136
  Number theory, classical   ix
  Numbers   6
  Over a field   228
Triangle inequality   224

Tubbs' constant   263
U-number   159
Vandermonde determinant   242
Vanilla   130
Wade, L.   234
Waldschmidt, M.   114
Warned, duly   47
Weierstrass
  σ-function   209, 246
  $\wp$-function   195, 250
  Form of an elliptic equation   188
Weierstrass, K.   44
Yu, J.   253